# Lecture Notes in Intelligent Transportation and Infrastructure

**Series Editors**

Janusz Kacprzyk, Systems Research Institute, Polish Academy of Sciences, Warsaw, Poland

Olegas Prentkovskis, Vilnius Gediminas Technical University, Vilnius, Lithuania

The series "Lecture Notes in Intelligent Transportation and Infrastructure" (LNITI) publishes new developments and advances in the various areas of intelligent transportation and infrastructure. Merging theoretical foundations, practical applications, and forward-looking insights, LNITI provides a comprehensive understanding of both the state-of-the-art and the future prospects within this dynamic field.

LNITI is designed to be an inclusive platform that covers an extensive array of topics including, but not limited to intelligent transportation systems, smart mobility, intelligent logistics, critical infrastructure, smart architecture, smart cities, intelligent governance, construction design, data security, operational analysis, optimal route planning, digitalization, autonomous vehicles, the evolution of transport systems as well as green and sustainable urban structures. The series contains monographs, conference proceedings, edited volumes, lecture notes and textbooks. Of particular value to both the contributors and the readership are the short publication timeframe and the world-wide distribution, which enable wide and rapid dissemination of high-quality research output.

Proceedings published in the series are indexed by INSPEC.

All books, including proceedings, published in the series are evaluated by Web of Science.

Henrikas Sivilevičius · Lijana Maskeliūnaitė

# Assessment of the Quality of Passenger Transportation by Train Using Multiple Criteria Decision Making Methods

Springer

Henrikas Sivilevičius
Faculty of Transport Engineering,
Transport and Logistics Competence Centre
Vilnius Gediminas Technical University
Vilnius, Lithuania

Lijana Maskeliūnaitė
Faculty of Transport Engineering,
Department of Mobile Machinery
and Railway Transport
Vilnius Gediminas Technical University
Vilnius, Lithuania

ISSN 2523-3440 ISSN 2523-3459 (electronic)
Lecture Notes in Intelligent Transportation and Infrastructure
ISBN 978-3-031-85879-6 ISBN 978-3-031-85877-2 (eBook)
https://doi.org/10.1007/978-3-031-85877-2



This Springer imprint is published by the registered company Springer Nature Switzerland AG
The registered company address is: Gewerbestrasse 11, 6330 Cham, Switzerland

# Preface

The study considers the system for assessing and improving the quality indicators of the rail passenger transport process. Multi-criteria decision-making (MCDM) methods are used to assess the quality of this process. The characteristics of a passenger train are presented, and road, rail and air transport alternatives are compared using the Analytical Hierarchy Process (AHP) method. As rail transport competes with air and road transport, a system of criteria has been developed that outlines the indicators by which rail passengers choose this mode of transport as an alternative to other modes of transport. The questionnaire was used to rank passengers' opinions on the importance of the criteria. The MCDM methods (ARTIW-L, ARTIW-N and IHAMCI) were employed to calculate the normalised weights of the hierarchically unstructured and structured criteria, indicating the relevance of each sub-criterion and the significance of the main criterion as a factor in the choice of train journey.

One of the authors of the monograph has worked through his career in the position of a train manager on the international train. His long practical experience and knowledge on the passenger rail transport have benefited the identification of the factors (criteria) that are critical to the quality of travel. The employment of MCDM methods has contributed to the hierarchical structure of the criteria, guided by scientific principles, to be explored at different levels. The AHP method was applied to assess the significance of the four key criteria and their constituent sub-criteria, as measured by passengers, train crew and management staff. The non-controversial nature of their individual subjective opinions allowed to use the average ranks and weights of the criteria as the outcome of the task, which comprehensively measures the significance of the main criteria and sub-criteria.

The study presents a concept and a new approach showing that it is advantageous to measure the quality of each train journey by a single number which is calculated from an additive mathematical model where the weights of the criteria, determined by expert research methods, are multiplied by the number of criteria variables. The variable number represents the actual value of the train journey sub-criterion, which depends on the ratio of the journey quality parameter to its normal, best, or worst value. It ranges from 0 (worst quality) to 1 (best quality). The adequacy of the mathematical model has been verified by conducting a study of the actual values of all sub-criteria of the International Passenger Train (IPT) and by calculating a Comprehensive Quality Index (CQI). This method provides a way to express the quality of the train under study in quantitative terms, when the result is not influenced by the number and quality of other alternatives (trains). The methodology and algorithm for the assessment of the quality of IPTs can be applied in practice to compare the compliance of trains of a rail passenger transport company to the required service quality, as well as to motivate high performing individuals and teams and improve their performance.

Vilnius, Lithuania

Henrikas Sivilevičius
Lijana Maskeliūnaitė

**Competing Interests** The authors have no competing interests to declare that are relevant to the content of this manuscript.

# Introduction

*Broliai seserys, imkit mane ir skaitykit*
*Ir tatai skaitydami permanykit*[1]
*Brothers and sisters, take me and read me*
*And when reading these, understand*
*Martynas Mažvydas*

In any country, the transport system is designed to provide military/defence mobility. It also performs the service function of transporting people (passengers) and freight within its territory and beyond. Passengers are transported by land (road and rail), air and water. It is necessary to take into consideration the fact that transportation is an international business that goes beyond the geographic boundaries of one country, irrespective of its size. It is also important to note that in the course of time individual transport companies and modes of transport taken separately cannot claim status of key players in international transport markets. In the global market, the toughest competition is going on within supply and transportation chains, the so-called transport corridors. Hence, a transport or logistics company, committed to success in international markets, must strive to infiltrate international transport corridors efficiently [1]. Different modes of transport have their advantages and disadvantages and are designed to increase the mobility of the public. Mobility has a major impact on the internal market and on the quality of life of citizens who enjoy the freedom of travel. Quality travel increases the attractiveness and competitiveness of a mode of transport among other modes of transport. The safety and other benefits of rail transport have been demonstrated by United States (US) President Joe Biden's 10 h train journey from Poland to Ukraine capital—Kyiv, on February 20, 2023, though the country is at war with Russia.

Train or rail transportation is favoured in many countries due to its high efficiency, large capacity, comfort, and safety. The numbers of passengers who choose trains as their preferred mode of transportation are regularly growing. For that reason, it

[1] A quote from the first Lithuanian book *Catechism* published in the Lithuanian language in 1547.

is important to improve passenger satisfaction on continuous basis. Ride comfort is a vital metric when evaluating the quality or experience of a passenger on board a train [2].

There are many good examples in some European Union (EU) Member States which demonstrate that rail transport can provide a quality service to satisfy the consumer. The internal market for rail transport services is an area which still faces numerous challenges, and which requires to be accomplished as a matter of priority in establishing a single railway area. To do this, it is necessary to remove the technical, administrative, and legal barriers that prevent the entry into national rail markets. By 2030, the EU expects to triple the total length of the existing high-speed rail tracks and network and to develop a dense rail network in all Member States. By 2050, the development of the European high-speed rail infrastructure network is expected to be finalised with most of the medium-distance passenger transport to be provided by rail.

The quality, accessibility and reliability of transport services will gain increasing importance in the coming years, inter alia due to the ageing of the population and the need to promote public transport. Attractive frequencies, comfort, easy access, reliability of services and inter-modal integration are the main characteristics of service quality. The availability of information over travelling time and routing alternatives is equally relevant to ensure seamless door-to-door mobility for passengers.

The reasons for choosing to travel by train can be divided into four levels depending on the degree of freedom of choice:

- *Zero level*—when a passenger chooses the place (country, city, town, village) where he/she wants to go by train, without being forced to do so and on the route that best suits his/her needs, in terms of timetable, price, safety and other indicators related to the quality of the train journey.
- *Level 1*—when a passenger travels on a route of his/her own choice and on a scheduled train in compliance with temporary restrictions on personal and other behaviour imposed by the national governmental authorities with a view to prevent the spread of viral COVID-19 or other infections, which reduce the quality of the train journey, but are inevitable for the benefit of the public.
- *Level 2*—when a passenger is forced by extraordinary abnormal circumstances or force majeure not attributable to him/her (hurricanes, floods, volcanic consequences, radiation pollution or anthropogenic pollution, acts of war) to leave his/her place of residence, work, vacation or other location, by train and on the route which facilitates, despite the quality parameters of the journey, the evacuation outside the area unsuitable for normal life (e.g. accidents at the Chernobyl and Fukushima nuclear power stations).
- *Level 3*—when people are boarded against their will onto train carriages that are not designed to carry people and transported by the power structures of a foreign country to areas of exile and imprisonment, where they are exposed to forced labour and a life in inhuman conditions, and often to death. For example, the deportations of the Baltic populations to Siberia and other territories in Russia

when the Soviet Union under Stalin occupied the Baltic States in 1940, the deportation of the Crimean Tatars in 1944, and the war in Ukraine launched by Putin's Russia on February 24, 2022. There are no criteria for the quality of this trip.

This work presents a zero-level study of rail travel, when passengers under normal (non-emergency) conditions choose their own itinerary, timing, and train, considering the convenience of a regular timetable, travel expenses, comfort, potential risks, quality of service and other factors.

The study establishes the significance of the required characteristics (criteria) for passenger transport and compares road, rail, and air transport alternatives against each of the ten criteria. The research data generated through the Analytic Hierarchy Process (AHP) method have shown that rail transport is the best alternative for this purpose. The passenger, given the opportunity to choose the most preferable mode of transport (road, rail, or air), evaluates the possible travel alternatives according to criteria that he/she understands. The development of the system of criteria for the advantages of travelling by train compared to travelling by air and by bus, and the analysis of the significance (weights) of these criteria, have resulted in the identification of the most important factors for passengers in their choice to travel by rail. The survey data from a sufficiently large sample of international train passengers were processed by applying popular rank correlation and AHP methods, as well as ARTIW-L and ARTIW-N (Average Rank Transformation into Weight Linear and Non-linear) multiple criteria decision making (MCDM) methods developed by the authors.

Given the reasons why passengers choose to travel by train, an original hierarchical framework of train journey quality factors was presented and fully explored during the second phase of the study. The study has adopted a principle that makes it possible to assess the object (the quality of travel by train) comprehensively through a single number, varying from 0 to 1. The hierarchy is structured around four key criteria that assess the following areas: technical condition of the train elements and track (16 sub-criteria), the organisation and technology of the train travel process (19 sub-criteria), the cost of the train journey (6 sub-criteria), and the safety of the train journey (8 sub-criteria). The significance of the main criteria and the constituent sub-criteria and the consistency of the respondents' opinions were estimated through the Analytic Hierarchy Process (AHP) and the rank correlation methods. The significance of the main criteria and sub-criteria has been assessed by passengers (P), train staff (TS) and administrative staff (AS). The calculation of the subjective normalised overall weights of the sub-criteria has been followed by the assessment of the variable for each sub-criterion, which is calculated from the factual value of the sub-criterion compared to the maximum or minimum allowed value, so that the calculated value of this variable is between 0 (worst quality) and 1 (best quality).

For each of the main criteria, the correctness (adequacy) of the model used to calculate the comprehensive quality index (CQI) has been verified by examining factual data on international passenger train travel. The findings show that the mathematical models provide a reliable quantification of the quality of the train journey and can be applied in the practice of rail transport companies.

The main objective of the monograph is a comprehensive study of the qualitative parameters of the rail passenger transport process through the application of quantitative mathematical methods. MCDM and mathematical statistical methods are the tools used to achieve this objective.

The study is aimed at researchers and strategic decision-makers in the development and quality improvement of rail passenger transport companies. The MCDM methods presented therein can be used by Master degree and Doctoral students in their research.

## References

1. Šakalys A (2013) Transportas ir komunikacija—vienas svarbausių šalies rūpesčių. Transp 2(21):8–9. (in Lithuanian)
2. Jiang Y, Chen BK, Thompson C (2019) A comparison study of ride comfort indices between Sperling's method and EN 12299. Int J Rail Transp 7(4):1–18. https://doi.org/10.1080/23248378.2019.1616329

# Contents

# About the Authors

**Prof. Habil. Dr. Henrikas Sivilevičius** serves as a professor in the Faculty of Transport Engineering at the Vilnius Gediminas Technical University—VILNIUS TECH. His research interests cover mathematical and decision-making methods for improving road and rail passenger transport. Author of one English textbook published by Vilnius: Technika (2000), 12 methodological publications and 11 normative-technical documents in Lithuanian, as well as 264 scientific articles in both English and Lithuanian, Henrikas Sivilevičius is also a member of the Joint Doctoral Committee in the field of Transport Engineering, organized by Vilnius Gediminas Technical University in partnership with Klaipėda University, Vytautas Magnus University and Kaunas University of Technology. He served the editorial board of five scientific journals. He has supervised or acted as scientific advisor for eight doctoral theses. He is currently a senior researcher at the VILNIUS TECH Transport and Logistics Competence Centre and the Civil Engineering Research Centre. The Ministry of Transport and Communications of the Republic of Lithuania has awarded him the Badge of Honorary Road Worker.

**Assoc. Prof. Dr. Lijana Maskeliūnaitė** is a Doctor of Technological Sciences in the field of Transport Engineering and Associate Professor at the Department of Mobile Machinery and Railway Transport, Faculty of Transport Engineering, Vilnius Gediminas Technical University (VILNIUS TECH). She has been working there since 2013. She also worked as a researcher at the VILNIUS TECH Transport and Logistics Competence Centre from 2016 to 2017. In 2013 she defended her Doctoral Thesis in the field of Transport Engineering (Technological Sciences), entitled "The Model for Multicriteria Evaluation of the Quality of Passenger Transport by International Trains". She was awarded the title of Associate Professor of Technological Sciences at VILNIUS TECH in 2020. Her research interests cover the application of multi-criteria decision-making and mathematical methods in transport. Author of one textbook (published in Lithuanian), and twenty-two research articles (published in English), Lijana Maskeliūnaitė has gained over 30 years' experience in the rail passenger transport sector. She has worked as a train manager on Lithuania's first high-speed train "Baltija", and on the international trains "Vilnius–Warsaw", "Vilnius–St. Petersburg" and "Vilnius–Moscow". In 2005, she was awarded a Certificate of Appreciation from the Ministry of Transport and Communications of the Republic of Lithuania for her work and long-term professional excellence in Lithuanian railway transport. In 2008, the Vilnius–Moscow international train team under her leadership was nominated as the "Best Passenger Train Team of the Year 2008".

# Abbreviations and Description

| | |
|---|---|
| A1...A16, B1...B19, C1...C6, D1...D8 | Abbreviation of sub-criterion |
| A, B, C, D | Abbreviation of key criterion |
| AA | Assortment abundance |
| ABAN | Axle-box and axle-neck |
| AFE | Aerosol fire extinguisher |
| AHP | Analytic Hierarchy Process method |
| APE | Automatic powder extinguisher |
| ARAS | Additive Ratio ASsessment |
| ARTIW-L | Average Rank Transformation Into Weight Linear method |
| ARTIW-N | Average Rank Transformation Into Weight Non-Linear method |
| AS | Administration staff |
| AT | Air transport |
| BT | Road transport |
| BTP | British Transport Police |
| BWM | Best-worst method |
| C | Compartment car |
| CILOS | Criterion Impact Loss method |
| CIS | Commonwealth of Independent States |
| $CO_2$ | Carbon Dioxide |
| COPRAS | Complex Proportional Assessment method |
| CORST | COmbining MCDM appRoaches using method Stability coefficient |
| COVID-19 | Coronavirus disease is an infectious disease caused by the SARS-CoV-2 virus |
| CQI | Comprehensive Quality Index |
| CRA | Communications Regulatory Authority |
| CSCC | Delhi State Consumer Commission |
| CVD | Cardiovascular disease |
| DB | Deutsche Bahn |

| | |
|---|---|
| DEA | Data Envelopment Analysis model |
| DOT | Department of Transportation |
| EBC | Emergency brake cock |
| ECF | European Cyclists' Federation |
| EFTA | European Free Trade Association |
| ERTMS | European Rail Traffic Management System |
| ETPS | Expenses of the trip and the provided services |
| EU | European Union |
| FAS | Fire alarm system |
| FCCS | Fire containment and control system |
| FDCU | Fire detectors, a fire detection and control unit |
| FED | Fire extinguishing device |
| FQ | Food (dish) quality |
| FRA | Federal Railroad Administration |
| GDP | Gross Domestic Product |
| GDPC | Gross Domestic Product per Capita |
| GHG | Greenhouse gas |
| GMT | Greenwich Mean Time Zone |
| GSM-R | Global system for mobile communications on railway |
| HABD | Hot axle-box detection system |
| HB | Handbrake |
| HIV | Human Immunodeficiency Virus |
| HSR | High-speed railway |
| HST | High-speed train |
| HVAC | Heating, ventilation and air conditioning system |
| IA | Internet access |
| ICT | Information and communications technology |
| IDOCRIW | Integrated Determination of Objective Criteria Weights method |
| IHAMCI | Inverse Hierarchy for Assessment of Main Criteria Importance method |
| IPT | International passenger train |
| JSC | Joint Stock Company |
| L | Luxury car |
| LTG Link | Railway passenger company of Lithuania |
| MCC | Meal change cycle |
| MCDM | Multiple Criteria Decision Making method |
| MGE | Multilevel Grey Evaluation method |
| ÖBB | Austrian Federal Railways |

| | |
|---|---|
| OECD | Organisation for Economic Co-operation and Development |
| OPS | Overall passenger satisfaction |
| ORR | Office of Rail Regulation |
| P | Passengers |
| PCM | Pairwise comparison matrix |
| PFE | Portable fire extinguisher |
| pkm | Passenger-kilometres |
| PLTS | Probabilistic linguistic term set |
| PM | Price of meals |
| PPE | Personal protective equipment |
| PROMETHEE | Preference ranking organization method for enrichment of evaluations |
| PT | Public transport |
| QLI | Quality level indicator |
| QTT | Quality of travel by train |
| RAILPOL | Railway Police |
| RPCB | Rail Passenger Convenience Benefit |
| RPI | Railway Performance Index |
| RSGDL | Railings, stairs, gangways, doors and locks |
| RT | Rail transport |
| RTQ | Railway trip quality |
| RTV | Rail transportation vehicle |
| S | Sleeping car |
| SAW | Simple Additive Weighting method |
| SEM | Structural equation modelling |
| SFTR | Spring Festival Travel Rush |
| specSE | Thermo-specific self-efficacy |
| SQ | Service quality |
| SWOT | Strengths, Weaknesses, Opportunities, and Threats |
| TEN-T | Trans-European Transport Network |
| TF | Ticket fare |
| TOPSIS | Technique for Order of Performance by Similarity to Ideal Solution method |
| TOWS | Threats, Opportunities, Weaknesses and Strengths |
| TQC | Track quality class |
| TQI | Track quality index |
| TRV | Track recording vehicle |
| TS | Train staff |
| UK | United Kingdom |
| US | United States |
| UTASTAR | An ordinal regression method for building additive value functions |

| | |
|---|---|
| VR | Virtual reality |
| WC | Toilet |
| WiFi | Wireless network |
| WT | Water transport |

## Symbols

| | |
|---|---|
| $\alpha$ | Significance level |
| $\beta$ | Coefficient depending on the number of intermediate stations on route *b* |
| $\gamma$ | Coefficient indicating the degree (level) of "fill-rate" of a train with passengers boarding at intermediate stations $B_e$ on the journey from stations A to C |
| $\Delta_i$ | Absolute error of the significance assessment for *i*-th criterion |
| $\Delta T_{ABAN}$ | Maximum permissible temperature difference (°C) |
| $\Delta T_{maxl}$ | Axle box of the train carriage the greatest temperature difference (°C) |
| $\lambda_{max}$ | Largest eigenvalue |
| $v$ | Degree of freedom |
| $\sigma_g$ | Standard deviations (mm) |
| $\sigma_g^2$ | Variance of track geometric parameters ($mm^2$) |
| $\sigma_{g,st}^2$ | Standard deviations of each of the seven geometric parameters (mm) |
| $\sigma_{Ri}$ | Standard deviation of the ranks $R_{ij}$ of the evaluated *i*-th criterion |
| $\sigma_T$ | Standard deviations of the temperature in carriage (°C) |
| $\sigma_{tot}^2 = \sum_{g=1}^{G} \sigma_{gfs}^2$ | Numerical value is equal to the total variance ($mm^2$) |
| $\chi^2$ | Pearson's chi-square test statistic |
| $\chi_{v,\alpha}^2$ | Critical Pearson's statistic, assuming the degree of freedom and the significant level |
| $x_{Ai}, x_{Bi}, x_{Ci}, x_{Di}$ | Variables of the *i*-th sub-criteria of the *f*-th key criterion used for calculating the real key criterion value |
| $x_{C2PM}, x_{C2FQ}, x_{C2AA}, x_{C2MCC}$ | Variable of the second key criterion C, showing the price of meals, the food quality, the assortment abundance, the meal change cycle |
| $\boldsymbol{\omega}$ | Column vector weights |
| $\omega_b$ | Sum of the global weights $\overline{\omega}_i$ for key criterion |
| $\tilde{\omega}_b$ | Weight of sub-criterion |
| $\tilde{\omega}_g$ | Weight of *g*-th key criterion |

| | |
|---|---|
| $\omega_i$ | Subjective normalised weight |
| $\overline{\omega}_i$ | Global weight of *i*-th criterion |
| $\omega_{i,a}$ | Normalised weight of the *a*-th alternative for the *i*-th criterion |
| $\omega'_{i,a}$ | Idealised weight of alternative *a*-th calculated for criterion *i*-th |
| $\omega_{i,amax}$ | Highest normalised weight of alternative *a*-th calculated for criterion *i*-th |
| $\omega_{iASj}$ | Normalised subjective normalised weight (eigenvector) of the *i*-th sub-criterion for the key criteria A, B, C, or D, respectively, identified by the *j*-th expert (administration staff), through the AHP method |
| $\omega_{iPj}$ | Normalised subjective weight (eigenvector) for the *i*-th sub-criterion of the key criteria A, B, C, or D, respectively, of the *j*-th respondent (passenger), as determined by the AHP method |
| $\omega_{iTSj}$ | Normalised subjective weight (eigenvector) of the *i*-th sub-criterion for the key criteria A, B, C, or D, respectively, identified by the *j*-th expert (train service staff), through the AHP method |
| $\omega_{PM}$, $\omega_{FQ}$, $\omega_{AA}$, $\omega_{MCC}$ | Weight price of meals, the weight of the food (dish) quality, the weight of the assortment abundance, the weight of the meal change cycle |
| $\mathbf{A}$ | Pairwise comparison matrix |
| $a$ | Vertical vibration acceleration of the carriage body ($cm/s^2$) |
| $a_{i,j}$ | Elements of the pairwise comparison matrix |
| $ACF_f$ | Factual assessment of the emergency ambulance call-out to trains option |
| $ACF_{max}$ | Maximum possible score of the emergency ambulance call-out to trains option |
| $AFE_f$ | Factual score of the availability and condition of the aerosol fire extinguishing equipment in carriage |
| $AFE_{max}$ | Maximum possible score of the availability and condition of the aerosol fire extinguishing equipment in each carriage |
| $APE_f$ | Score of factual number and condition of automatic powder extinguishers in carriage |
| $APE_{max}$ | Maximum possible score of number and condition of automatic powder extinguishers in each |
| $ATS_{fp}$ | Actual estimate in points exterior the appearance of train service staff members |
| $ATS_{max}$ | Highest possible estimate |
| $b$ | Number of intermediate stations on the route |

| | |
|---|---|
| $B_e$ | Intermediate stations |
| $BD_f$ | Estimate in points of the berth structure in the *c*-th carriage and the adaptation of a passenger carriage for the disabled passengers |
| $BD_{max}$ | Highest possible estimate of the berth structure and the adaptation of a passenger carriage for the disabled passengers |
| $BL_{fc}$ | Actual estimate in points delivery of bedclothes, their condition and making up the bed for passengers |
| $BL_{max}$ | Highest possible estimate of the considered service |
| $BS_f$ | Means the presence or absence of a special carriage or compartment for carrying bicycles and scooters in the train and the evaluation of transportation conditions in points |
| $BS_{max}$ | Highest estimate of the conditions for carrying bicycles and scooters in the train |
| $BT_f$ | Actual estimate in points possibility of buying a ticket on the train (from the train manager) |
| $BT_{max}$ | Highest possible service estimate |
| $CC_f$ | Estimate in points of the possibility to call the conductor to a passenger compartment of the *c*-th carriage in emergency |
| $CC_{max}$ | Highest possible estimate of the possibility to call the conductor to a passenger compartment in emergency |
| $CE_f$ | Real condition (cleanness, damages) of the *c*-th carriage expressed in points |
| $CE_{max}$ | Highest estimate value of the carriage in points |
| $C.I.$ | Consistency index |
| $CI_{f,c}$ | Real estimate of the *c*-th carriage interior in points |
| $CI_{max,p}$ | Maximum possible carriage interior estimate value in points |
| $COFG_{fp}$ | Actual estimate in points competence, objectivity and communication skills of customs officers and frontier guards |
| $COFG_{max}$ | Highest possible sub-criterion estimate |
| $CPO_f$ | Factual score of the possibility of calling the police in passenger train |
| $CPO_{max}$ | Maximum possible score of possibility of calling the police in train |
| $C.R.$ | Consistency ratio |
| $D$ | Rate of ordering (serving) the dishes of all categories offered in the dining car, i.e. the total number of all served dishes |

| | |
|---|---|
| $D_d$ | Rate of the *d*-th dish ordering, i.e. the number of the served (sold) dishes of this kind |
| $DC_f$ | Estimate of the dining car availability and the provided services in points |
| $DC_{max}$ | Highest possible estimate of the dining car availability and the provided services |
| $DCR_f$ | Actual estimate in points possibility to reserve a table in the dining carriage |
| $DCR_{max}$ | Highest possible service estimate |
| $DM_f$ | Highest possible estimate of the delivery of meals included in the ticket price |
| $DM_{max}$ | Denotes the real estimate (in points) of the delivery of meals (or its absence), included (not included) in the ticket price of the passengers travelling in the first-class double compartment |
| $E$ | Number of respondents and expert categories |
| $EA_f$ | Estimate of the possibility for passengers to use electrical appliances (hairdryer and an iron) on the train expressed in points |
| $EA_{max}$ | Highest estimate of the possibility for passengers to use a electrical appliances (hairdryer and an iron) on the train |
| $EBC_f$ | Factual score condition and action of emergency brake cock in the carriage |
| $EBC_{max}$ | Maximum possible score of the condition and action of emergency brake cock lever in carriage |
| $EE_f$ | Factual score of emergency exits in carriage |
| $EE_{max}$ | Maximum possible score of emergency exits in carriage |
| $F$ | Number of key criteria describing the quality (characteristics) of the considered object |
| $F(f_v)$ | Modifying coefficient (frequency weighting function) of the vibration frequency that expresses human vibration sensitivity |
| $f_v$ | Vertical vibration frequency of the carriage body (Hz) |
| $FAK_f$ | The factual assessment of the first aid kits on board the train |
| $FAK_{max}$ | Maximum possible score given to the assessment of the availability of first aid kits on-board the train |
| $FAS_f$ | Score of the factual availability and condition of the automatic fire alarm system in carriage |
| $FAS_{max}$ | Highest possible score of the availability and condition of the automatic fire alarm system in the carriage |

| | |
|---|---|
| $FD_f$ | Actual estimate in points possibility of ordering food and drinks from the dining car to be delivered to the compartment (via car attendant) |
| $FD_{max}$ | Highest possible estimate of the considered service in points |
| $FED_f$, $FED_{max}$ | Actual and maximum scores of the types of fair extinguishing devices in the carriage |
| $FLTS_{fa}$ | Actual estimate in points knowledge of foreign languages by the train service staff |
| $FLTS_{max}$ | Highest possible estimate |
| $FQ_0, FQ_1, \ldots, FQ_4$ | Denote the estimate of a food expressed in points |
| $G$ | Number of geometric parameters in the Track Quality Index |
| $g$ | Number of key criteria describing the research object |
| $GV_f$ | Actual service estimate in points possibility to get a visa at the frontier post |
| $GV_{max}$ | Highest possible service estimate |
| $HB_f$ | Factual score condition and action of handbrake in carriage |
| $HB_{max}$ | Maximum possible score of the condition and action of handbrake |
| $HIP_f$ | Actual estimate in points possibility to get a foreign health insurance policy on the train |
| $HIP_{max}$ | Highest possible service estimate |
| $IA_f$ | Actual estimate in points possibility to access the internet |
| $IA_{max}$ | Highest possible service estimate |
| $K$ | Value of the train journey quality level indicator |
| $k$ | Number of sub-criteria in the criterion |
| $K_A$ | Comprehensive quality index related to the train elements and the technical state of railway track |
| $K_B$ | Comprehensive quality index related to the railway trip planning and technology |
| $K_C$ | Comprehensive quality index related to the expenses of a trip by train and the provided services |
| $k_c$ | Consistency coefficient |
| $K_D$ | Comprehensive quality index related to the internal safety of railway trip |
| $L$ | Total length of the railway track (km) |
| $l_l$ | Denotes the lengths of the sections of the track (km) |
| $m$ | Number of criteria or sub-criteria |
| $m_0, m_1, \ldots, m_4$ | Denote the number of points assigned by the passengers, evaluating the food |

| | |
|---|---|
| $MCC_f$ | Denotes the real estimate in points awarded to the meal change cycle |
| $MCC_{max}$ | Highest estimate given to the cycle |
| $N$ | Number of comparable alternatives |
| $n$ | Number respondents or experts |
| $N_{A+C}$ | Number of passengers who boarded at the departure station A and travelled to the destination station C |
| $n_{AAf}$ | Factual number of dishes offered in the dining car during the investigated trip |
| $n_{AAmax}$ | Maximum number of dishes ever offered in the dining car |
| $n_{AS}$ | Number of administration staff |
| $n_b$ | Number of the types of braking measures in the carriage of the train |
| $n_C$ | Number of the categories of dishes served in the dining car or number of the second-class 4-berth compartment cars |
| $n_{car}$ | Number of carriages in the train |
| $n_e$ | Number of assessors (respondents or experts) who have measured the significance of the main criteria |
| $n_{ex}$ | Number of fire extinguisher types in the carriage |
| $n_f$ | Number of sub-criteria forming the $f$-th key criterion |
| $N_{km}$ | Number of kilometre-long sections of railway track |
| $n_L$ | Number of the first-class luxury two-berth cars |
| $n_{min}$ | Minimum number of respondents or experts |
| $n_P$ | Number of passengers |
| $n_{ps}$ | Number of the sub-criteria describing services |
| $n_S$ | Number of ordinary second-class sleepers |
| $n_{stage}$ | Number of railway track sections between stations |
| $n_T$ | Number of values of $t$-measurements of the carriage interior temperature |
| $N_{total}$ | Total number of passengers on the train |
| $n_{TS}$ | Number of train service staff |
| $n_v$ | Number of visas required for the trip |
| $n_y$ | Number of cars of $y$-th category in the train |
| $NL_{f,c}$ | Factual audio pressure level in the $c$-th carriage (dBA) |
| $NL_{max,\ a}$ | Maximum permissible noise level in the $c$-th carriage (dBA) |
| $NL_{min}$ | Minimal audio pressure level (dBA) |
| $NM_f$ | Actual service estimate in points distribution of popular newspaper and magazines on the train |
| $NM_{max}$ | Highest possible service estimate |
| $num$ | Numerator |

| | |
|---|---|
| $PC_f$ | Actual estimate in points possibility of setting for onboard services by payment card |
| $PC_{max}$ | Highest possible service estimate |
| $PFE_f$ | Factual score of number and condition of portable fire extinguishers in carriage |
| $PFE_{max}$ | Score of maximum necessary number and condition of portable fire extinguishers in each carriage |
| $PH_f$, $PH_{min}$, $PH_{max}$ | Denote the real, the lowest and the highest price of health insurance abroad (€) |
| $PL_f$ | Actual estimate in points protection of passenger luggage and other valuable things on the train |
| $PL_{max}$ | Highest possible service estimate |
| $PM_{average}$ | Average price of the dish (€) |
| $PM_d$ | Price of the dish of $d$-th category (€) |
| $PM_{min}$ | Lowest price of the dish served in the dining car of the train (€) |
| $PNM_f$ | Shows the estimate of the criterion describing the price of newspapers and magazines offered in the train in points |
| $PNM_{max}$ | Highest possible estimate given to the price of newspapers and magazines in the train |
| $PP_{fc}$ | Actual evaluation in points of delivery rations provision for passengers in the $c$-th car |
| $PP_{max}$ | Maximum number of points given to rations provision |
| $PPE_f$ | Score of factual supplying carriage $c$ with manufactured personal protective equipment and its condition |
| $PPE_{max}$ | Maximum possible score of supplying each carriage $c$ with manufactured personal protective equipment and its condition |
| $Q_a^{DM}$ | Global weight of the $a$-th alternative is calculated using the distribute mode of AHP method |
| $Q_a^{IM}$ | Global weight of the $a$-th alternative calculated by the ideal mode of AHP method |
| $Q_{fi}$ | Mean local weights of the $i$-th sub-criteria for key criterion $f$-th |
| $\overline{Q}_{fi}$ | Normalised total (local) weight of the $i$-th sub-criterion of the key criterion $f$-th |
| r | Coefficient of correlation |
| $R^2$ | Coefficient of determination |
| $\overline{R}_f$ | Average rank of the $f$-th key criterion |
| $\overline{R}_i$ | Average $i$-th criterion rank |
| $\left\|\overline{R}_{1-2}\right\|_i$ | Modulus of differences (or the absolute value of the ranks), assigned to the key criteria (A, B, C or D) or |

| | |
|---|---|
| | to $i$-th sub-criteria by the respondents or by experts and experts |
| $R_{ij}$ | Rank given to the $i$-th criterion of the $j$-th expert |
| $\overline{R}_{ip}$ | Average of the population set |
| $RB_f$ | Estimate in points of the operation of the radio broadcasting unit in the $c$-th carriage and its centralized on/off switching |
| $RB_{max}$ | Highest possible estimate of the radio broadcasting unit operation and centralized on/off switching |
| $R.I.$ | Random index |
| $RM_{fp}$ | Actual estimate in points music broadcast and information in conformity with passenger requests |
| $RM_{max}$ | Highest possible service estimate |
| $RSGDL_f$ | Factual score of railings, stairs, gangways, doors and locks in carriage |
| $RSGDL_{max}$ | Maximum score of railings, stairs, gangways, doors and locks in carriage |
| $S$ | Sum of the squares of the sum of $i$-th criterion rank $R_{ij}$from the overall mean $\overline{R}$ |
| $S_{max}$ | Maximum possible value |
| $SE_f$ | Estimate of shower equipment availability and its work hours in the $c$-th carriage in points |
| $SE_{max}$ | Highest possible estimate of shower equipment availability and its work hours |
| $SP_f$ | Estimate of smoking places in the train in points |
| $SP_{max}$ | Highest possible estimate of smoking places in the train |
| $SS_f$ | Actual estimate in points sale of souvenirs on the train |
| $SS_{max}$ | Highest possible service estimate |
| $t$ | Value of $t$ distribution, which depends on the probability taken to assess the importance of the criterion in deciding to go by train as an alternative to the bus |
| $T_{ABAN\,f}$ | Factual temperature of the measured axle-box and axle-neck (°C) |
| $T_{ABAN\,f\,max}$ | Maximum measured real temperature of all train axle-boxes and axle-necks (°C) |
| $T_{air}$ | Temperature of environment (ambient) air on the measured site (station) (°C) |
| $t_{arr,B}$ | Train arrival delay time at intermediate station $B_e$ (min) |
| $t_{arr.C}$ | Train arrival delay time at the destination station C (min) |

| | |
|---|---|
| $t_{arr.p} = t_{dep.p}$ | Tolerance of delay time arriving at or departing from the departure, destination or intermediate station (5 min) |
| $\overline{T}_{f,c}$ | Average temperature in the $c$-th carriage measured $n_T$ times at a height of 1.10 m above the floor of the carriage during the trip (°C) |
| $t_{dep.A}$ | Train departure delay time from station A (min) |
| $t_{dep.B}$ | Train departure delay time from intermediate station $B_e$ (min) |
| $T_{ft}$ | Factual temperature obtained in the $t$-th measurement (°C) |
| $T_L$ | Lowest permissible carriage temperature (°C) |
| $T_{lim}$ | Maximum permissible axle-box and axle-neck temperature limits (°C) |
| $T_{LS}$ | Lowest permissible temperature value in summer (°C) |
| $T_{LW}$ | Lowest permissible temperature value in winter (°C) |
| $T_o$ | Optimum temperature of the carriage compartment (interior) (°C) |
| $T_{OS}$ | Temperature of the carriage compartment (interior) in the warm season (°C) |
| $T_{OW}$ | Temperature of the carriage compartment (interior) in the cold season (°C) |
| $T_U$ | Maximum permissible carriage temperature (°C) |
| $T_{US}$ | Highest permissible temperature value in summer inside a passenger carriage (°C) |
| $T_{UW}$ | Highest permissible temperature value in winter inside a passenger carriage (°C) |
| $TCF_f$ | Actual estimate in points possibility of ordering a taxi |
| $TCF_{max}$ | Means its highest possible service estimate |
| $TF_f$, $TF_{min}$, $TF_{max}$ | Factual, the lowest and the highest ticket fare, respectively, for a car of $y$-th category |
| $TF_{fC}$, $TF_{minC}$, $TF_{maxC}$ | Regular, the lowest and the highest fare in the car of C category (€) |
| $TF_{fL}$, $TF_{minL}$, $TF_{maxL}$ | Factual, the lowest and the highest fare in the car of L category (€) |
| $\underline{TF_{fS}}$, $\underline{TF_{minS}}$, $\underline{TF_{maxS}}$ | Regular, the lowest and the highest fare in the car of S category (€) |
| $\overline{TQI}_f$ | Average track quality index of evaluated railway track the length of which is $N_{km}$ (mm$^2$) |
| $TQI_{fmax}$ | Largest value for track length (mm$^2$) |
| $TQI_{fmin}$ | Smallest value obtained for 1 km track length (mm$^2$) |

| | |
|---|---|
| $TQI_{fs}$ | Quality index of *s*-th one-kilometre track length, the numerical value of which is equal to the total variance $\sigma^2_{gfs}$ of the particular *g*-th geometric parameters, ($mm^2$) |
| $u_i$ | Normalized value of each criterion |
| $v_{f,l}$ | Real (measured) and the maximum permissible speed difference of a train at the *l*-th section of the track (km/h) |
| $v_{max}$ | Highest permissible train speed on the route (railway track) (km/h) |
| $v_{max,a}$ | Highest permissible speed on all railway track sections (km/h) |
| $v_{max,la}$ | Highest permissible speed of a train at the *l*-th section of the track (km/h) |
| $VCL_f$ | Estimate (in points) of the operation of the *c*-th carriage systems |
| $VCL_{max}$ | Highest possible estimate of the systems operating in the carriage |
| $VF_f$, $VF_{min}$, $VF_{max}$ | Denote the real, the lowest and the highest prices of *v*-th visa's fees |
| $W$ | Concordance coefficient |
| $W_{min}$ | Smallest value of the concordance coefficient |
| $W_s$ | Sperling index |
| $WC_f$ | Estimate of the real condition of lavatories in the *c*-th carriage, expressed in points |
| $WC_{max}$ | Highest possible estimate (in points) for the *c*-th carriage |
| $\overline{Z}_A$ | Mean values of normalized subjective weight of key criterion A |
| $Z_{Ae}$ | Weight given by the respondents or experts of the *e*-th category to key criterion A |
| $\overline{Z}_B$ | Mean values of normalized subjective weight of key criterion B |
| $Z_{Be}$ | Weight given by the respondents or experts of the *e*-th category to key criterion B |
| $\overline{Z}_C$ | Mean values of normalized subjective weight of key criterion C |
| $Z_{Ce}$ | Weight given by the respondents or experts of the *e*-th category to key criterion C |
| $\overline{Z}_D$ | Mean values of normalized subjective weight of key criterion D |
| $Z_{De}$ | Weight given by the respondents or experts of the *e*-th category to key criterion D |
| $Z_f$ | Significance indicator of criterion |

$\overline{Z}_f^{\star}$ Subjective normalised weight for $f$-th the key criterion, obtained by interviewing train passengers, train service staff and administrative staff

# Chapter 1
# Impact of the Rail Passenger Traffic System on Mobility

*When passengers start thinking about the train driver, it means he has made a mistake. "Hey, you're not hauling firewood!" If all is well, the train driver is as good as gone.*
Oleg Divov, Maks Rubljov "Ne prislonjat'sja"

Over the last 30 years, the European Union (EU) and its Member States have worked together to develop a common rail transport policy. EU rail policy contributes to the broader policy objectives for the transport sector as a whole - ensuring sustainable, safe, and reliable mobility. In the rail sector, EU policy is focused on the core objective of creating a single EU rail network system that will enable the rail sector to grow with reference to competition, technical harmonisation, and the common development of cross-border connections. On October 17, 2016, the European Council approved new rules to encourage more rail travel in Europe. The reform is to open and expand national rail passenger markets in the Member States and create a fair playing field for railway companies by ensuring non-discriminatory access to the network. The reform of EU rail passenger rights is to enforce the rights of all passengers, including passengers with disabilities or reduced mobility.

Rail transport is the most sustainable mode of transport. According to the European Environment Agency, in 2017 the rail sector accounted for just 0.5% of the total EU greenhouse gas emissions. In the EU, rail accounts for 6.9% of total passenger mobility. Consumer satisfaction with rail transport across the EU is high. In 2017, it accounted on average for 74.9%. In 2021, the European Council adopted conclusions on the need to further develop rail transport in the European Union and to invest in the interoperability and better connectivity of national systems [1].



H. Sivilevičius and L. Maskeliūnaitė, *Assessment of the Quality of Passenger Transportation by Train Using Multiple Criteria Decision Making Methods*, Lecture Notes in Intelligent Transportation and Infrastructure,
https://doi.org/10.1007/978-3-031-85877-2_1

## 1.1 Regulations Governing the Rail Passenger Traffic Process

The rail passenger traffic process is governed by the appropriate documentation: EU Directives, regulations, codes, provisions, standards, rules, and instructions. A schematic diagram of the documents regulating the passenger transport process in the Republic of Lithuania is shown in Fig. 1.1. The main EU regulatory instruments governing the rail passenger traffic process are the EU Directives and the Regulation of the European Parliament and the Council. *Directive (EU) 2016/797 of the European Parliament and of the Council on the interoperability of the rail system within the EU* [2] lays down the conditions that must be met to achieve the interoperability of the EU rail system, with a view to achieving an optimal level of technical harmonisation. The document also focuses on facilitation, improvement, and development of rail transport services within the Union and with third countries, as well as contributes to the creation of a single European railway area and internal market. These conditions relate to the design, construction, commissioning, upgrading, renewal, and maintenance of parts of the system, and to the professional qualifications and health and safety conditions applicable to the personnel who operate and maintain it.

*Directive (EU) 2016/798 of the European Parliament and of the Council on railway safety* [3] lays down provisions to ensure the development and improvement of the safety of the EU railway system and to improve access to the market for rail transport services. The purpose of *Regulation (EC) No 1370/2007 of the European Parliament and of the Council on public passenger transport services by rail and by road and repealing Council Regulations* [4] is to set out how the competent authorities may act in the field of public passenger transport, in accordance with the provisions of Community law, with a view to securing the provision of a service in the general economic interest in a way that is more extensive, safer, of higher quality and at lower cost than in the case of a service provided under normal market conditions. *Regulation (EC) No 1371/2007 of the European Parliament and of the Council on rail passengers' rights and obligations* [5] lays down rules on:

(a) the information to be provided by railway companies, the conclusion of traffic contracts, the issue of tickets and the implementation of a computerised rail transport information and reservation system.
(b) the liability of railway companies and their insurance obligations in respect of train passengers and their luggage.
(c) the obligations of railway companies to train passengers in the event of delays.
(d) protection and assistance for disabled persons and persons with reduced mobility travelling by rail.
(e) setting and monitoring service quality standards, management of risks to passengers' personal security, handling of complaints.
(f) general rules for implementation.

*The Railway Transport Code of the Republic of Lithuania* establishes the terms and conditions of railway transport activities as well as the requirements for railway

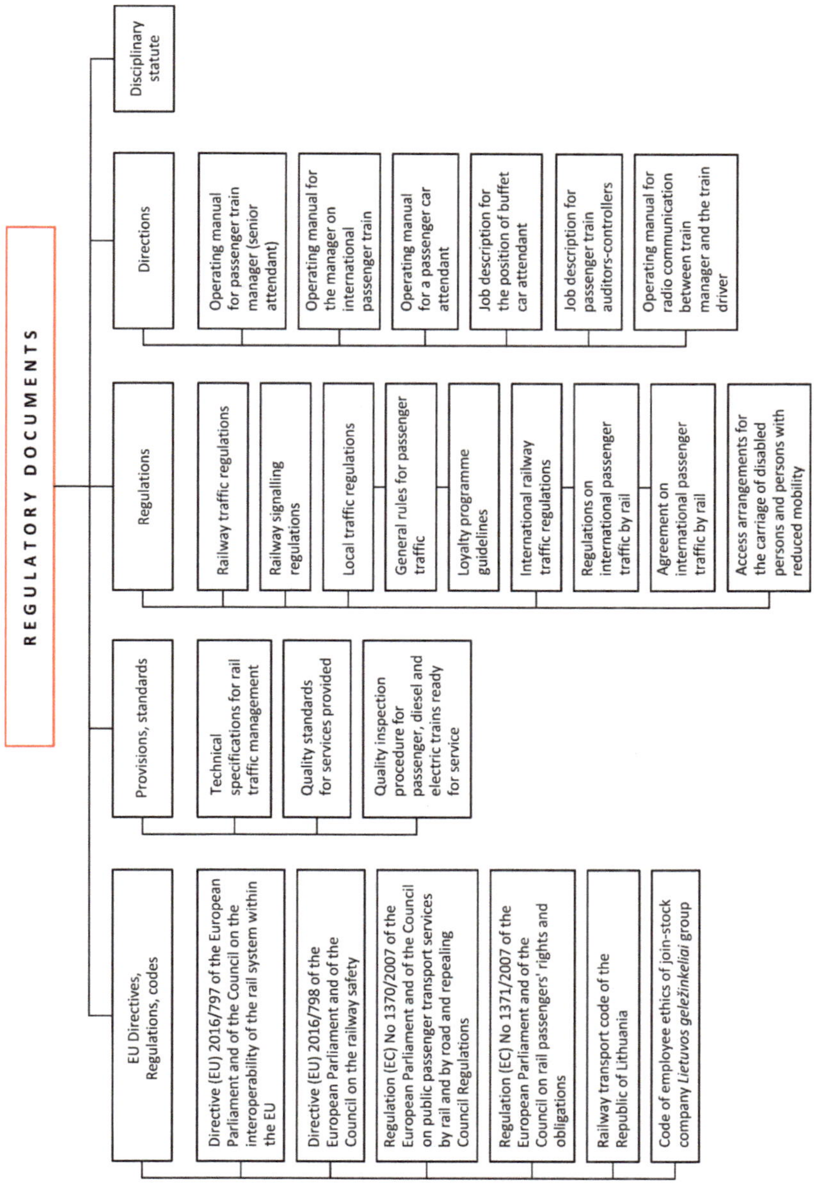


**Fig. 1.1** Schematic diagram of documents regulating the rail passenger transport process in the Republic of Lithuania

transport employees. It also regulates the relations arising from the transportation of passengers, luggage and/or cargo by railway transport in the Republic of Lithuania [6]. *The Code of Employees Ethics of Joint-stock Company Lietuvos geležinkeliai Group* establishes the basic principles of employee conduct within the company, which are aimed at ensuring honesty, transparency, and reliability. The company expects ethical behaviour from its employees in the performance of their duties. Outside working hours, the employees of *Lithuanian Railways* are to avoid situations in which their unethical behaviour could be associated with the company and could damage the reputation of the *Lithuanian Railways* group of companies. The *Code of Conduct* applies to the employees of the *Lithuanian Railways* group of companies and to members of the collegiate management bodies, trainees or interns. Employees are expected to adhere to the principles of the *Code of Conduct* in their communication with each other, as well as with clients, contractors, partners, employees of state and municipal authorities, organisations, the media, etc. [7].

*The Technical Specifications for rail traffic management* lay down the operating framework for 1520 mm gauge railways, the dimensions, requirements, maintenance standards for the main facilities, equipment and rolling stock, as well as the principles for the organisation of train traffic and signalling [8]. The quality standards for services provided by joint-stock company *Lietuvos geležinkeliai UAB LTG Link* have been prepared in accordance with *Regulation (EC) No 1371/2007 of the European Parliament and of the Council on rail passengers' rights and obligations dated October 23* [5], and the *European Standard EN 13816:2002 of the European Committee for Standardisation* (CEN/TC 320), which was adopted by the Lithuanian Standards Department by way of approval as a Lithuanian standard and which has been in force since December 15, 2002 [9]:

1. Tickets and information.
2. On-time performance of international services and the basic principles of passenger service in the event of service disruption.
3. Cancellation requirements for international transport services.
4. Cleanliness of rolling stock and station premises.
5. Regular customer surveys on service quality.
6. Processing of complaints, payments, and compensation for non-compliance with quality standards.
7. Assistance for people with disabilities and reduced mobility.

The rolling stock cleaning process is organised and carried out in accordance with the *Quality Inspection Procedure for Passenger, Diesel, and Electric Trains Ready for Service*. This procedure establishes technological requirements and addresses the preparation of the train before it is diverted to the departure track to pick up passengers. It aims to ensure their safe transportation and a civilised service during the train journey, and the design, layout, installation and maintenance of passenger train interior audiovisual information measures (rules for the transportation of passengers and luggage by rail, information signs, information posters, diagrams) [10].

*The Railway Traffic Regulations* lay down the principles of train traffic management, the procedure for the admission, release and shunting of trains at a station.

This procedure also specifies the actions to be taken by rail transport staff in the event of an emergency or abnormal situation and in the event of a malfunction of the traffic control equipment system [11]. *The Railway Signalling Regulations* lay down the signalling system for rail transport, specifying the purpose of signalling installations, their types, the meaning of signals and the procedure for their application [12]. The terms and conditions of the contractual relationship between the carrier and the passenger when the carrier carries passengers, hand luggage, bicycles and animals on local transport routes are laid down in the *General Rules for the Carriage of Passengers.* These rules make up an integral part of the contract of carriage and are issued in accordance with the current versions of the following legislation [13]:

- Regulation (EC) No. 1371/2007 of the European Parliament and of the Council of October 23, 2007, on rail passengers' rights and obligations.
- Civil Code of the Republic of Lithuania.
- The Law on Transport Concessions of the Republic of Lithuania.
- Other legislation.

The terms and conditions for participation in the special loyalty programme run by UAB "LTG Link" (the railway passenger company) are determined by the *Loyalty Programme Guidelines.* They must be adhered to by all *loyalty programme* members. *Loyalty programme*—is a special programme run by the Company which entitles the holder of a loyalty card to receive a discount on train travel on local routes at a rate set by the Company. A *loyalty programme* member is entitled to receive other discounts offered by the Company and can also take advantage of the Company special offers, such as: "Business loyalty", "Family card", "Fixed-term e-ticket", "Group ticket", "Single ticket", "Birthday discount", "5% loyalty programme", "Electronic teacher's certificate", "Discount coupon", "Student", "Eurail" [15].

On international routes, the *Rules for Passengers on International Trains* apply. They apply to passengers travelling on international trains of the Commonwealth of Independent States (CIS), the Republic of Latvia, the Republic of Lithuania, and the Republic of Estonia. They set out the procedure, rights, and responsibilities for the carriage of passengers, luggage, and load luggage in international rail traffic [16]. *The Agreement on International Passenger Traffic by Rail (SMPS – Soglashenie o mezhdunarodnom passazhirskom soobshhenii),* which considers the peculiarities of the application of the individual provisions in the CIS countries, Georgia, the Republic of Latvia, the Republic of Lithuania, and the Republic of Estonia, establishes the procedure for the carriage of passengers, luggage, and load luggage in international rail traffic [17]. Special terms and conditions for the carriage of disabled persons or persons with reduced mobility on the territory of Lithuania and on international routes are laid down in the *Access Arrangements for the Carriage of Disabled Persons and Persons with Reduced Mobility.* The purpose of the rules is to ensure equal and non-discriminatory access to rail travel for people with mobility and other disabilities, whether on grounds of age or for any other reason. Particular attention is paid to the provision of information on the accessibility of rail services, access to trains and their facilities. *Lithuanian Railways* and the *Lithuanian Disability Forum,* which acts in the interests of people with disabilities in Lithuania, were involved in the process of development, harmonisation, and approval of Access Rules [18].

The instructions for train crews set out crew duties, the proper performance of which ensures the safe operation of the train and a good passenger service. *The Position Description for Passenger Train Manager (Senior Attendant)* sets out the duties of the passenger train manager before and during the train journey, at the return and shunting points. The description also sets out the actions, rights, and responsibilities of the train manager in the event of an accident and in other emergency situations, at border and customs control stations [19]. *The Position Description of the Manager on the International Passenger Train* is the main document regulating the duties of a train manager in the railways of the CIS countries, the Republic of Latvia, the Republic of Lithuania, the Republic of Estonia and other countries the train passes through [20]. *The Position Description for Passenger Car Attendant* sets out the duties of the passenger car attendant before and during the train journey, at the return and shunting points. The description also specifies a car attendant's actions in emergency and other exceptional cases, at border and customs control stations. The description outlines the activities, rules of conduct, rights and responsibilities of a passenger car attendant in international rail traffic [21]. *The Post Regulations for the Position of a Buffet Car Attendant* of the Passenger Transport Directorate Catering Subdivision at *Lithuanian Railways* specify the duties and functions, rights, responsibilities, and relations of buffet car attendants [22]. *The Job Description for the Position of Auditor-Controller on Passenger Trains* specify the duties of auditors-controllers in the inspection of passengers, buffet cars and baggage cars, the procedure for documenting the inspection outcomes, and the actions they take before and after the train journey [23]. *The Instruction for the Use of Radio Communication Between the Manager of the Train and the Train Driver* lays down the procedure for the installation of radio communication between the head manager of the passenger train and the locomotive operator, the rules for its use and the regulation of conversations [24]. Radio communication is used for formal communication between the train manager and the train driver to improve traffic safety, passenger service culture and quality.

Rail transport is one of the most essential parts of the transport system of the Republic of Lithuania. For rail transport to work in harmony, for passengers and freight to reach their destination on time, and for rail traffic safety to be ensured, rail transport employees are required to work honestly and with integrity, to exercise work discipline, to carry out the orders of their superiors in a timely and punctual manner, to adhere to the requirements of technological and industrial discipline, to comply with the rules of safety at work, and to refrain from leaving the place of work until they are succeeded by other employees.

*The Disciplinary Statute for Employees of Lithuanian Railways* establishes the particularities of the employment contract and disciplinary liability in prevention the activities that may lead to disruptions in the work of the company in connection with large-scale accidents and their serious consequences for people and the natural environment. The statute also provides for the procedure of incentives for employees of *Lithuanian Railways* [25].

Currently, in the Lithuania most of the job descriptions are not applicable. For example, the positions of train manager, car attendant, buffet car attendant and auditor-controller are no longer available. Local trains are serviced by travel attendants. The performance of duties of a passenger and train crew is monitored by the Carrier's authorised staff. They must have special certificates or written instructions (for one-off checks) [14].

An overview of the principal documents regulating rail passenger transport in the Republic of Lithuania provides that the focus is on the rights, duties and responsibilities of passengers and carriers, the principles, and rules of conduct of employees, and the rules governing the organisation, management, and safety of train traffic.

## 1.2 Quality Indicators in Rail Passenger Traffic

The quality of rail passenger traffic is a set of characteristics of the rail traffic process based on its potential to meet passenger needs. Passenger rail traffic performance is measured by the Railway Performance Index (RPI) across three dimensions [26]:

- Traffic volume which indicates how many passengers travel by rail.
- Quality of service which indicates whether trains are fast and running on time and whether rail travel is affordable?
- Safety which indicates whether the rail system meets the highest safety standards?

European railways are used by passengers to get to work, to school, to leisure or holiday destinations. The annual rail passenger traffic in the EU is 9.5 billion. 4 out of 5 Europeans (80%) are train users. Occasional travellers account for 67% and those who travel frequently, i.e., at least once a week, account for 13%. More than 90% of European passenger trains run without delays. Rail traffic provides energy-efficient mobility and lower emissions. In the EU, railways account for only 2% of total energy consumption in the entire transport sector [27].

Quality is a much-used term when assessing the services offered. It is usually measured in terms of its relevance to the interests of the parties concerned. However, there is no definitive definition of the quality of rail traffic timetables, as the parameters for assessing it are not yet established clearly. High-quality timetables, which are crucial for improving railway performance, must ensure the sometimes-conflicting interests of different parties concerned, including government authorities, infrastructure managers, railway undertakings, shippers, and passengers. The expectations of the parties involved are analysed to determine the quality of the timetables, and their evaluation is discussed in the timetable development [28] Infrastructure managers focus on timetable implementation and capacity, while carriers and passengers are more concerned about convenient journey times. The aim of the study [29] was to develop and test the design of a methodological procedure to assess the quality of transport connectivity on a given transport network. Its practical application consists in comparing available and recommended timetables. The quality of passenger traffic is one of the key factors from the perspective of the public. The performance of rail systems in Europe has been assessed with a focus on rail passenger transport [30]. It is recommended to look at passenger charging services on international trains [31].

Long-distance railway transport is more energy efficient and generates fewer environmental impacts than air and automobile travel. One strategy for increasing rail use is to introduce high-speed rail systems and other railway improvements. However, even with these capital improvements, rail travel may take too long for certain trips. Therefore, another strategy is to increase the use of rail for at least part of a trip (e.g., by replacing a short flight), which could be done by improving intermodal connections (e.g., flight-to-train connections at airports) or by using rail on the outbound trip and air on the return trip (or vice versa). Sauter-Servaes and Nash [32] research found that an effective internet-based multimodal transport information and reservation system is needed to successfully increase rail travel. This system should provide accurate and complete information about all travel alternatives in a simple package and should enable customers to purchase a single ticket for the entire trip (including local public transport).

The results of a passenger survey at the Joydebpur railway station in Bangladesh are provided [33]. The responses of 1000 rail passengers on 24 selected services and 5 demographic parameters were analysed. Critical consideration from the perspective of a developing country revealed six unique factors, including level crossing facility, illegal establishments, illegal shops, floating people, arrival performance and departure performance, which have never been explored in any previous studies. The regression analysis has revealed that 13 service quality factors significantly affected the level of passenger satisfaction, in particular the ones in relation to food and beverages, road connectivity, sanitation, and departure facilities at the station. Of the five demographic factors, passenger age, occupation and frequency of travel contributed significantly to the overall passenger satisfaction (OPS). The study findings have identified a need for policy makers to focus on senior citizens who are financially well-off and frequent travellers. Priority should be given to lounges, road connectivity, sanitation, and departure area facilities.

One of the quality indicators is the on-time running of trains. The impact of winter climate on high-speed train delays is assessed [34]. In the period of 25 years between 1994 and the end of 2019, there has been a significant increase in rail travel in the UK. This has resulted in increasing train delays and dissatisfied passengers. The focus is on quantitative approaches [35]. The aim is to quantitatively assess passengers' views towards rail service disruptions [36]. An attempt is made to identify the situations that cause the greatest dissatisfaction. The study has shown that passengers reacted negatively to more than 30 min of delay. Dissatisfaction increased when passengers were made to stand during the journey and/or received insufficient information, or when trains were cancelled. Train operators should therefore: (1) cancel trains only in the most extreme cases; (2) give priority to trains that are up to 10 min late; (3) minimise delays for trains carrying large numbers of standing passengers; (4) improve the quality of information and the mechanisms for providing it to the greatest extent possible.

Rail constitutes a vital part of the transport system and transport is a vital function of any modern economy. Dynamic pricing is a modern instrument for rail, airline, and bus companies. It aims to increase revenue through timely passenger monitoring and consistent ticket fare adjustments. The passenger decides when to buy a ticket to save money. This makes it very difficult to predict ticket fares. This issue is generally addressed through the analysis of previous ticket fares and selected train/airplane/bus departure options (e.g., number of days before departure and which day of the week the departure takes place) as predictor variables. The aim of the study [37] is to investigate whether open data from a search request can improve the quality of ticket fare prediction for models based on statistical and artificial neural networks. Rail Passenger Convenience Benefits (RPCBs) can be assumed as a special generalised fare. The results of a questionnaire survey [38] conducted on two trains in China show that there is a linear relationship between RPCBs and the fare, with a correlation coefficient of 0.9616. The aim [39] is to develop a dynamic rail passenger ticket pricing model that would maximise revenue for the Railway Operation Department. A seating management strategy is put forward. It tests whether dynamic pricing and optimal seat selection method can be used to maximise profits. With the view to ticket fares and quality of service, the EU aims to explore the provision of services to rail passengers and to assess how this has been influenced by national policies across different Member States [40].

The EU rail system has one of the highest safety levels in the world. When compared to the multimodal EU transport system, rail makes the safest mode of land transport. Analysis of fatal accidents reveals that safety performance has improved steadily since 1990. The number of accidents has fallen by more than 5% per year on average. Major accidents with 5 or more fatalities are rare, with only 2 such accidents in the last two years. While safety levels have improved in recent years, the pace of improvement has slowed down, and in 2016–2017 was almost at a standstill. Behind the overall positive trends, there are problems that require the attention of railway managers and politicians. Despite the reduction in major accidents since 2010, collisions, derailments and fires on rolling stock still occur. Progress in the EU Member States also varies considerably, although safety levels remain high. Since 2014, the number of fatalities of railway employees has remained high. Around 60 rail employees are seriously injured each year. While the number of passenger fatalities (casualties per volume of traffic) has been decreasing since 2014, this trend is less pronounced for the number of rail employees killed. Over the last decade, due to the introduction of cutting-edge systems, safety at level crossings has improved: Between 2010 and 2018, the annual average rate of accidents was 3% lower and that of fatalities 4% lower. The level of rail infrastructure safety varies considerably between EU Member States. It is obvious that safety requires further improvement. There is also a crucial imperative to share information, knowledge, and experience across the EU [41].

High quality transport services, i.e., improvement of the attractiveness of rail transport, are essential for attracting more passengers. It is important to know your passengers' needs. For this purpose, passenger questionnaires are conducted. Structural equation modelling (SEM) is used to assess service quality and passenger loyalty

[42]. It describes quality indicators for rail passenger traffic and offers a new methodology to assess them [43]. This methodology has been tested on a selected route and relates to monitoring the quality of traffic routes on the rail network.

It establishes quantitative significance of quality indicators [44, 45]. Structural observations are conducted. Structural observation is one way to measure how public transport passengers spend their journey time. Researchers in New Zealand interviewed around 812 adult passengers in Wellington [46, 47]. They observed passengers and their behaviour on different routes for 4 min. The observation of passengers has shown that on buses more often than on trains, most passengers (65.3%) were more likely to "look ahead/out the window". Around a fifth of all passengers observed were reading and this referred predominantly to rail passengers. Other passengers were wearing headphones and listening to something, were chatting, texting, or sleeping, or just had their eyes closed. Passenger behaviour was compared by gender, age group, mode, and time of day. In transport planning and transport economics, there is an assumption that travel time is inefficient and should be minimised. Passengers do not always "do nothing" while travelling. They can use this time in productive ways, such as working on the computer, reading, or knitting. Transport planners and economists should consider this point. In Germany, a train journey from Berlin to Munich turned into a visit to a performance, when two friends, artists, took advantage of a short stretch of track where the high-speed train slows down due to the wave-like countryside. When the train slows down, people turn away from their phones and computers and start looking through the windows. Passengers watched up to 50 short performances through the windows [48].

Passengers on the Shanghai–Nanjing high-speed rail (HSR) train ranked staff attitudes, ticketing, and comfort of travel as the most important quality indicators [49].

Austrian Federal Railways (ÖBB) is planning to launch new, ultra-modern and comfortable trains. Seat reservation and ride quality are being improved. The train will have WiFi, new apps such as bedtime stories to read or listen to, music to fall asleep, etc. The berths on the new night trains will be slightly longer even though the European track gauge does not allow berths of 2 m. In the buffet car, aluminium beverage cans will be replaced by eco-friendly glass bottles. This means that in the future, ÖBB will release around 1.4 tonnes less aluminium waste [50].

## 1.3 Impact of Rail Traffic Dynamics on the National Economy

In the EU, demand for passenger traffic increased steadily between 2015 and 2019. (Fig. 1.2). A peak of 414 billion passenger-kilometres (pkm) was documented in this period (up 10.4%). This upward trend ended in 2020 with the onset of the COVID-19 pandemic. Due to strict precautionary operational measures and a reduction in the number of trains as of March 2020, rail passenger traffic has fallen significantly in

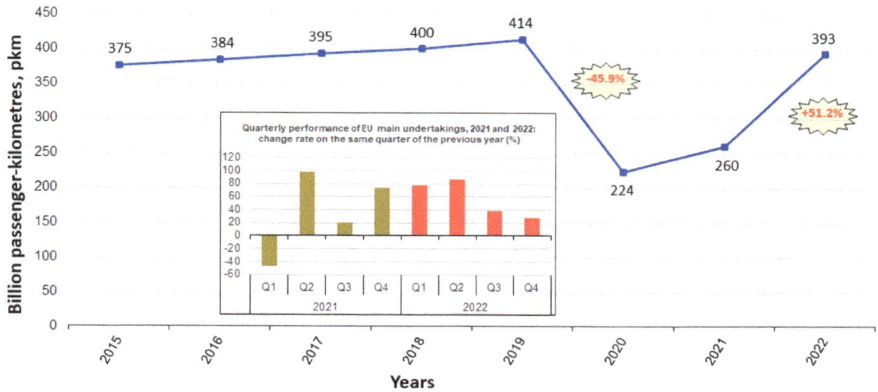


**Fig. 1.2** Rail passenger transport for main undertakings, EU, 2015–2022 (billion passenger-kilometres) (adapted from Ref. [51]). *Note* The y-axis is cut. *Source* Eurostat (online data code: rail_pa_typepas)

all EU Member States. Compared to 2019, rail passenger traffic in the EU almost halved (-45.9%) to 224 billion pkm. In 2021 only a partial recovery can be observed, with a 16.7% increase compared with 2020. However, in 2022, for most European countries, the rail passenger transport offer has already returned to a level close to that of 2019 and for few countries even higher. Rail passenger transport increased by 51.2% in the EU compared with 2021, reaching a performance of 393 billion pkm. The three following quarters of 2021 and all quarters of 2022 recorded again substantial increases (compared with the same quarters of the previous year) [51, 52].

In 2022, six EU Member States fully recovered after the massive drop of rail passenger transport performance in 2020 and even surpassed the pre-COVID level (Fig. 1.3). The highest increase in 2022 compared with 2019 among the EU Member States was recorded by Slovenia (+ 12.6%), followed by Croatia (+ 12.2%), France and Lithuania (both + 6.5%), Bulgaria (+5.3%) and Denmark (+ 3.3%). The highest decrease in 2022 compared with 2019 among the EU Member States, for which data are available, was observed in Ireland (– 27.1%), followed by Slovakia (– 20.0%), Italy (– 17.8%), Latvia and Luxembourg (both – 16.0%), Czechia (– 13.5%), Portugal (– 12.6%), Sweden (– 11.9%) and Greece (– 10.8%). A similar trend was observed in Norway (– 17.1%), Switzerland (– 11.3%), North Macedonia (– 25.5%) and Montenegro (– 22.7%). When comparing 2022 with 2020, all countries recorded a substantial increase in their rail passenger performance. The highest growths were recorded by Spain, Italy and Ireland. The three countries more than doubled from 2020 to 2022. In 2022, France and Germany were the largest contributors to the rail passenger transport performance in the EU, with 102.8 billion pkm (or 26.1% of the EU total rail passenger transport performance) and 92.3 billion pkm (or 23.5%), respectively. At the other end of the scale, nine EU Member States recorded less than 2 billion pkm in 2022. The candidate countries Bosnia and Herzegovina, Montenegro and North Macedonia recorded less than 60 million pkm in 2022 [51].

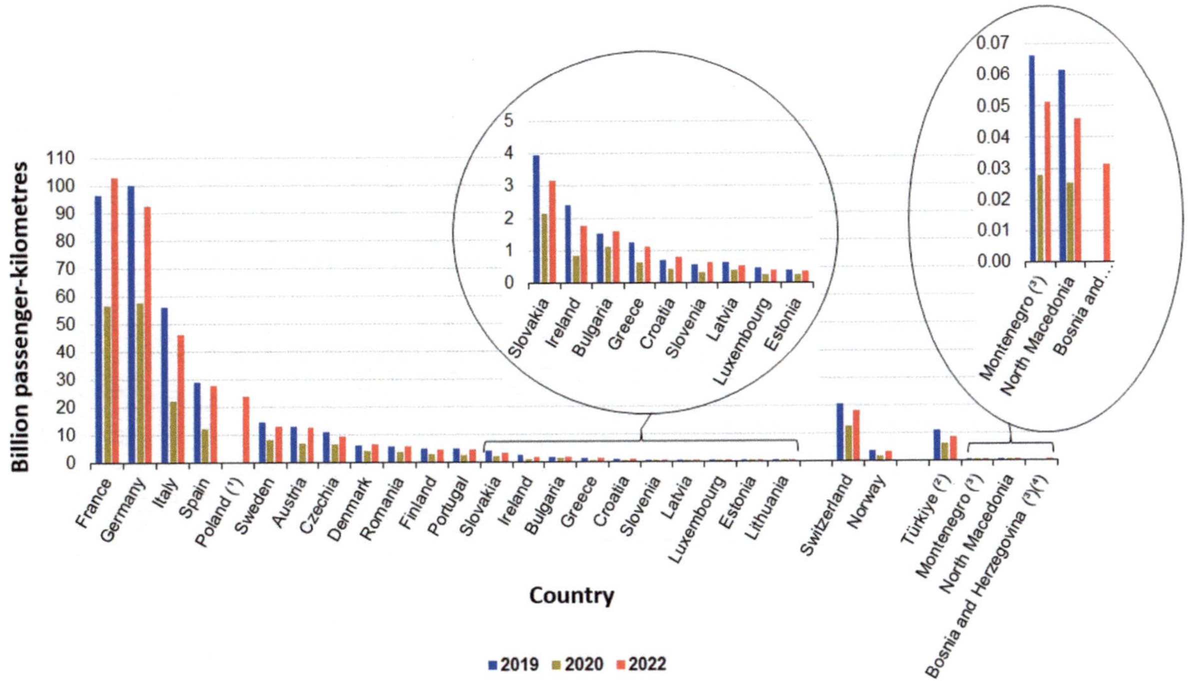


**Fig. 1.3** Rail passenger transport for main undertakings, 2019, 2020 and 2022 (billion passenger-kilometres) [51]. *Note* Countries are ranked based on 2022 data. Data for Belgium, Hungary and the Netherlands are confidential. Cyprus and Malta have no railways. (1) 2019 and 2020 data are confidential. (2) 2021 data instead of 2022. (3) Based on quarterly data. (4) 2019 and 2020 are not available

From the perspective of the share between national and international transport, national transport is usually predominant. It represented more than 90% of the total rail passenger transport performance for all EU Member States in 2022, with the exception of Luxembourg and Czechia where national transport represented 69.4% and 83.4%, respectively (Fig. 1.4). Only national passenger transport was reported for 2022 by Estonia, Greece and Latvia [51].

With the exception of four EU Member States, in all countries reductions in the number of passengers transported by rail in 2022 (Fig. 1.5) compared with 2019. The largest decrease was recorded in Ireland, by 28.4%, followed by Italy (− 22.8%), Greece (− 21.3%), Finland (− 18.0%), Slovakia (− 17.3%), Latvia (− 15.7%), Estonia (− 14.9%), Germany (− 14.7%), Luxembourg (− 11.9%), Spain (− 11.7%) and Lithuania (− 10.6%). Four EU Member States recorded growths in 2022 compared with 2019. The highest rebound among the EU Member States was recorded by Denmark (+ 46.0%). This phenomenon can be mainly attributed to an increase in the number of companies offering international rail passenger services. Increases in passengers transported can also be observed in Bulgaria (+ 8.1%), Slovenia (+ 4.9%) and Romania (+ 2.1%). It should be noted that France, Croatia and Lithuania recorded decreases in the number of passengers transported by rail in 2022 compared with 2019 while they reported increases in the rail transport performance in passenger-kilometres. Conversely, Romania reported an increase in the number of passengers transported by rail in 2022 while a decrease was recorded in the rail transport performance. Regarding the number of passengers transported by rail, Germany was by far the largest contributor in the EU, recording 2.5 billion passengers (or 34% of the EU total) in 2022. France, Italy and Spain followed with 1.1, 0.7 and 0.6 billion passengers, respectively. At the other end of the scale, Lithuania and Estonia registered less than 8 million passengers in 2022. In the candidate countries Bosnia and Herzegovina, Montenegro and North Macedonia, less than one million passengers were recorded in 2022 [51].

Figure 1.6 presents the number of passengers transported in relation to the population of the reporting countries. At EU-level, an average of 16.3 passengers per capita travelled by rail within their country in 2022, increasing by 5.7 passengers per capita compared with 2020, but dropping by 2.1 passengers per capita compared with 2019. Denmark and Luxembourg had the highest rates in 2022, with 51.1 and 33.7 passengers per capita, respectively. The EFTA country Switzerland also registered a high rate, with 49.4 passengers per capita. In 2019, Switzerland's rate was even higher, with 60.6 passengers per capita. Germany, Austria and Sweden were the other EU Member States with more than 20 passengers per capita (29.9, 28.9 and 23.3, respectively). For seven EU Member States, the rate stood between 17 and 11 passengers per capita, for five others between nine and five passengers per capita, and another five countries had a rate between five and one passengers per capita. Greece registered the lowest rate with 1.5 rail passengers per capita in 2022. The candidate country North Macedonia recorded even less than 1 passenger per capita transported by rail in 2022. When comparing the rates of 2022 with 2019, increases were observed for four EU Member States. The most significant increase was recorded by Denmark (+ 15.6 passengers per capita) [51].

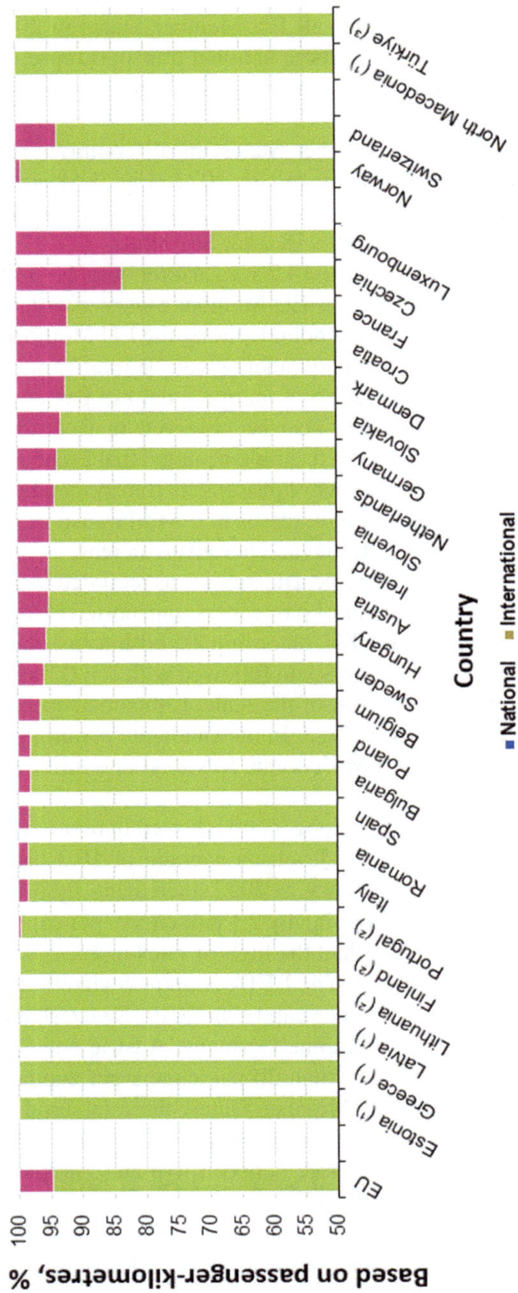


**Fig. 1.4** Rail passenger transport by type of transport for main undertakings, 2022 (%, based on passenger-kilometres) [51]. *Note* Cyprus and Malta have no railways. Data are not available for Montenegro and Bosnia and Herzegovina. (1) No international transport. (2) International transport is less than 0.5 %. (3) 2021 data instead of 2022

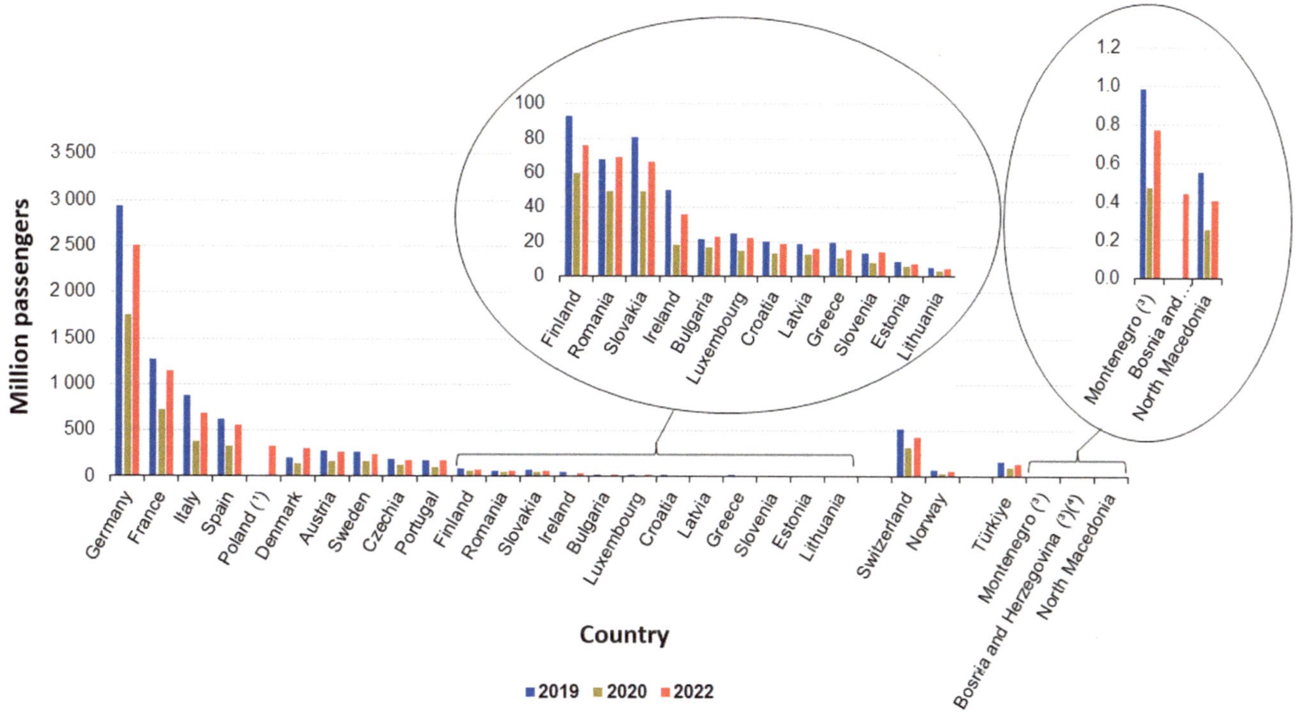


**Fig. 1.5** Rail passenger transport for main undertakings, 2019, 2020 and 2022 (million passengers) [51]. *Note* Countries are ranked based on 2022 data. Data for Belgium, Hungary and the Netherlands are confidential. Cyprus and Malta have no railways. (1) 2019 and 2020 are confidential. (2) 2021 data instead of 2022. (3) Based on quarterly data. (4) 2019 and 2020 data are not available

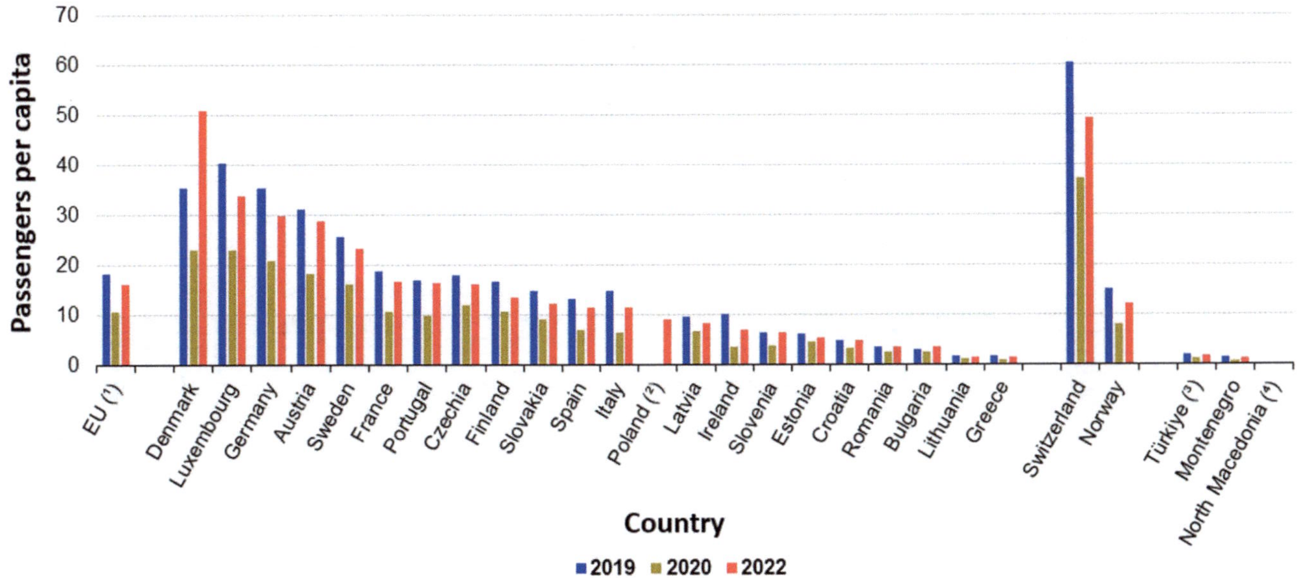


**Fig. 1.6** Rail passenger transport for main undertakings, 2019, 2020 and 2022 (passengers per capita) [51]. *Note* Countries are ranked based on 2022 data. Data for Belgium, Hungary and the Netherlands are confidential. Cyprus and Malta have no railways. Data are not available for Bosnia and Herzegovina. (1) Includes only national transport. (2) 2019 and 2020 data are confidential. (3) 2021 data instead of 2022. (4) Number of passengers based on quarterly data. Values: 0.3 in 2019, 0.1 in 2020 and 0.2 in 2022.

Transport is of great importance for social, economic, and environmental development. The link between the main modes of transport, related investment, specific air pollutants and sustainable economic growth is examined [53]. Fixed-effects regression results for the EU-28 from 1990 to 2016, based on Eurostat, Organisation for Economic Co-operation, and Development (OECD) and World Bank data, show that road, inland waterway, maritime and air transport infrastructure have a positive impact on Gross Domestic Product per Capita (GDPC). Meanwhile, the impact of rail transport on GDPC is negative. The empirical results on transport infrastructure investment show a positive impact on economic growth for all transport modes except inland waterways. In addition, $CO_2$ emissions from all modes of transport, along with other specific air pollutants, have a negative impact on Gross Domestic Product (GDP).

Nonetheless, transport is an important economic factor in the EU: around 5% of the population is employed in transport and transport accounts for 7% of GDP [54]. The results show [55] that the elasticity of the one-year lagged GDP variable is positive and statistically significant in all specifications considered. This reflects a dynamic growth pattern. In general, the value of transport investment over a one-year lag shows a positive and significant statistical relationship with regional GDP across several dynamic model specifications. On the other hand, the value of rail accessibility has a positive and significant impact on GDP in two years lag. The study shows that the impact of the monetary value of transport investment was generally felt immediately the following year. However, improvements in rail accessibility require a certain time frame to achieve the benefits.

The development of rail passenger traffic in EU countries is analysed [56], based on criteria related to the traffic process and the level of economic development of countries. The study offers a methodology based on a combination of multi-criteria methods. First, the indicators that will be used to measure rail passenger numbers are identified. The main assessment criteria include social, economic infrastructure factors as well as factors associated with travel and technological development. Second, the weights of the criteria and sub-criteria are calculated using the Fuzzy AHP method. Third step involves the assessment of the countries under study using the PROMETHEE method. The results show that in the ranking of countries, travel-related and technological factors play the most important role. Taking all factors into account, the most developed rail passenger traffic is identified in Germany, Great Britain and France.

Until the end of the twentieth century, European cities were connected by night trains. Over the last two decades, the numbers of night trains have fallen sharply. The climate crisis and the COVID-19 pandemic have forced a change in travel habits. Night trains are a good example of more sustainable mobility. The pandemic has disrupted rail services, including night trains. Just before the summer holidays, night train operators have resumed cancelled routes. Cross-border night trains face more challenges than border closures. The challenges may include different legal requirements in EU Member States, different signalling and electrification systems, and different track and loading gauges. In addition, like all railway companies, night trains must pay infrastructure access charges, which is an important disadvantage when competing with flights. Night trains in Europe are one of the most sustainable and convenient means of transport for travelling long distances. However, they are not yet the most frequent choice of passengers. Demand for night trains can be assessed on the basis of a potential analysis, considering the target market, travel time and the competitive situation with other modes of transport. The development of new night train services requires comfortable rolling stock, which is currently in short supply in Europe. Only a few night train carriages have been built in recent years. Some companies prefer to destroy their old rolling stock rather than sell it to other operators who can refurbish it and give it a second life. An EU-supported rolling stock pool for all carriers could be one way to address this problem. Another way would be to force all carriers to offer the surplus rolling stock they no longer want to use to other carriers (incumbents and new entrants) [50].

Efforts are being made to reduce the risk of coronavirus infection when travelling by train. The Dutch national government and the Dutch railways operator have launched some policy measures. Train passengers have been instructed to wear masks, and home working and remote training instructions have been developed. Other policy measures, such as the introduction of a reservation system, have also been considered. The experiment was conducted to investigate whether train passengers support these policy measures and how the implementation of these changes their awareness of infection when travelling by train [57].

Recommendations to suspend travel have led to a reduction in rail passenger numbers of around 80% in all countries during the quarantine. The closure of borders has led to an almost 100% reduction in the number of international rail passengers for all operators [58]. The coronavirus outbreak has been the biggest humanitarian and economic challenge the EU has faced. It was difficult to estimate how long the pandemic would last and how long the restrictions would last. The potential impact of COVID-19 on the transport sector and specifically on railways has been analysed. The pandemic has had consequences not only for carriers and passengers, but also for the environment, leading to changes in the supply chain. The health crisis has halted passenger mobility and severely constrained air and sea freight capacity, resulting in positive environmental impacts. The removal of the capping measures will reverse the downward trend in greenhouse gas (GHG) emissions. Greenhouse gas emissions are projected to increase. It is therefore essential to develop new ecological mobility that is environmentally friendly. The five necessary actions to be taken by the rail sector to deliver future services have been outlined. They include resilience, return, rethinking, reform, and research. New research areas are highlighted that could be instrumental in making railways more competitive and resilient to future crises. The pandemic, it is noted, can be seen as a testing ground for future crises and as an opportunity to introduce a debate on a new paradigm for green and public mobility. The COVID-19 crisis has confirmed the resilience and stability of the European rail system. Passenger and freight rail transport provides a permanent territorial cohesion between regions [59].

The history of the railway led to the construction and use of the so-called Russian 1520 mm gauge railway track in Lithuania [60]. Over time, with the changing geopolitical situation, the rapid development of technology, the changing needs of passengers and the increasing mobility of goods, it became clear that old railways have become morally obsolete. With the view to create a single European railway area, ERTMS (*European Rail Traffic Management System*) was launched in the late 1990s, with the aim of replacing all existing different signalling systems in Europe with a single system. The system would ensure interoperability between national rail networks and cross-border rail transport. Trains would pass EU countries without any disruptions, the competitiveness of railway transport would be promoted, traffic safety would be ensured. The current total length of Lithuanian railway network lines is 1910.7 km.

*Rail Baltica* is a rail transport infrastructure project that aims to integrate the Baltic States into the European railway network. The budget of the *Rail Baltica* project increased four times from 2017 to 2023: from 5.8 billion Eur to 23.8 billion Eur, and it is the largest investment aimed at the enhancement of mobility in the region, development of business, tourism, and trade. The project involves 5 EU countries: Poland, Lithuania, Latvia, Estonia and Finland which is involved indirectly. The line will connect Helsinki, Tallinn, Pärnu, Riga, Panevėžys, Kaunas, Vilnius, and Warsaw. 870 km of high-speed electrified railway double track will be built in Lithuania, Latvia, and Estonia. The projected speed of passenger trains is to be 249 km/h. According to preliminary calculations, from 2030 to 2040, *Rail Baltica* railway line via Kaunas will annualy be used by more than 4.5 million passengers. *Rail Baltica* will act as an extensive catalyst for the continued economic development of the Baltic States and during the construction phase it will boost hundreds of new jobs and contribute to the region's GDP. It will also reduce the distance between the Baltic countries and bring the rest of Europe closer. A modern high-speed rail link will provide a convenient, safe, and environmentally friendly alternative for passenger movement. Passengers in the Baltic States will have access to travel experiences in terms of quality, speed, convenience, and passenger expertise that have long been available in Western Europe.

In the summer of 2015, *Lithuanian Railways* cancelled the rout "Vilnius–St. Petersburg–Vilnius", which had a low passenger interest [61], and in December of the same year the company decided to discontinue the rout "Vilnius–Moscow–Vilnius" [62]. The decision to abolish the routes was not only due to a reduction in the number of people travelling by train. Low-cost airline flights had been launched. Besides, the geopolitical context had also played a role: Lithuanians started refraining from travelling to Russia [61]. For some time, several Lithuanian carriages were attached to the Russian railway company train "Kaliningrad–Moscow" transiting via Lithuania. However, due to the low number of passengers, these carriages were abandoned in the spring of 2017 [63]. Due to a quarantine situation in Lithuania, the "Vilnius–Minsk–Vilnius" international train route has been suspended since March 16, 2020 [64]. Due to the epidemiological and geopolitical situation, this route has not been renewed to date [65].

## 1.4 Understanding the Benefits/Priorities of Travelling by Train

Anyone can exercise choice in terms of the means of transport they use to travel. For example, Americans are much less likely to travel by train than people in other developed countries. This is a logical choice, as far-away destinations can be reached much faster by plane than by train or car. On the other hand, for shorter distances, the train is a good choice: there is more space in the carriage, there is no need to arrive at the railway station much earlier before departure of the train, there is no rigorous personal screening of each passenger, the railway stations are mostly located in the central part of the city, while airports are mainly located on the city outskirts. The train also offers more benefits, such as more spacious seats, free Wi-Fi, lower baggage fees and restrictions, and greater price flexibility when it comes to rescheduling your journey. Besides, flying by plane will not offer you the beautiful scenery you can appreciate on a train. For nature lovers, travelling by train is less environmentally polluting than travelling by plane or car (Fig. 1.7). Rail transport is one of the most sustainable, energy-efficient, and safe modes of transport and will play an essential role in the future EU mobility system.

None of the modes of transport is perfect. Some people like trains, others avoid them. Whether a train is the best way to reach your destination depends on the route of travel as well as the personal needs and preferences of each passenger.

Studies conducted in Germany and Switzerland have shown [67] that passengers consider rail transport to be preferable to road transport. Why is this the case? Passenger choices on the Beijing–Shanghai high-speed railway (HSR) channel were identified by using random sampling [68]. The aim was to survey passengers travelling by rail, air and coach. A total of over 60,000 completed questionnaires were collected. The questionnaires were used to identify passengers' preferred means of transportation, travel purposes and intentions, occupation and income level. The questionnaire data was used to analyse the alternatives for HSR passengers in terms of the speed of the journey, the quality of the HSR service and the quality of the service of other means of transportation, the convenience of the journey, the comfort of the journey and the safety. The results reveal that people with higher incomes

**Fig. 1.7** Specific $CO_2$ emissions per passenger transport mode [66]

and frequent travellers are more willing to travel by HSR. The benefits of high-speed railway rely on the distance travelled. Compared to ordinary trains and buses, high-speed railway has no clear advantages on short-distance routes. In addition, passengers who like to travel by plane value speed first and foremost. HSR enjoys the benefits of high running speeds, short waiting times and a comfortable travel environment.

Trains are gaining popularity as a means of transport because of their comfort and environmental friendliness. Recent statistics show that over the last decade, the number of train passengers has increased by 40%. New high-speed trains have made rail transport more attractive. They are required to have a high level of comfort and a high degree of safety. Methods for long-term ride comfort assessment are being developed [69]. They aim to assess passengers' awareness of comfort in rail vehicles. Factors that are assessed include vibration (level, frequency and duration), temperature, noise and the space per passenger on the train. The reduction in accidents has slowed down research into passive safety. Therefore, recent works focus on the reduction of second impact effects and minimizing biomechanical damage criteria in the event of a crash, using new rolling stock interior design solutions.

The quality of rail passenger transport is difficult to quantify and measure directly by one single criterion. When travelling by train, a passenger is surrounded by a certain environment (Fig. 1.8). The environmental factors of a passenger can be grouped into the following categories: microclimate and psychological climate inside the carriage, visual environment, physical and mechanical pollution. Ride comfort is characterised by indoor temperature, humidity, lighting, noise level, dustiness and vibration.

The ventilation needs to be at a level where at least half of the air in the room changes within an hour. A minimum of 14.4 $m^3/h$ of outdoor air per adult is required to keep indoor $CO_2$ concentrations within the permissible limit. Poorly ventilated rooms accumulate carbon dioxide, moisture and pollutants and increase microbiological contamination. If the environment is full of viral pathogens, poorly ventilated rooms are more likely to get them [70].

A good indoor climate on a passenger train is crucial, not only because it will affect energy consumption and thus its sustainability, but also because good comfort is essential for long journeys. A survey [71] that assessed heat and air quality has been conducted in the sleeping cars of Chinese long-distance trains. The results are intended to be used to optimise the management of the ventilation system. This study looked at seven key parameters including thermal comfort, relative humidity, air quality, clothing, etc. A total of 91 questionnaires containing passenger responses were returned. Most respondents said the air quality in the carriage was average or good. Most of the passengers in the carriage could sense some smell, but it was not strong. The results of this study have shown that ventilation, temperature, humidity and clothing constitute the main factors that influence the thermal comfort of a passenger. There is a growing interest in the role of scent in passenger transport. Scents have an incredible effect on the energy of a space and the people in it. Malodours in the carriage have a negative impact on a passenger's well-being. For example, anxiety/stress levels are often higher when travelling (especially when

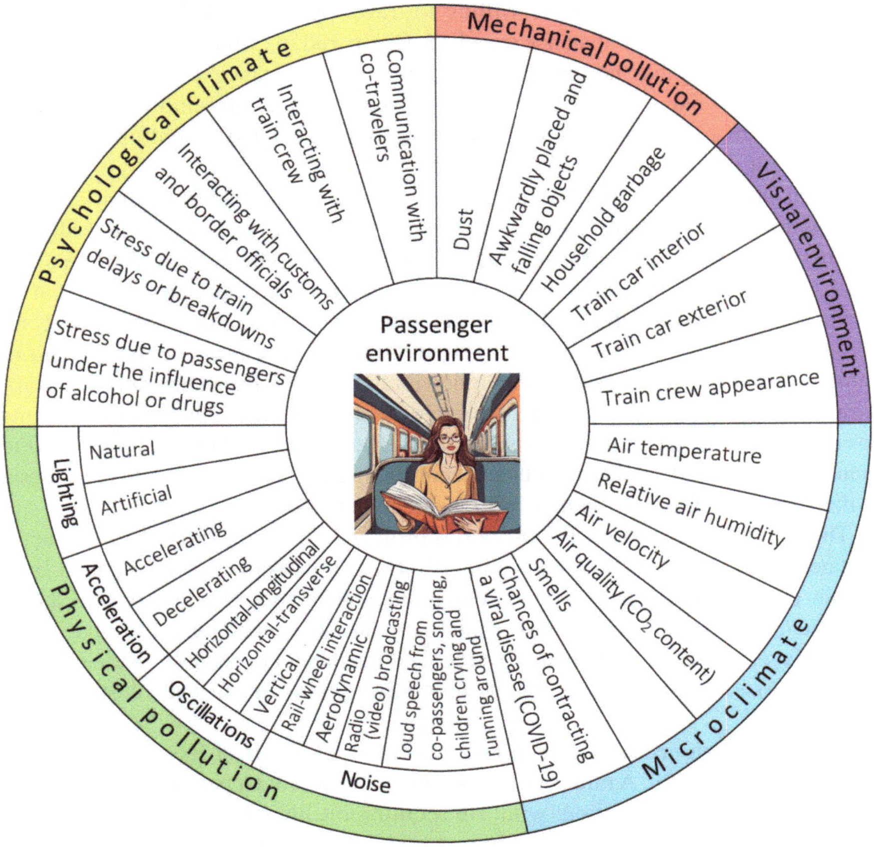


**Fig. 1.8** Factors affecting a train passenger comfort during the journey

travelling by plane). Odours often violate a passenger's personal space. The possibility of using scent to reduce passenger stress and anxiety, or to give the impression of cleanliness, is being considered [72]. The impact of environmental scent on the emotions and behaviour of public transport passengers has been assessed [73]. Research shows that olfactory marketing interventions in public transport need to be given more attention by transport managers.

Due to its low environmental impact, rail is the preferred mode of transport for mass passenger mobility. One of the key elements that make rail travel more appealing is the thermal comfort of the passenger carriage. A new approach to improving thermal comfort in railway passenger carriages is based on the idea of personalised comfort zones. Individual climate control is assumed to ensure the thermal comfort of passengers. Studies [74] have further explored this assumption considering the notion of thermo-specific self-efficacy (specSE) as a psychological factor in the context of thermal comfort in railcars. Two studies were carried out, involving 11 test drives with

172 respondents. A mock-up of a passenger railway carriage was used. The ambient climatic conditions in the mock-up were monitored and measured. It was established that specSE could be considered as a separate psychological factor that had a significant impact on the prediction of thermal comfort and climate satisfaction. In addition, it moderated the effects of available and exercised control. The results presented here are complementary to previous findings on the concept of personal control and support the role of specSE in predicting thermal comfort. Thermal comfort conditions are assessed not only on trains, but also at train stations [75].

Due to the high hygrothermal comfort requirements, the energy consumption of modern trains for heating/cooling can be up to 30% of the total related electricity demand. The energy savings of heating, ventilation and air-conditioning systems on the train can be duly estimated using dynamic modelling techniques. Weather conditions are to be assessed dynamically, considering the actual location and orientation of the moving train. This methodology can be applied to analyse a wide range of innovative actions for energy efficiency, reduction of environmental impacts and comfort improvement, and assess their economic feasibility. A new study carried out on a medium-distance train in Southern Italy demonstrates the potential of this approach. The loads and efficiency of the heating/cooling system of a standard and an upgraded train are dynamically assessed, along with the related electricity consumption. In the context of the heating, ventilation and air-conditioning (HVAC) improvements, various energy saving factors are considered. The results are beneficial in terms of energy savings, reduction of $CO_2$ emissions and comfort improvements. The study also provides useful design and operational criteria for train manufacturers and operators [76].

Trains, like other places of large-scale assembly, such as shopping centres and public institutions, pose a risk factor for illness. Staphylococci, fungal agents, Human Immunodeficiency Virus (HIV), Hepatitis B and C agents, noroviruses, rotaviruses and now the COVID-19 virus are all dangerous to passengers. Public transport is an ideal environment for the transmission of micro-organisms. People can contract them by touching the mucous membranes of the eyes, nose and mouth with contaminated hands. Researchers around the world are regularly conducting studies on the pathogen problem in public transport [77, 78].

The psychological climate can be affected by several characteristics: interaction with other passengers on the train, with train staff and on an international train, with customs and border guards. It is important for the train manager to be aware of the psychological climate on board the train, as many of the factors influencing this climate can be changed. For example: to discipline a passenger who fails to comply with the norms of behaviour (e.g., under the influence of alcohol, drugs); to eliminate or reduce service quality failures and extend a helping hand in case of unforeseen situations. A significant factor in establishing good relations with passengers and building a healthy psychological climate at work is the match between the personal qualities of employees and the psychological requirements for the job.

The success of a trip and the satisfaction of travelling depends greatly on the mood of the traveller. Delays on public transport have become an integral part of the daily routine of many commuters. The study [79] was conducted with the aim of investigating the relationship between the various psychological feelings that Malaysian rail passengers may experience due to delays and their satisfaction in transport. Passengers on these rail services are often delayed for more than an hour due to breakdowns or other technical problems. 396 passengers completed the questionnaires at departure areas. The aim was to identify passengers' views on how delays affect the quality of their journey. The ordered mixed logic model was employed to perform the analysis. It was decided that the hybrid model would offer more flexibility and protect against biased evaluations. The findings of the study have provided important insights into the different feelings experienced by passengers that affect the quality of rail transport services.

Although people are often encouraged of using public transport, it is not always a convenient option. The study [80] employs service elements to measure passenger anxiety. The authors apply a conceptual model based on a rail passenger service chain perspective. Passengers' anxiety related to train travel is measured using a modern psychometric method, the Rasch model. The study analysed the interviews of 412 rail passengers. The results of the analysis have shown that the elements that cause anxiety for passengers when travelling by train include congestion, delays, station accessibility, finding the right train on the platform and changing trains. The empirical results obtained using the Rasch method can be useful for the development of an effective strategy to reduce train passenger anxiety. This empirical study also shows that anxiety varies according to the passenger's gender, age, frequency and type of journey. This information is useful for transport planners and policy makers to create a comfortable rail travel environment and to assess the special needs of specific passenger groups.

Despite operating one of the safest railways in Europe, operators in the United Kingdom (UK) are still facing serious consequences of alcohol abuse [81]. Recent studies have shown that 60% of passenger deaths and assaults in the last decade were alcohol related. Technological advances, improved passenger information and enhanced guidance and regulation have made the UK the safest rail network in Europe. However, there are still some alcohol-related problems that pose a risk to travellers and workers, especially during the holiday season, when more drunk passengers travel by train. To manage the risk posed by drunk passengers, staff numbers are being increased, CCTV cameras are being installed, station lighting, stair railings and floor surfaces are being upgraded. In 2012, Scottish rail operator ScotRail banned alcohol on its trains between 9 pm and 10am in response to an increase in antisocial behaviour reported by passengers. Paul Leach, human factors specialist at RSSB, does not believe that a total ban of alcohol on trains is the solution. He justifies his opinion by saying that it is preferable for people to take the train rather than their own car when drunk. One approach that has worked well has been to ban alcohol at specific popular events involving large crowds, such as football matches. An application for data sharing, such as SmartGUARD, is one tool that could be applied in railways. Information about anti-social behaviour is shared with other staff. In this way, a person who has been banned from entering a station or train because of his/her abusive or threatening behaviour would not be admitted by another staff member.

Physical pollution in the train carriage includes inadequate lighting, excessive vibration, noise, deceleration and acceleration (Fig. 1.8). As lighting is an energy distribution that improves the visibility of objects, improper lighting can cause discomfort to train carriage occupants, e.g., when reading. The overall level of artificial lighting on the table surface and, where there is none, on the surface of the passenger seating or sleeping area, shall not be < 100 lx [82]. So far, only functional parameters have been included in the specifications for interior lighting in trains. Lighting (level and uniformity) and standards are more focused on safety rather than comfort. The visual comfort research programme aims to understand the customer's preferences and expectations regarding the lighting environment and to improve the specifications of both new and existing trains. Experimental results on the combined effects of light parameters and seat colours for passengers are presented [83]. It has been found that lighting levels below 150 lx are considered relaxing and soft. Illumination > 200 lx is blinding, aggressive and tiring. This is the lighting that is mostly inside the train.

The vibration of the carriage impairs the well-being of the passenger and the efficiency of the staff. The movement of a train is the result of the various forces at work. A train, and therefore a carriage, can move smoothly (at a steady speed), decelerate, or accelerate. One of the main characteristics of braking is deceleration and one of the main characteristics of gaining speed is acceleration. Braking and acceleration efficiency determines the dynamic performance of the carriage and thus the passenger's ride comfort. The rapid development of around the world is making high-speed trains (HSTs) a competitor to aeroplanes for long-distance travel. To improve ride quality, studies on aerodynamic noise reduction in high-speed trains [84] and vibration analysis of the suspension system [85] have been conducted. The influence of car suspension parameters on ride comfort [86] and the influence of accelerations and road roughness on the ride quality of a railway passenger car [87] have been assessed. Compared to aeroplanes, HST carriages have much larger windows. Large windows provide better lighting and views of the landscape. The sunlight coming through the windows has a significant effect on the temperature inside the carriage. The study presents a numerical investigation of solar thermal comfort in HST. The effect of sunlight was assessed in terms of airflow model, temperature distribution and thermal comfort indexes. Parametric tests were carried out at seven different times of the day. The effect of using a roller window shade has also been explored. The overall carriage air temperature was found to have risen significantly due to sunlight, especially in the vicinity of passengers. It has been documented that passengers sitting near the windows experienced a marked change in thermal comfort at different times of the day. To improve passenger comfort and reduce energy consumption at high outdoor temperatures, the use of window shades would be effective [88]. Carriage cabins with sleeping berths shall be equipped with protection against direct sunlight (shades, blinds, etc.) [82].

Excellent passenger comfort should once again become a competitive advantage for railways. For a long time, railways have been losing their advantage over other modes of transport such as cars. The "Next Generation Train" project on high-speed trains has considered many aspects of passenger comfort, such as smoothness of the ride, vibration, acoustic noise, pressure fluctuations, lighting and air conditioning. The study carried out by Winter et al. [89] provides a comprehensive overview of passenger comfort improvement measures performed at the German Aerospace Centre since 2007. It presents the findings of laboratory studies regarding ventilation and lighting that have effect on a human being. Two studies were conducted with the focus on the thermal comfort: one on the sliding ventilation with air outlets installed under the passenger seats and the other on the infrared heating through a panel installed in front of the first row. In addition, two tests evaluated the impact of the new lighting concept with OLEDs on passenger comfort.

External sounds can also adversely affect one's hearing. Snoring, children crying or running around the carriage is a common problem, especially in the sleeping carriages. Around 45% of people snore occasionally and 25% snore regularly. In the US, around 37 million people snore regularly. Moderate snoring is between 50 and 65 dB(A) but can reach around 90 dB(A). In the UK, 111.6 dB(A) snoring has been registered [90]. Trains are considered the best mode of transport for travelling with children, as there is more space to play and move around. Toilets are also available. Having a buffet car on board the train makes it easy to feed your child. Children aged between 9 months to 6 years old can have a crying level between 99 and 120 dB(A) [91]. It must be admitted that snoring train passengers and those travelling with small children negatively affect the quality of the journey for other passengers. However, just because a person snores or travels with a child, he or she cannot be banned from travelling. Rail companies are the ones who are supposed to find solutions to these problems. On the French International (Intercities) train (Paris–Limoges–Tolouse line) children can play in the "Espace Kid" area [92]. It has everything a child needs: pillows, a play rug, special seats for children aged 1–4 to keep them safe. There are also baby changing tables, sockets for feeding bottle warmers and areas to store prams, pushchairs or strollers. Train staff could, in an exceptional case, provide the snoring passenger with a free compartment (if available).

When designing new rolling stock, a lot of attention is paid to interior styling. It is very important that the increase in seating capacity does not compromise the comfort of the ride. In the past, trains used to have a variety of seating configurations, but in recent years the seating arrangement in the carriage has become more like an aeroplane. The rows of chairs are often arranged to face each other. The idea is to change this to some extent by creating a more dynamic layout. Quiet zones are being created to allow passengers working on their computers to concentrate. At the same time, the more active zones are also offered so that passengers can talk on the phone or with each other in groups, depending on their destination. There is also a greater need for open areas that allow passengers on short journeys to stand [93].

One of the most appreciated amenities by passengers are armrests and tables. Everyone wants to have their own personal space, i.e., not to feel too close to someone sitting next to or in front. Sitting opposite each other feels different than when sitting next to each other. It is very popular to install some of the seats higher up in the carriage. A large desk is handy for working at. This is a very attractive option for young people, where you can sit and play a game or chat. This is a great way to improve the interior of the carriage. In addition to the physical furnishing of the interior, ways are being explored to create a more comfortable environment for passengers. A lot of attention is paid to interior design: colours, seat covers, interior fittings [93].

Dust, falling objects (e.g., passengers' luggage falling from overhead racks, crockery falling from a table, another passenger not wearing a seatbelt falling from a berth) can all be classified as mechanical pollution of the carriage (Fig. 1.8). The harmful effects of dust depend mainly on its physical and chemical properties, particle size, shape and concentration in the air, as well as the length of time spent in dusty environments. Most dust is harmful to the respiratory system. Some can even lead to skin and other diseases. Various bacteria move with the dust. Dust from the environment enters the interior of the carriage through open or leaky windows, where coal is used to heat the train carriages. The carriage compartment and toilet must be clean throughout the journey [82]. Waste is collected in dedicated containers. Waste collection containers are emptied when full, washed with water and disinfected if they become contaminated.

All of the above factors affect the comfort of the passenger's journey in the carriage.

## References


1. Bendros Europos geležinkelių erdvės kūrimas (2022) Europos vadovų taryba. https://www.consilium.europa.eu/lt/policies/single-eu-railway-area/ (in Lithuanian)
2. Directive (EU) 2016/797 of the European Parliament and of the Council on the interoperability of the rail system within the EU (2016). Official Journal of the European Union (OJ L138/44, 26.5.2016). https://eur-lex.europa.eu/legal-content/EN/TXT/PDF/?uri=CELEX:32016L0797&rid=9
3. Directive (EU) 2016/798 of the European Parliament and of the Council on railway safety (2016). Official Journal of the European Union (OJ L138/102, 26.5.2016). https://eur-lex.europa.eu/legal-content/EN/TXT/PDF/?uri=CELEX:32016L0798
4. Regulation (EC) No 1370/2007 of the European Parliament and of the Council of 23 October 2007 on public passenger transport services by rail and by road and repealing Council Regulations. Official Journal of the European Union (L 315/1, 3.12.2007). EUR-Lex. https://eur-lex.europa.eu/legal-content/EN/TXT/HTML/?uri=CELEX:32007R1370&from=EN
5. Regulation (EC) No 1371/2007 of the European Parliament and of the Council of 23 October 2007 on rail passengers' rights and obligations. Official Journal of the European Union (L 315/14, 3.12.2007). EUR-Lex. https://eur-lex.europa.eu/LexUriServ/LexUriServ.do?uri=OJ:L:2007:315:0014:0041:en:PDF
6. Lietuvos Respublikos geležinkelių transporto kodeksas (2021) Teisės akto redakcija įsigaliojo nuo: 2021–06–16. https://www.infolex.lt/ta/76168:str1 (in Lithuanian)
7. AB "Lietuvos geležinkeliai" įmonių grupės darbuotojų etikos kodeksas (2021) Patvirtinta AB "Lietuvos geležinkeliai" generalinio direktoriaus 2021 m. AB "Lietuvos geležinkeliai", Vilnius. https://www.litrail.lt/documents/10279/0/Darbuotoj%C5%B3%20etikos+kodeksas+2021-02-03 (in Lithuanian)

8. Techninio geležinkelių naudojimo nuostatai (1996) Lietuvos Respublikos susisiekimo ministerija. Patvirtinta Lietuvos Respublikos susisiekimo ministro 1996 m. rugsėjo 20 d. įsakymu Nr.297. Informacijos ir leidybos centras, Vilnius, 126 p (in Lithuanian)
9. Teikiamų paslaugų kokybės standartai (2011) AB "Lietuvos geležinkeliai" Keleivių vežimo direkcija. Vilnius. 24 p. https://www.litrail.lt/documents/10279/101738/LG_kokybes_standartai.PDF/0d6b51f8-7ea7-4c67-9865-b4ddb0156c4d (in Lithuanian)
10. Keleivinio, dyzelinio ir elektrinio traukinio, parengto reisui, kokybės tikrinimo tvarka (2004) Patvirtinta Keleivių vežimo valdybos viršininko 2004 m. gegužės 7 d. įsakymu Nr. Į – 158. Vilnius: Gelspa, 16 p (in Lithuanian)
11. Geležinkelių eismo taisyklės (2000) ADV/003. Lietuvos Respublikos susisiekimo ministerija. Patvirtinta Lietuvos Respublikos susisiekimo ministro 1999 m. gruodžio 30 d. įsakymu Nr. 452. Informacijos ir leidybos centras, Vilnius, 263 p. ISBN 9986-537-53-3 (in Lithuanian)
12. Geležinkelių signalizacijos taisyklės (1998) ADV/002. Lietuvos Respublikos susisiekimo ministerija. Patvirtinta Lietuvos Respublikos susisiekimo ministro 1997 m. gruodžio 30 d. įsakymu Nr. 483. Informacijos ir leidybos centras, Vilnius, 133 p. ISBN 9986-537-23-1 (in Lithuanian)
13. Bendrosios keleivių vežimo taisyklės (2011) Patvirtinta AB „Lietuvos geležinkeliai" generalinio direktoriaus 2011 m. gruodžio 13 d. įsakymu Nr. Į-966. AB „Lietuvos geležinkeliai",Vilnius (in Lithuanian)
14. Bendrosios keleivių vežimo taisyklės (2017) UAB „LTG Link". https://ltglink.lt/bendrosios-uab-ltg-link-keleiviu-vezimo-taisykles (in Lithuanian)
15. Lojalumo programos taisyklės. UAB „LTG Link" (2017) https://www.traukiniobilietas.lt/portal/info/lojalumoprogram (in Lithuanian)
16. Keleivių važiavimo tarptautinio susisiekimo traukiniais taisyklės (1997) (in Lithuanian)
17. Tarptautinio keleivių vežimo susitarimas (Soglashenie o mezhdunarodnom passazhirskom soobshhenii (SMPS) (1997) (in Lithuanian)
18. Prieigos taisyklės, taikomos vežant neįgalius ir ribotos judėsenos asmenis (2009) Sukurtos remiantis 2007 m. spalio 23 d. Europos Parlamento ir Tarybos reglamentu (EB) Nr. 1371/2007 „Dėl geležinkelių keleivių teisių ir pareigų". Atmintinė, priedas Nr. 2. Vilnius, 14 p (in Lithuanian)
19. Keleivinio traukinio viršininko (Vyresniojo palydovo) pareiginė instrukcija (2001) (VL/119A). Patvirtinta Keleivių vežimo valdybos viršininko 2001 m. gruodžio 20 d. Antrasis pataisytas ir papildytas leidinys. Gelspa, Vilnius, 26 p (in Lithuanian)
20. Tarptautinio susisiekimo keleivinio traukinio viršininko instrukcija (2003) Gelspa, Vilnius, 16 p (in Lithuanian)
21. Keleivinio vagono palydovo pareiginė instrukcija (1999) (VL/120) Patvirtinta Komercijos direktoriaus 1999 m. rugsėjo 3 d. Informacijos ir leidybos centras, Vilnius, 20 p (in Lithuanian)
22. Maitinimo poskyrio palydovo bufetininko pareigybės nuostatai (2006) AB „Lietuvos geležinkeliai" Keleivių vežimo direkcija. Vilnius, 5 p (in Lithuanian)
23. Keleivinių traukinių revizoriaus kontrolieriaus pareiginė instrukcija (2005) AB „Lietuvos geležinkeliai". Vilnius, 5 p (in Lithuanian)
24. Radijo ryšio tarp traukinio viršininko ir lokomotyvo mašinisto priemonių naudojimo instrukcija (KL/42) (1999) Patvirtinta Riedmenų ūkio direktoriaus 1998 m. gruodžio 28 d. Informacijos ir leidybos centras, Vilnius, 12 p (in Lithuanian)
25. Lietuvos geležinkelių transporto darbuotojų drausmės statutas (1995) Lietuvos Respublikos susisiekimo ministerija. Informacijos leidybos centras, Vilnius, 12 p (in Lithuanian)
26. Duranton S, Audier A, Hazan J, Langhorn MP, Gauche V (2017) The 2017 European Railway Performance Index. Boston Consulting Group. https://www.bcg.com/publications/2017/transportation-travel-tourism-2017-european-railway-performance-index
27. The Voice of European Railways (2018) Community of European Railway and infrastructure companies (CER). http://cer.be/topics/passenger
28. Lenze W, Nießen N (2021) Quality in rail timetabling—a detailed review of stakeholders' interests compared with current practices in Germany. J Rail Transp Plan Manage 20(2021):1–14. https://doi.org/10.1016/j.jrtpm.2021.100282

29. Ľupták V, Droździel P, Stopka O, Stopková M, Rybicka I (2019) Approach methodology for comprehensive assessing the public passenger transport timetable performances at a regional scale. Sustainability 11(13):35321–18. https://doi.org/10.3390/su11133532
30. Fraszczyk A, Lamb T, Marinov M (2016) Are railways really that bad? An evaluation of rail systems performance in Europe with a focus on passenger rail. Transp Res Part A Policy Pract 94:573–591. https://doi.org/10.1016/j.tra.2016.10.018
31. IRG (2020) Overview of international passenger services 20(9):1–33. Independent Regulators' Group—Rail. IRG—Rail Charges Working Group. file:///C:/Users/40507/Downloads/IRG-Rail__20__9_-_Overview_of_International_Passengers_Services%20(1).pdf
32. Sauter-Servaes T, Nash A (2009) Increasing rail demand by improving multimodal information and ticketing. Results of the night and flight case study. Transp Res Record J Transp Res Board 2117(1):7–13. https://doi.org/10.3141/2117-02
33. Islam MR, Ahmed MT, Anwari N, Hadiuzzaman M, Amin S (2022) The aspiration for happy train journey: commuters' perception of the quality of intercity rail services. CivilEng 3(4):909–945. https://doi.org/10.3390/civileng3040052
34. Wang J, Granlöf M, Yu J (2021) Effects of winter climate on delays of high speed passenger trains in Botnia-Atlantica region. J Rail Transp Plan Manage 18:100251, 1–8. https://doi.org/10.1016/j.jrtpm.2021
35. Bešinović N (2020) Resilience in railway transport systems: a literature review and research agenda. Transp Rev 40(4):457–478. https://doi.org/10.1080/01441647.2020.1728419
36. Monsuur F, Enoch M, Quddus M, Meek S (2021) Modelling the impact of rail delays on passenger satisfaction. Transp Res Part A Policy Pract 152:19–35. https://doi.org/10.1016/j.tra.2021.08.002
37. Stavinova E, Chunaev P, Bochenina K (2021) Forecasting railway ticket dynamic price with Google Trends open data. Proc Comput Sci 193:333–342. https://doi.org/10.1016/j.procs.2021.10.034
38. Huang W, Shuai B (2018) A methodology for calculating the passenger comfort benefits of railway travel. J Mod Transp 26(2):107–118. https://doi.org/10.1007/s40534-018-0157-y
39. Liu Y, Jing Y (2019) Research on railway passenger ticket pricing and seat management based on revenue management. In: 19th COTA International Conference of Transportation Professionals. https://doi.org/10.1061/9780784482292.273
40. Cartmell J (2016) Study on the prices and quality of rail passenger services. European Commission Directorate General for Mobility and Transport, Brussels, 236 p
41. Report on Railway Safety and Interoperability in the EU (2020) European Union Agency for Railways (ERA). Publications Office of the European Union, Luxembourg, 116 p. https://doi.org/10.2821/30980
42. Wonglakorn N, Ratahavaraha V, Karoonsoontawong A, Jomnonkwao S (2021) Exploring passenger loyalty and related factors for urban railways in Thailand. Sustainability 13(10):1–15. https://doi.org/10.3390/su13105517
43. Dedík M, Gašparik J, Záhumenská Z (2017) Quality assessment in the logistics of rail passenger transport. MATEC Web Conf 134:1–7. https://doi.org/10.1051/matecconf/201713400009
44. Maskeliūnaitė L, Sivilevičius H, Podvezko V (2009) Research on the quality of passenger transportation by railway. Transport 24(2):100–112. https://doi.org/10.3846/1648-4142.2009.24.100-112
45. Sivilevičius H, Maskeliūnaitė L (2010) The criteria for identifying the quality of passengers' transportation by railway and their ranking using AHP method. Transport 25(4):368–381. https://doi.org/10.3846/transport.2010.46
46. Russell ML (2012) Travel time use on public transport: what passengers do and how it affects their wellbeing. University of Otago, Dunedin, 325 p. https://ourarchive.otago.ac.nz/bitstream/handle/10523/2367/RussellMarieL2012PhD.pdf?sequence=1&isAllowed=y
47. Russell M, Price R, Gerring Z, Signal L, Cumming J, Stanley J (2011) What do passengers do during travel time? Structured observations on buses and trains. J Public Transp 14(3):123–146. https://digitalcommons.usf.edu/cgi/viewcontent.cgi?article=1133&context=jpt

48. Traukinio keleiviams pro langą rodo spektaklius (2017) LRT televizijos laida „Labas rytas, Lietuva“, LRT.lt. https://www.lrt.lt/naujienos/veidai/14/185752/traukinio-keleiviams-pro-langus-rodo-spektaklius (in Lithuanian)
49. Zhen F, Tang J, Cao J (2018) Exploring correlates of passenger satisfaction and service improvement priorities of the Shanghai-Nanjing high speed rail. J Transp Land Use 11(1):559–573. https://doi.org/10.5198/jtlu.2018.958
50. Cerny P (2021) 10 questions about night trains that you always wanted, but never dared, to ask. Heinrich Böll Stiftung, Brussels. European Union. https://eu.boell.org/en/10-questions-night-trains
51. Railway passenger transport statistics—quarterly and annual data (2023). Eurostat statistics explained. https://ec.europa.eu/eurostat/statistics-explained/index.php?title=Railway_passenger_transport_statistics_-_quarterly_and_annual_data#In_2020.2C_rail_passenger_transport_almost_halved_in_the_EU_compared_to_2019
52. IRG (2021) Impacts of the COVID-19 crisis and national responses on European railway markets in 2020. Independent Regulators' Group—Rail. file:///C:/Users/Home/Downloads/IRG-Rail__2021__5_____Covid_publication_2021.pdf
53. Gherghina ŞC, Onofrei M, Vintilă G, Armeanu DŞ (2018) Empirical evidence from EU-28 countries on resilient transport infrastructure systems and sustainable economic growth. Sustainability 10(8):2900, 1–34. https://doi.org/10.3390/su10082900d
54. European Transport Policy. Mobility (2020) 07 02. Federal Ministry for Digital and Transport. https://www.bmvi.de/SharedDocs/EN/Dossier/european-transport-policy/european-transport-policy.html
55. Alotaibi S, Quddus M, Morton C, Imprialou M (2022) Transport investment, railway accessibility and their dynamic impacts on regional economic growth. Res Transp Bus Manage 43(1–12):100702. https://doi.org/10.1016/j.rtbm.2021.100702
56. Stoilova S (2018) Study of railway passenger transport in the European Union. Tehnički vjesnik 25(2):587–595
57. Molin E, Kroesen M (2022) Train travel in corona time: safety perceptions of and support for policy measures. Transp Res Part A Policy Pract 158:196–209. https://doi.org/10.1016/j.tra.2022.03.005
58. UIC Covid-19 Task Force (2020) Management of Covid-19. First estimation of the global economic of Covid-19 on rail transport. International Union of Railways (UIC)—Paris, 27 p. https://uic.org/IMG/pdf/economic_impact_v2.pdf
59. Tardivo A, Zanuy AC, Martin CS (2021) COVID-19 impact on transport: a paper from the railways' systems research perspective. Transp Res Record 2675(5):367–378. https://doi.org/10.1177/0361198121990674
60. Maskeliūnaitė L (2021) Railways in Lithuania: from tsarist Russia to Rail Baltica. Transport 36(4):364–375. https://doi.org/10.3846/transport.2021.16086
61. Seniausias traukinio maršrutas Lietuvoje iškeliavo į paskutinį reisą (2015) 15 min.naujienos. Available from Internet: https://www.15min.lt/pasaulis-kiseneje/naujiena/kelioniu-istorijos/seniausias-traukinio-marsrutas-lietuvoje-iskeliavo-i-paskutini-reisa-639-506749 (in Lithuanian)
62. „Lietuvos geležinkeliai“ nutraukia traukinio Vilnius–Maskva–Vilnius maršrutą (2015) LRT radijas, LRT.lt. Available from Internet: https://www.lrt.lt/naujienos/verslas/4/122427/lietuvos-gelezinkeliai-nutraukia-traukinio-vilnius-maskva-vilnius-marsruta (in Lithuanian)
63. Navakas N (2017) „Lietuvos geležinkelių“ vagonai neberiedės į Maskvą. Verslo žinios. https://www.vz.lt/transportas-logistika/2017/05/12/lietuvos-gelezinkeliu-vagonai-neberiedes-i-maskva (in Lithuanian)
64. Dėl koronaviruso stabdomas tarptautinių keleivinių traukinių eismas (2020) Delfi žinios. https://www.delfi.lt/news/daily/lithuania/del-koronaviruso-stabdomas-tarptautiniu-keleiviniu-traukiniu-eismas.d?id=83777479 (in Lithuanian)
65. Jockus A (2021) „Lietuvos geležinkeliai“: keleivinis traukinys į Minską nevyksta nuo pernai, kroviniai į Baltarusiją gabenami įprastai. Vakarų ekspresas. Alfa.lt. https://ve.lt/aktualijos/lietuva/lietuvos-gelezinkeliai-keleivinis-traukinys-i-minska-nevyksta-nuo-pernai (in Lithuanian)

66. CER (2022) The voice of European Railways. Next generation on track. Activity Report 2022. 28 p. https://www.cer.be/images/CER_ActivityReport_2022.pdf
67. Scherer M, Dziekan K (2012) Bus or rail: an approach to explain the psychological rail factor. J Public Transp 15(1):75–93. https://doi.org/10.5038/2375-0901.15.1.5
68. Zhou Y, Au Z-W, Wang H-X (2017) On the passenger travel behavior of high-speed railway in China—with Beijing-Shanghai high-speed railway Channel as an example. Transp Res Cong. https://doi.org/10.1061/9780784482513.027
69. Pinheiro da Silva PF, Mendes J (2020) Passengers comfort perception and demands on railway vehicles: a review. In:International Congress on Engineering—Engineering for Evalution, pp 257–270. KnE Engineering. https://doi.org/10.18502/keg.v5i6.7039
70. Patalpų oro kokybės gerinimas naudojant rekuperacinę sistemą (2023) Ekspertai.lt. Rekuperatoriai. Parengta pagal Namams24.lt medžiagą. https://www.ekspertai.lt/rekuperatoriai/straipsniai/patalpu_oro_kokybes_gerinimas_naudojant_rekuperacine_sistema (in Lithuanian)
71. Ye X, Lu H, Li D, Sun B, Liu Y (2016) Thermal comfort and air quality in passenger rail cars. Int J Vent 3(2):183–192. https://doi.org/10.1080/14733315.2004.11683913
72. Spence C (2021) Scent in motion: on the multiple uses of ambient scent in the context of passenger transport. Front Psychol 12(702517):1–19. https://doi.org/10.3389/fpsyg.2021.702517
73. Silva J, Sá ES, Escados M, Carvalho J (2021) The influence of ambient scent on the passengers' experience, emotions and behavioral intentions: an experimental study in a public bus service. Transp Policy 106:88–98. https://doi.org/10.1016/j.tranpol.2021.03.022
74. Maier J, Zierke O, Hoermann H-J, Goerke P (2021) Effects of personal control for thermal comfort in long-distance trains. Energy Build 247:1–9. https://doi.org/10.1016/j.enbuild.2021.111125
75. Nakano J, Tanabe S (2019) Thermal comfort condition of passengers in naturally ventilated train stations. E3S Web Conf 111:02069, 1–6. https://doi.org/10.1051/e3sconf/201911102069
76. Barone G, Buonomano A, Forzano C, Palombo A (2020) Enhancing trains envelope—heating, ventilation, and air conditioning systems: a new dynamic simulation approach for energy, economic, environmental impact and thermal comfort analyses. Energy 204(117833):1–14. https://doi.org/10.1016/j.energy.2020.117833
77. Hernández AM, Vargas-Robles D, Alcaraz LD, Peimbert M (2020) Station and train surface microbiomes of Mexico city's metro (subway/underground). Sci Rep 10(8798):1–10. https://doi.org/10.1038/s41598-020-65643-4
78. Cho H-W, C, Chu (2011) Is the public transportation system safe from a public health perspective? Osong Public Health Res Perspect 2(3):149–150. https://doi.org/10.1016/j.phrp.2011.11.037
79. Rezapour M, Ferraro FR (2021) The impact of commuters' psychological feelings due to delay on perceived quality of a rail transport. Human Soc Sci Commun 8(190):1–8
80. Cheng Y-H (2010) Exploring passenger anxiety associated with train travel. Transportation 37:875–896. https://doi.org/10.1007/s11116-010-9267-z
81. Grey E (2016) 12 20. Tackling alcohol consumption on Britain's railways. Railway Technology. https://www.railway-technology.com/features/featuretackling-alcohol-consumption-on-britains-railways-5693290/
82. HN 128:2013. „Keleiviniai geležinkelių riedmenys. Sveikatos saugos reikalavimai". Lietuvos higienos norma, patvirtinta Lietuvos Respublikos sveikatos apsaugos ministro 2013 m. birželio 14 d. įsakymu Nr. V-628. https://e-seimas.lrs.lt/portal/legalAct/lt/TAD/TAIS.451868?jfwid=-9dzqnu8op (in Lithuanian)
83. Talotte C, Ségrétain S, Gonac'h J, Le Rohellec J (2011) Improvement of lighting ambience on board trains: experimental results on the combined effect of light parameters and seat colours on perception. In: 9th Word Congress Railway Research 2011, May 22–26, pp 1–7. http://www.railway-research.org/IMG/pdf/d2_talotte_corinne.pdf
84. Zhang Y, Zhang J, Li T, Zhang L, Zhang W (2016) Research on aerodynamic noise reduction for high-speed trains. Shock Vib 2016(4):1–21. https://doi.org/10.1155/2016/6031893

85. Prithvi C, Srinidhi R, Karthik Hebbar A (2021) Vibration analysis of railway wagon suspension system for improved ride quality using MATLAB simulink. Oper Manage Syst Eng. https://doi.org/10.1007/978-981-15-6017-0_6
86. Dižo J, Blatnický M, Gerlici J, Leitner B, Melnik R, Semenov S, Mikhailov E, Kostrzewski M (2021) Evaluation of ride comfort in a railway passenger car depending on a change of suspension parameters. Sensors 21(8138):1–16. https://doi.org/10.3390/s21238138
87. Haji Abdulrazagh P, Hendry MT, Gül M, Roghani A, Toma E (2021) Use of measured accelerations from a passenger rail car to evaluate ride quality and track roughness—a case study. Proc Inst Mech Eng Part F J Rail Rapid Transit 236(6):733–742. https://doi.org/10.1177/09544097211041459
88. Yang L, Li X, Tu J (2020) Thermal comfort analysis of a high-speed train cabin considering the solar radiation effects. Indoor Built Environ 29(8):1101–1117. https://doi.org/10.1177/1420326X19876082
89. Winter J, Windemut I, Kevlishvili N, Schmeling D, Maier J (2016) Passenger comfort for the next generation train: subject trials for ventilation and lighting. In: Proceedings of the Third International Conference on Railway Technology: Research, Development and Maintenance. Civil-Comp Press, Stirlingshire, pp 1–12. https://doi.org/10.4203/ccp.110.196
90. Rakicevic M (2020) 36 alarming snoring statistics and facts for. https://disturbmenot.co/snoring-statistics/
91. Calderon LE, Carney LD, Kavanagh KT (2016) The cry of the child and its relationship to hearing loss in parental guardians and health care providers. J Evid Inf Social Work 13(2):198–205. https://doi.org/10.1080/23761407.2015.1018031
92. Travelling by train with our baby (2023) SNCF. https://www.sncf.com/en/passenger-offer/travel-for-everyone/travel-by-train-with-a-baby
93. Burroughs D (2020) Capacity vs comfort: balancing the interior design. Int Railway J. https://www.railjournal.com/in_depth/capacity-vs-comfort-balancing-the-interior-design

# Chapter 2
# Passenger Choice of Travelling by Train as an Alternative to Other Modes of Transport

*Trains are wonderful.... To travel by train is to see nature and human beings, towns and churches and rivers, in fact, to see life*
Agatha Christie

## 2.1 Conceptualising the Choice of Public Transport as a Mode for Travel

Nowadays, for travelling people can use different modes of transport, and they usually choose the most suitable and reliable one of those that are available. The choice of one mode of transport as an alternative to another is subjective. It is usually built upon passenger attitude toward the advantages and disadvantages of using some travelling mode. All modes of transport and their elements interact not just in multi-modal passenger traffic, but also in competition with each other [1]. Alternatives for passenger traffic include road transport, rail transport and air transport. Several alternative transport technologies are available for hilly regions [2].

In Sweden, there are several competing railway companies. They compete in an attempt to introduce innovations and new services for passengers. Competition has been a key factor behind the current growth in passenger numbers in Sweden since 2010 (+45%). This is one of the highest rates in Europe. Traditional sleeping carriages meet the needs of passengers. However, significant fare increases are essential to make them profitable and sustainable. Currently, the demand for a much higher fare (for a traditional sleeper carriage) is very low. Therefore, the focus is on improving passenger comfort: new and more comfortable mattresses, blankets and pillows. Passengers who want privacy can book a separate compartment [3].

In a rapidly changing world, with increasing competition in all sectors of transportation, railways are in a period of restructuring their management and technology. New methods of organization are introduced, commercial and tariff policies change



H. Sivilevičius and L. Maskeliūnaitė, *Assessment of the Quality of Passenger Transportation by Train Using Multiple Criteria Decision Making Methods*, Lecture Notes in Intelligent Transportation and Infrastructure,
https://doi.org/10.1007/978-3-031-85877-2_2

radically, a more entrepreneurial spirit is required. At the same time, new high-speed tracks are being constructed and old tracks are renewed, high comfort rolling stock vehicles are being introduced, logistics and combined transport are being developed. Awareness of environmental issues and the search for greater safety give a new role to the railways within transportation system. Meanwhile, methods of analysis have significantly evolved, principally due to computer applications and new ways of thinking and approaching old problems. Thus, it becomes necessary to come up with a new scientific approach to tackle management and engineering aspects of railways, to understand in-depth the origins and inter-relationships of the various situations and phenomena, and to suggest the appropriate methods and solutions to solve the various emerging problems [4].

The sustainable development of the EU is strongly influenced by transport, which is polluting from an environmental point of view. Transport infrastructure accessibility criteria, accessibility distance have a positive effect on sustainable development [5]. To mitigate transport sector greenhouse gas emissions, it is necessary to assess the efficiency of transport policy [6]. One of the most effective ways of improving the sustainability of the transport sector is the choice to use a less polluting mode of transport. Passenger satisfaction is an important factor in choosing a mode of transport to travel in municipalities and especially in big cities [7].

Mobility is crucial for the development of the country internal market and for maintaining the desired life quality of citizens, as it is important for people to exercise their freedom to travel. This means more frequent travel by bus, rail, and air. Viable options can only be available through better integration of modal networks, which means that airports, railway stations, metro and bus stations should be increasingly interconnected and transformed into multimodal passenger transport platforms [8].

There will be more possibilities for passengers to choose a particular mode of transport when various systems of transportation become more closely integrated. Experience has shown that passengers usually choose the mode of transport that best suits their personal needs, habits and understanding of the expected quality of travelling. The mode of transport which is the most acceptable to a passenger can be determined by developing a system of criteria and calculating their weights using multiple criteria decision making (MCDM) methods [9–13]. Passenger loyalty development increases the profits of airline, rail, and road transport companies. This is undoubtedly important for the success of the institution [14].

Travelling allows us to learn more about the world. However, it brings about some inconveniences and discomfort. People who travel by bus for long distances spend much time on a bus and get tired.

For many people it causes stress unless they are inveterate travellers. Therefore, when planning a trip, people should learn more about particular transport modes, trip duration, possibilities to have a rest and the cost of the trip. Various transport modes (bus, train, or airplane) have advantages and disadvantages [15]. It is important to inform urban public transport passengers on the estimated time of arrival as it influences the forecast of travelling by bus [16]. Train traffic has increased over the last decade and is likely to continue to do so, as more passengers and freight are transported by train rather than by car. This will reduce energy consumption and pollutant emissions [17].

Travel by train is still popular in the world. The railways, rolling stocks, services provided to passengers as well as safety are being constantly improved, while the environ-mental pollution is being reduced. This helps rail transport to compete with other modes of transport. Matuška [18] examines the accessibility of rail transport by providing ways to assess the accessibility of railway infrastructure and trains. He applies a two-step model to assess the availability of departure halls. Train services are barrier-free for passengers, especially those traveling long distances, but they are still less accessible to disabled passengers in the suburbs and regions.

Given that a passenger may travel by several modes of transport on the chosen route, it is necessary to investigate which factors are the most influential in the choice of a transport mode. The choice of a transport mode depends on a passenger's personal preferences, awareness and experience, as well as his or her knowledge of the journey to be made. A study of the opinions of a large group of passengers on the preferences regarding public transport can help to identify public expectations about different modes of transport. The studies have identified positive and negative characteristics of different modes of transport that meet or fail to meet passenger needs. The findings can be used to develop national and inter-country strategies for passenger transport.

Each mode of transport can be compared with other modes of transport by building a system of factors (criteria) and identifying their significance. MCDM methods are best tailored for this purpose. They are validated by taking the average opinion of experts or respondents on the basis of the importance given to the criteria in the survey questionnaires, and expressed in terms of ranks, scores, weights or other numerical estimates. Various MCDM methods are known to produce close or different results, however none of them has a theoretical advantage over the other methods. The general principle of all these algorithms is basically the same: the most important criterion must be given the highest weight. The numerical values of the weights have to be consistent with the ranks of the criteria and the sum of the weights $\omega_i$ of all criteria has to be equal to one, i.e., the weights have to be normalised: $\sum_{i=1}^{m} \omega_i = 1$.

In theory and in practice, there are a few ways to determine the importance of the indicators (criteria). One of the most popular ways to determine the importance of criteria is to rank them. The most simplistic is the rank correlation method [19]. Often direct weighting in percentages or unit fractions is used [20]. The number of criteria describing an object determines the results of the assessment, i.e. their weights and precision. The large number of criteria makes it difficult for an expert to assess the unit as a whole. To this end, improved but more sophisticated mathematically based methods of criteria weighting have been proposed, where the criteria are compared two by two. This determines which criterion is more important than the other in each pair, and by how much. However, when using more sophisticated methods to determine the weights of the criteria, it is useful to start by ranking the criteria, i.e. determining their order of importance. This makes the application of these methods much easier. One such method is the AHP method [21]. The AHP is a theory of measurement through pairwise comparisons and relies on the judgements of experts to derive priority fundamental scales [22]. While the Saaty method is inherently powerful as a questioning device and easy to use, there is a degree of imprecision in the specification of what factors should be considered when determining the weights [23]. This approach, like other MCDM approaches, has its pros [24, 25] and cons [7, 26–29]. One of the major drawbacks is that not every expert is able to fill in the pairwise comparison matrix adequately to satisfy the transitivity condition and keep the consistency ratio (C.R.) below 0.1. This method is rational when 5–9 criteria characterise the object of study [30]. When there are more criteria, it is reasonable to divide them into a hierarchical structure consisting of main criteria and sub-criteria [31–34].

## 2.2 Selecting the Best Alternative for a Passenger Transport Mode

### *2.2.1 The Hierarchy Model for Criteria and Alternatives*

Public transport for passenger traffic involves three main modes: road transport (BT), rail transport (RT) and air transport (AT). Where there are navigable rivers or large lakes within a country or between countries, passengers can travel by water transport (WT). These modes of transport have their pros and cons, which are weighed up by a passenger when choosing the most appropriate mode of travel, i.e., the best alternative of the available modes of transport. With the view of the importance of all the factors (criteria) that have the greatest impact on the quality of the passenger journey, it is possible to rationally optimise them so as to increase the overall attractiveness of each mode of transport. Each criterion compares individual modes of transport against each other to determine their normalised weights, i.e., the weights of the alternatives. The normalised weights of the criteria and the normalised weights and idealised weights of the transport modes benchmarked against each criterion are then calculated to give the normalised global weights of the alternatives.

The best alternative for passenger transport modes can be selected using the Analytic Hierarchy Process (AHP) approach in the distributive mode and the ideal mode [35]. The hierarchy model for the selection of the best alternative for passenger transportation is presented in Fig. 2.1. The first level of the hierarchical structure contains the goal, the second level is made up of 10 factors (criteria) and the third level comprises 3 alternatives for the passenger transport modes.

Currently, there are more than 20 different methods that can be used to derive the priority weights of alternatives for pairwise comparison matrix (PCM) [36].

Other MCDM methods can also be employed to select the best alternative. Other popular approaches include Technique for order of performance by similarity to ideal solution (TOPSIS) method, COmplex PRoportional ASsessment (COPRAS) method, Simple Additive Weighting (SAW) method and Multilevel grey evaluation (MGE) method. The MGE and TOPSIS methods are employed by Chen et al. [37]

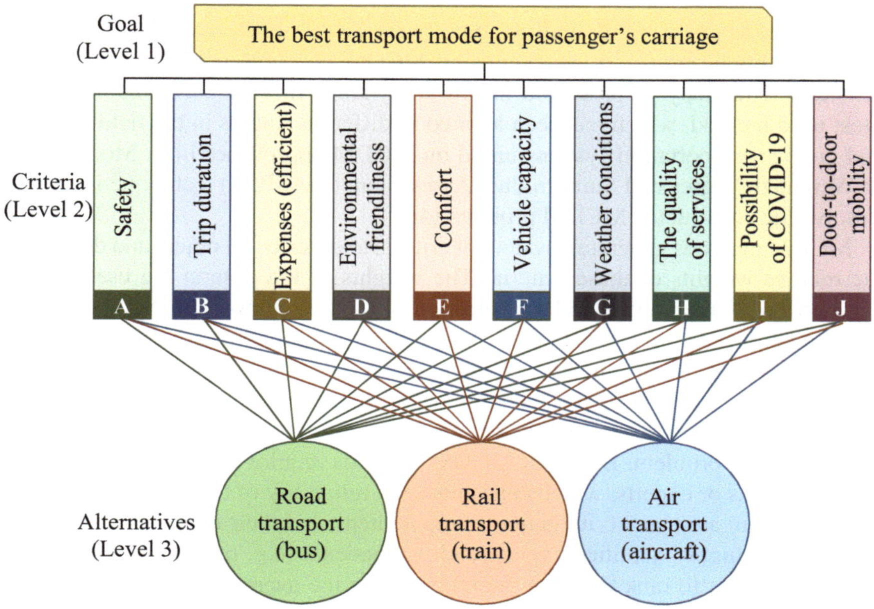


**Fig. 2.1** A three-level hierarchy for selection of the best transport mode of passenger transportation

to evaluate the overall performance of passenger transfer at large transport terminals in different alternatives. The integrated weighting method is adopted to overcome the biased weight set from individual subjective or objective perspectives in the evaluation.

MCDM methods are regarded as the main decision-making methods that consider more than one criterion in the decision-making process [38]. MCDM techniques are mathematical tools that help decision makers evaluating and ranking in an automatic way many possible alternatives over multiple conflicting criteria in highly complex situations [39]. MCDM methods have applications in transport, civil engineering, economics, environmental, social and other societal issues. Šikšnelytė-Butkienė et al. [40] aim to overview the application of MCDM techniques for solving sustainability issues in the transport sector and to provide the main insights for methods and sustainability criteria selection. Alternative MCDM methods applied to the same problem often produce different results. In the search for a more robust tool, a framework of MCDM methods is proposed to address one problem [41]. Researchers often use multiple models to achieve their goal, e.g. combining PLTS and UTASTAR and BWM approaches when selecting a battery for a new energy vehicle [42]. Four different MCDM methods—ARAS, SAW, TOPSIS and PROMETHEE—are used to evaluate stair safety alternatives. Since the selection of the most suitable MCDM method is full of uncertainties, the novel combined-decision-making approach formally called CORST is proposed in Zavadskas et al. [43]. A new integrated MCDM-SWOT-TOWS model is applied to the analysis of strategic decisions of a transport company [44]. A bibliographic overview of some recent evaluations of urban and public transport systems using MCDM decision-making approaches is provided. The analyses presented in this study show that the AHP method is the most used method, which has been applied to different studies in the field of urban and public transportation systems based on MCDM approaches [45]. Modification of Weighted Aggregated Sum Product Assessment (WASPAS) technique is applied to solve the constructed MCDM problem [46].

MCDM methods generate a system of criteria to describe an object and determine the relative weights of these criteria. The weights of the criteria are used for the comparison of alternatives and the selection of the best alternative. Sometimes the problem is solved by determining only the relative weights of the criteria. To increase the reliability of the results, several methods are used to determine the weighting of the criteria, as none of the methods has a theoretical advantage. The average of the weights from several methods can reasonably be taken as the result of the solution to the problem. Different MCDM methods require different qualifications of respondents or experts, which determines the reliability of the results.

When there are many criteria describing an item, and their importance is judged by both non-highly qualified experts and respondents (e.g. train passengers), it is most convenient to rank the criteria and calculate the averages of the ranks of the group of raters. For this purpose, the ARTIW-L and ARTIW-N methods proposed by the authors of the monograph are used to calculate the relative weights of the criteria from the averages of their ranks. A new IHAMCI method is also proposed to calculate the global relative weights of the key criteria (groups of criteria) from the local weights of many hierarchically unstructured sub-criteria.

### *2.2.2 Characteristics of the Transport Mode Required for Passenger Traffic*

A framework of 10 factors relevant to each mode of transport has been developed, based on the study of research papers, a scientific discussion with transport professionals and developers as well as personal experience. The definitions (titles) and brief comments on these ten criteria, which facilitate a comprehensive assessment of the required characteristics of any transport mode, are provided in Table 2.1.

Some criteria are maximised (max) (A, D, E, F, H, J) while others (B, C, G, I) are minimised (min). The definition of each criterion and the comments render the content of the criterion. This makes it easier for the expert to make an assessment decision by prioritising all criteria. The expert assigns ranks to all the criteria and completes the pairwise comparison matrix for the AHP method criteria. Each passenger transport mode is then compared with other modes for each criterion.

**Table 2.1** Criteria for the selection of the best mode of passenger transport

| Criterion | Definition and comments |
|---|---|
| A<br>Safety<br>(*max*) | Safety is related to the likelihood of not being involved in an accident on a journey, as indicated by the statistics and analysis of accidents involving the mode of transport, as well as passenger hunch and personal experience. It is not possible to predict the location and timing of accidents in advance. Safety is enhanced by the availability of emergency drop-off points, fire-fighting and medical equipment, seat belts, good running condition of the vehicle, and high driver qualification |
| B<br>Trip duration<br>(*min*) | Journey times are reduced by increasing the speed of the vehicle, choosing the right route, and shortening border checks and customs clearance times. Travel time is often considered as lost time because a passenger is unable to do work other than IT technology. Travel time can be used for partial relaxation |
| C<br>Expenses<br>(efficient)<br>(*min*) | Travel costs are borne both by the carrier and passengers. The passenger travel costs include the ticket fare, booking of special seats for passengers, baggage, transit visas, meals. Acquisition and maintenance of vehicles, fuel, maintenance of staff, infrastructure costs are paid by a carrier. A passenger wants a low price and a carrier pursues profit. It must therefore be acceptable to both sides: competitive and cost effective |
| D<br>Environmental friendliness<br>(*max*) | Different types of vehicle engines (internal combustion, electric and hydrogen) have different emissions. Therefore, the vehicles powered by polluting or non-polluting engines as well as vehicles with low pollution are different ecologically. The use of eco-friendly vehicles is a strategic tool for the development of the European transport area, with a view to reducing greenhouse gas emissions. A less polluting mode of transport is more beneficial for society. It is also reflected in vehicle noise emissions and potential spillage of consumables |
| E<br>Comfort<br>(*max*) | Passengers must feel comfortable, not tired and stress-free during the boarding, travelling and disembarking process. This is linked to the air temperature inside the vehicle as well as the air conditioning, lighting, accelerations, comfort of the seats, the procedure and location of border and customs checks (at the terminal or on board the vehicle), the behaviour of other passengers, and the secure storage of baggage |
| F<br>Vehicle capacity<br>(*max*) | Capacity is defined as the maximum number of passengers that a vehicle can carry when all seats or berths are occupied. A carrier pursues the highest degree of capacity utilisation while a passenger does not care about this or prefers to travel in an uncrowded vehicle. Manufacturers tend to increase the capacity of the vehicle. Capacity is not equal to the total number of passengers carried by the transport means |
| G<br>Wheater conditions<br>(*min*) | Weather conditions (clear weather, rain, snow, fog, hail, storm clouds, freezing, heating, icy roads, volcanic ash, strong winds) affect the speed, safety and fuel consumption of the vehicle. Adverse weather conditions may lead to the cancellation or delayed start of the journey |
| H<br>The quality of services<br>(*max*) | On public transport, passengers are attended by staff (conductors, stewards, attendants) from the company operating the journey. Their helpful and friendly approach towards travel-related requests of each passenger increases the quality of service |

(continued)

**Table 2.1** (continued)

| Criterion | Definition and comments |
|---|---|
| I<br>Possibility of COVID-19<br>(*min*) | Travelling is a risk of contracting the viral disease COVID-19. Close contact between passengers or passengers and the crew members, inadequate personal protection, and disregard of established rules and guidelines increase the likelihood of this. The density (number) of occupants per unit area in the vehicle and the efficiency of the ventilation system are also critical |
| J<br>"Door-to-door" mobility<br>(*max*) | A passenger starts the journey from his/her home or workplace and finishes the journey when he/she reaches the destination. It is not always possible to travel from the point of departure to the point of arrival by a single mode of transport, i.e., "door-to-door". Often several modes of transport are used—multimodal transport. This necessitates transfers between different modes of transport, resulting in inconvenience for passengers |

### 2.2.3 Alternatives to Modes of Transport

Road transport (BT) is the least safe in terms of statistics on injuries and fatalities. A bus journey on a short route takes the least time. Railway transport and especially air transport become more advantageous as distances increase. Bus travel costs are moderate to low; bus and train fares are almost the same. Road vehicles are more polluting than other types of transport when driven closer to where people live (especially in urban areas), and where internal combustion engines are still dominant. Bus passengers experience moderate comfort, as there is limited opportunity to rest while lying down. In comparison to train and aircraft capacity, bus capacity is the smallest. However, this is not of great importance for a passenger. The capacity of the vehicle and the degree to which it is being filled by passengers is relevant for a passenger carrier. The drivability of road vehicle is almost independent of the weather conditions (rain, fog, snow, wind), but it does depend on the condition of the road surface (icy, snow-covered). The quality of passenger service on a bus is lower than on a train or aircraft, where there are attendants and stewards. It also has a moderate to higher risk of contracting the viral disease COVID-19, depending on the effectiveness of the preventive measures on board. Road transport is the most consistent with the principle of door-to-door mobility, as the bus station can be located in the city centre or close to it. Passengers can embark and disembark at intermediate stations and public transport stops. A passenger can get on and off the bus at the closest point to where he or she lives, works or is at leisure.

Rail transport is the safest compared to other modes of transport. Derailments or collisions with other trains are extremely rare. The journey time by non-express train is longer than by bus or aircraft. Train fares are similar to bus fares, but lower than air fares. Rail transport is the most environmentally friendly as many rail routes are electrified and this trend is being developed. Train passengers experience the greatest comfort when sitting or lying comfortably in their assigned seatings. They can walk on the train. Besides, a train can carry more passengers than an average aircraft and far more than a bus. Train travel is the least dependent on weather conditions, which rarely results in cancellations or delays in the arrival time at the destination station. The quality of travel/service, as measured by frequency, convenience, accessibility, reliability and the integration of different modes of transport, is average in rail transportation. There is a low risk of contracting COVID-19 on board the train if personal protection is worn and the requirements are adhered to. The lowest number of passengers per unit area also contributes to the prevention of contracting the infection. Consequently, this reduces contact between passengers and staff. Railway stations can also be located in city centres or close to them, which means that trains pick up passengers closer to where they live, although not as close as buses do. Accordingly, rail transport on average satisfies the principle of door-to-door mobility.

Air transport is superior to road transport in terms of safety, although it falls behind rail transport. The ratio of the number of aircraft crashes to the number of passengers carried is decreasing. Long-distance air travel is less time-consuming than train or coach travel. Air fares are higher than bus and train fares. Air transport is a major source of exhaust emissions and noise pollution. However, engine emissions are released into the atmosphere as the aircraft moves away from the airport and people's residences, thereby reducing sensible pollution. On board the aircraft, passenger comfort is lower than on other modes of transport, due to the higher density of passengers per unit area. The mandatory baggage and passport control as well as customs clearance procedures also add to the reduction of comfort. A medium-sized aircraft carries more passengers than a coach and fewer than a train. A large aircraft can carry more passengers than a train. Aircraft capacity and capacity load factor are of little importance to passengers, but they do matter to the carrier. Adverse weather conditions have a major impact on air transport, leading to flight cancellations or delays. On board the aircraft, passengers receive the highest quality of service compared to any other mode of transport, as during the flight they are attended by stewards. Because of the strict travel rules and effective personal protection measures, there is little chance of contracting COVID-19 in air travel. Close contact between passengers and staff increases the likelihood of contracting an infection. Air transport is the least satisfactory in terms of door-to-door passenger mobility, as airports are usually located outside the city. Airports are usually accessible by road (bus, trolleybus, taxi or private car) or rail (metropolitan train, tram).

### 2.2.4 Analytic Hierarchy Process Method for the Analysis of Criteria Weights and the Study of Transport Mode Alternatives

The normalised weights of the criteria and the weights of the alternatives for each criterion are calculated by applying the popular MCDM method, which was introduced by Saaty [21]. The MCDM method is known as the AHP method.

The AHP method is used to compare individual factors (criteria) in pairs against each other, prioritising them in relation to one another. It is used to determine the weights of indicators at one level of the hierarchy in relation to a higher level, or the weights of a system of unstructured criteria. The core of the method is the pairwise comparison matrix **A**. The expert compares all of the criteria that make up the study object: $R_i$ and $R_j$ $(i, j = 1, 2, \ldots, m)$, where $m$ is the number of criteria to be compared (characteristics of the mode of transport to carry passengers or comparable alternatives).

The AHP approach is convenient because it is easier to compare the required characteristics of passenger transport or transport alternatives in pairs rather than all at one time. The comparison of criteria is straightforward and of high quality for an expert with a good knowledge of the AHP method. One criterion is stated to be more important than the other. In a matrix of pairwise comparisons of criteria, the values $a_{ij}$ indicate the dominance of the criterion on the left over the criterion on the top. The method allows the qualitative (linguistic) assessment of criteria to be represented by a quantitative (numerical) assessment.

The result of a pairwise comparison of the criteria is a square matrix $\boldsymbol{A} = \|a_{ij}\|$ $(i, j = 1, 2, \ldots, m)$. To assess the importance of the criteria Saaty [21, 47] and Kou et al. [36] proposed the application of the commonly used odd integer scale of 1, 3, 5, 7 and 9. Intermediate even estimates $(a_{ij} = 2, 4, 6, 8)$ can also be used to complete the pairwise comparison matrix when the expert considers the state under consideration to be different from the representative state (Table 2.2).

Suppose you need to compare the significance or impact on the quality of passenger traffic of the required passenger transport characteristics $R_1, \ldots, R_m$. The situation is analysed under the assumption that the significance of the features to be compared is ideally equal. Significance is indicated by estimates $\omega_1, \omega_2, \ldots, \omega_m$. The ratio

$$a_{ij} = \frac{\omega_i}{\omega_j}, \qquad (i, j = 1, 2, \ldots, m) \tag{2.1}$$

shows how many times the $i$-th characteristic $R_i$ is more significant than the $j$-th characteristic $R_j$. The matrix $\boldsymbol{A}$ is reciprocal because its elements $a_{ij} = 1/a_{ji}$.

The elements of matrix $\boldsymbol{A}$ shall be completed in accordance with the following requirements. First, the elements of the matrix $\boldsymbol{A}$ are $a_{ij} = 1$ when the significance of the two criteria to be compared is the same with respect to the phenomenon/object under study. For this case, all elements of the main diagonal are $a_{ij} = 1$ $(i = 1, 2, \ldots, m)$, since each criterion is compared against itself. Second, the elements of matrix $\boldsymbol{A}$ are $a_{ij} = 3$, where criterion $R_i$ is slightly more significant than criterion $R_j$. Third, the elements of matrix $\boldsymbol{A}$ are $a_{ij} = 5$, where criterion $R_i$ is much more important than criterion $R_j$. Fourth, the elements of matrix $\boldsymbol{A}$ are $a_{ij} = 7$, where criterion $R_i$ is very much more important than criterion $R_j$. Fifth, the elements of matrix $\boldsymbol{A}$ are $a_{ij} = 9$ when criterion $R_i$ is absolutely and incomparably more important than criterion $R_j$.

The numbers in this scale are used to show the number of times an element with a higher priority score dominates against an element with a lower priority score in terms of a criterion or characteristic they share.

**Table 2.2** The Saaty's 9-points fundamental rating scale

| Intensity of importance | Definition | Explanation |
|---|---|---|
| 1 | Equal importance | Two activities contribute equally to the objective |
| 3 | Weak importance of one over another | Experience and judgment slightly favour one activity over another |
| 5 | Essential or strong importance | Experience and judgment strongly favour one activity over another |
| 7 | Very strong or demonstrated importance | An activity is favoured very strongly over another; its dominance demonstrated in practice |
| 9 | Absolute importance | The evidence favouring one activity over another is of the highest possible order of affirmation |
| 2, 4, 6, 8 | Intermediate values between adjacent scale values | When compromise is needed |
| Reciprocals of above nonzero | If activity *i* has one of the above nonzero numbers assigned to it when compared with activity *j*, then *j* has the reciprocal value when compared with *i* | A reasonable assumption |
| Rationals | Ratios arising from the scale | If consistency were to be forced by obtaining *m* numerical to span the matrix |

The elements $a_{ij}$ of the matrix $\boldsymbol{A}$ can be viewed as ratios of the values of the criteria $R_i$ and $R_j$ and, after normalisation of these values, as ratios of the weights of the criteria. In this case, the scale of the elements $a_{ij}$ would be the set of all rational numbers. The square matrix of the pairwise comparison of the criteria evaluated by the expert looks like this (*i*—rows, *j*—columns):

$$\boldsymbol{A} = \begin{bmatrix} a_{11} & a_{12} & \cdots & a_{1j} & \cdots & a_{1m} \\ a_{21} & a_{22} & \cdots & a_{2j} & \cdots & a_{2m} \\ \vdots & \vdots & \vdots\vdots\vdots & \vdots & \vdots\vdots\vdots & \vdots \\ a_{i1} & a_{i2} & \cdots & a_{ij} & \cdots & a_{im} \\ \vdots & \vdots & \vdots\vdots\vdots & \vdots & \vdots\vdots\vdots & \vdots \\ a_{m1} & a_{m2} & \cdots & a_{mj} & \cdots & a_{mm} \end{bmatrix}$$

$$
= \begin{bmatrix}
\frac{\omega_1}{\omega_1} & \frac{\omega_1}{\omega_2} & \ldots & \frac{\omega_1}{\omega_j} & \ldots & \frac{\omega_1}{\omega_m} \\
\frac{\omega_2}{\omega_1} & \frac{\omega_2}{\omega_2} & \ldots & \frac{\omega_2}{\omega_j} & \ldots & \frac{\omega_2}{\omega_m} \\
\vdots & \vdots & \vdots & \vdots & \vdots & \vdots \\
\frac{\omega_i}{\omega_1} & \frac{\omega_i}{\omega_2} & \ldots & \frac{\omega_i}{\omega_j} & \ldots & \frac{\omega_i}{\omega_m} \\
\vdots & \vdots & \vdots & \vdots & \vdots & \vdots \\
\frac{\omega_m}{\omega_1} & \frac{\omega_m}{\omega_2} & \ldots & \frac{\omega_m}{\omega_j} & \ldots & \frac{\omega_m}{\omega_m}
\end{bmatrix}. \tag{2.2}
$$

For the ideal case, where the elements of matrix $\boldsymbol{A}$ are ratios of unknown weights, the matrix is reciprocal, i.e., $a_{ij} = 1/a_{ji}$. In fact, $a_{ij} = \frac{\omega_i}{\omega_j}$ and $a_{ji} = \omega_j/\omega_i$. Therefore, it is possible to complete the part of the matrix above or below the main diagonal.

The reciprocal symmetry of the matrix $\boldsymbol{A}$ is in the ideal case understood in this way, e.g. if one object is five times heavier than another, then the latter is five times lighter or 1/5 lighter than the former. Therefore, for the ideal case where the elements $a_{ij}$ are ratios of unknown weights, the corresponding elements of any two adjacent columns of matrix $\boldsymbol{A}$ will be proportional. This means that the column elements will have the same proportions. For example, when taking the ratio of the first and second columns of matrix $\boldsymbol{A}$.

$$
\frac{a_{i1}}{a_{i2}} = \frac{\frac{\omega_i}{\omega_1}}{\frac{\omega_i}{\omega_2}} = \frac{\omega_2}{\omega_1}. \tag{2.3}
$$

Similarly, the elements of any two rows of the matrix will be proportional ($i = 1, 2, \ldots, m$).

The expert compares the criteria of the research object in pairs, and applies the linguistic description of the intensity of the mutual importance given in Table 2.2. Then the expert assigns the corresponding level of importance from 1 to 9. It is convenient to assign ranks to the object criteria at the beginning and then to compare the most significant criterion with the second most significant criterion, the third most significant criterion, etc., and at the end with the last ($m$-th) most significant criterion (the row of the most significant criterion is filled with integers) [48]. The elements of the most significant criterion column are fractional: they vary from 1/9 to 1/2. The second most significant criterion is then compared in the same sequence against all the other criteria (completing the row for the second most significant criterion of $m - 1$ integers and one fractional number) and so on. The consistency of the pairwise comparison matrix completed by the expert is of great relevance, since it is an indication of the non-contradictory logic of its completion. A study of scientific publications reveals that the matrix is not always completed correctly. A prerequisite for the consistency of a comparison matrix is the transitivity of the significance of its elements. If an element (criterion) A of matrix $\boldsymbol{A}$ is more significant (greater) than element B and element B is more significant than element C, then element A is more significant than element C. In practice, the transitivity condition makes it easy to identify the contradictory elements of the matrix that do not satisfy this condition (they either need to be corrected or not used in the study).

The consistency condition of the matrix can be checked mathematically. For the ideal case, Eq. (2.2) is used to multiply the matrix $\boldsymbol{A}$ by the column vector weights, i.e., the transposed row is $\boldsymbol{\omega} = (\omega_1, \omega_2, \ldots, \omega_m)^T$:

$$\boldsymbol{A\omega} = \begin{bmatrix} a_{11} & a_{12} & \cdots & a_{1m} \\ a_{21} & a_{22} & \cdots & a_{2m} \\ \vdots & \vdots & \vdots & \vdots \\ a_{m1} & a_{m2} & \cdots & a_{mm} \end{bmatrix} \begin{bmatrix} \omega_1 \\ \omega_2 \\ \vdots \\ \omega_m \end{bmatrix} = m \begin{bmatrix} \omega_1 \\ \omega_2 \\ \vdots \\ \omega_m \end{bmatrix} = m\boldsymbol{\omega}. \tag{2.4}$$

This is the solution to the largest eigenvalue and eigenvector problem in mathematics, which is well known to experts:

$$\boldsymbol{A\omega} = \lambda\omega \tag{2.5}$$

with eigenvalue $\lambda = m$, it is the row of the matrix $\boldsymbol{A}$ or the number of the research object (passenger transport characteristics or alternatives).

Equation (2.5) shows that the weights $\omega_i$ of Saaty's [21] pairwise comparison matrix $\boldsymbol{A}$ are the normalised values of the eigenvector corresponding to the largest eigenvalue $\lambda_{max}$, which sum to one: $\sum_{i=1}^{m} \omega_i = 1$.

The most accurate of the algorithms for determining the weights (relative significance) $\omega_i$ of the criteria is based on the calculation of the product of geometric means of the rows of the elements $a_{ij}$ of the pairwise comparison matrix [37, 49–51]:

$$\omega_i = \frac{\sqrt[n]{\prod_{j=1}^{n} a_{ij}}}{\sum_{i=1}^{m} \sqrt[n]{\prod_{j=1}^{n} a_{ij}}} \quad (i = 1, 2, \ldots, m). \tag{2.6}$$

The normalisation of the products of the rows of the matrix results in the calculation of eigenvector, which can reasonably be taken as criteria weights $\omega_i$, representing the opinion of one expert. The consistency of the pairwise comparison matrix $\boldsymbol{A}$ is checked by calculating the largest eigenvalue $\lambda_{max}$.

The product of the pairwise comparison matrix $\boldsymbol{A}$ is found from the weights of the column $\boldsymbol{\omega} = (\omega_1, \omega_2, \ldots, \omega_m)^T$.

To do this, the sum of the products of each criterion $\sum_{i=1}^{n} a_{ij}\omega_i$ is divided by the estimated value of the eigenvector (the normalised weight) $\omega_i$ of the corresponding criterion, to obtain the arithmetic mean of all the criteria of the object:

$$\lambda_{max} = \frac{1}{m} \times \sum_{i=1}^{m} \frac{\sum_{j=1}^{n} a_{ij}\omega_i}{\omega_i}. \tag{2.7}$$

It has been found [21] that the largest eigenvalue of the $m$-th order reciprocal matrix is $\lambda_{max} \geq m$. For the ideal case, if the pairwise comparison matrix is fully aligned, then the elements of its columns are also proportional, $\lambda_{max} = m$. The consistency of the matrix is shown by the margin $\lambda_{max} - m$ and the row n of matrix $\boldsymbol{A}$ $m \times m$. The consistency index (*C.I.*) is calculated as follows:

$$C.I. = \frac{\lambda_{max} - m}{m - 1}. \tag{2.8}$$

It is extremely uncommon in practice to obtain a matrix with *C.I.* = 0, even when the condition of transitivity of its elements is fully satisfied. The consistency degree of a reciprocal matrix can be quantified by comparing the *C.I.* of the matrix being evaluated with the random index (*R.I.*) of a randomly generated reciprocal matrix of the same order on a scale of 1–3–5–7–9 [30, 52]. Refer to Table 2.3.

When the pairwise comparison matrix of the research object is greater than 15 rows, the value of the *R.I.* can also be calculated according to formula (2.9) [53, 54]:

$$R.I. = 1.98\frac{m - 2}{m} = 1.98\left[1 - \frac{m - 1}{m(m - 1)/2}\right]. \tag{2.9}$$

The ratio of the matrix consistency index (*C.I.*) to the random index (*R.I.*) is called the consistency ratio (*C.R.*), which indicates the consistency degree of the pairwise comparison matrix $\boldsymbol{A}$:

$$C.R. = \frac{C.I.}{R.I.} = \frac{\lambda_{max} - m}{(m - 1)R.I.}. \tag{2.10}$$

A matrix can reasonably be assessed as consistent when its calculated *C.R.* is less than or equal to 0.1 or 10% [21, 36]. The resulting vector is accepted if *C.R.* is about 0.10 or less (0.20, and not more may be tolerated). For $m = 3$, it is 0.05, and for $m = 4$, it is 0.08. If it is not less than 010, study the problem and revise the judgements [55]. The inconsistency of 10 percent or less implies that the adjustment is small compared to the actual values of the eigenvector entries [47].

For instance, Aguarón and Moreno-Jiménez [56] proposed different thresholds for the matrices with different orders to provide an interpretation of the inconsistency threshold analogous to the *C.R.* = 10%, that is, 0.3147 for $m = 3$, 0.3526 for $m = 4$ and 0.370 for $m > 4$.

The *C.R.* offers a way to determine the consistency of each expert's assessment. The opinions of individual experts may be conflicting. Consistency of the opinions of a group of experts is possible if the whole group of experts agree by compromise

**Table 2.3** Randomindex *R.I.* values depending on matrica row (criteria) *m*

| *m* | 3 | 4 | 5 | 6 | 7 | 8 | 9 | 10 | 11 | 12 | 13 | 14 | 15 |
|---|---|---|---|---|---|---|---|---|---|---|---|---|---|
| *R.I* | 0.58 | 0.90 | 1.12 | 1.24 | 1.32 | 1.41 | 1.45 | 1.49 | 1.51 | 1.48 | 1.56 | 1.57 | 1.59 |

(discussion) to form a single harmonised matrix ***A***. The process of completing this common matrix of expert group views is long, challenging and not always achievable. It takes a long time for two experts to complete a single pairwise comparison matrix, because each of them has a personal opinion that they feel is more in line than that of the other.

### 2.2.5 Synthesis Method for Common Priorities

Once the priorities of the required characteristics (criteria) of passenger transport expressed in terms of the normalised weights $\omega_i$ of these criteria have been identified, as well as the priorities of the alternatives to the modes of transport expressed in terms of the weights $\omega_{i,a}$ or $\omega'_{i,a}$, the global priorities of the alternatives can be calculated. The global priorities for passenger transport are calculated using the two modalities of the AHP method. One of these is the distributive-mode relative AHP and the other is the ideal-mode relative AHP [35, 57].

The distributive-mode calculates the global preference of the alternative (BT, RT, AT) compared by the expert and indicated by the global normalised weight.

$$Q_a^{DM} = \sum_{i=1}^{m} \omega_i \cdot \omega_{i,a}, \tag{2.11}$$

where $Q_a^{DM}$ is the global normalised weight of the $a$-th passenger transport mode alternative calculated in a distributive mode; $\omega_i$ is subjective normalised weight ($i$ = 1, 2, ..., $m$) of the required passenger transport characteristic (criterion); $m$ is the number of criteria; $\omega_{i,a}$ is the normalised weight ($a$ = 1, 2, ..., $N$) of the $a$-th alternative for the $i$-th criterion; $N$ is the number of comparable alternatives.

The correctness of all pairwise comparison matrices for criteria and alternatives is checked by calculating their $C.R.$

In a distributive approach, the number and quality of other alternatives can affect the result. This approach, known as the dominance method, is used when the decision-maker assesses the dominance of each alternative over all other alternatives.

Therefore, the ideal approach is used for the selection of the best alternative, which focuses on an ideal benchmark without assessing the number and quality of other alternatives against this benchmark.

The ideal way of prioritising alternatives starts by calculating the idealised weights of the $a$-th alternatives:

$$\omega'_{i,a} = \frac{\omega_{i,a}}{\omega_{i,amax}}, \tag{2.12}$$

where $\omega_{i,a}$ is the normalised weight of alternative $a$-th calculated for criterion $i$-th $\left(\sum_{a=1}^{N} \omega_{i,a} = 1\right)$; $\omega_{i,amax}$ is the highest (most significant, best) normalised weight of alternative $a$-th calculated for criterion $i$-th.

This is followed by the calculation of the global priority of each alternative $a$-th as determined by the expert and quantified by a global normalised weight:

$$Q_a^{IM} = \frac{\sum_{i=1}^{m} \omega_i \cdot \omega'_{i,a}}{\sum_{a=1}^{N} \sum_{i=1}^{m} \omega_i \cdot \omega'_{i,a}}, \tag{2.13}$$

where $Q_a^{IM}$ is the global normalised weight of the $a$-th alternative for passenger transport mode calculated in an idealised way; $\omega_i$ is subjective normalised weight for the required characteristic (criterion) of passenger transport; $\omega'_{i,a}$ is the idealised weight of alternative $a$-th calculated for criterion $i$-th $\left(\sum_{a=1}^{N} \omega'_{i,a} > 1\right)$.

The sum of the global weights of the passenger transport alternatives (road transport—BT, rail transport—RT, air transport—AT) equals:

$$Q_{BT}^{DM} + Q_{RT}^{DM} + Q_{AT}^{DM} = \sum_{a=1}^{N} Q_a^{DM} = 1 \tag{2.14}$$

and

$$Q_{BT}^{IM} + Q_{RT}^{IM} + Q_{AT}^{IM} = \sum_{a=1}^{N} Q_a^{IM} = 1. \tag{2.15}$$

The average opinion of a group of experts can be considered when applying the distributive mode and the ideal mode of the AHP method to compare transport alternatives and make the best choice between them. The average of the consistent (non-contradictory) opinions of a group of experts is usually more reliable than the opinion of a single expert.

### 2.2.6 Significance of the Required Characteristics for Passenger Transport

The opinion of a single expert or the average of the opinions of a group of experts can be used to determine the weights of the criteria and the best alternative. In the examples where the AHP approach was used [35] to compare the priority of the amount of beverages consumed in the US and the priority of the amount of protein in foods and compare them with the actual amounts, the opinion of a single expert was used. In this book and in the article by Saaty [47], the opinion of family members is used to choose the best alternative for the house to be purchased. Key indicators affecting synchromodality were assessed through a four-expert discussion [58]. The opinion of five experts was used to prioritise pavement maintenance works [57]. In addition, 5 experts assessed critical success factors for construction projects [33].

The significance of the required characteristics of any mode of transport, expressed in normalised weights, has been determined on the basis of data from a single expert survey. The expert assigned ranks to the required characteristics (criteria) and completed the pairwise comparison matrix (Table 2.4).

The ten most important criteria for the expert are safety (A), trip duration (B) and comfort (E). Their ranks equal to 1, 2 and 3 respectively and their normalised weights are $\omega_i = 0.341, 0.171$ and $0.114$. The least significant in the expert's opinion are environmental friendliness (D), possibility of COVID-19 (I) and vehicle capacity (F). Their ranks are 8, 9 and 10 respectively. The weights $\omega_i$ for these least significant criteria are 0.043, 0.038 and 0.034 respectively. The other 4 criteria are of medium significance.

The consistency ratio of the criteria pairwise comparison matrix (Table 2.4) indicates a sufficiently high degree of consistency $C.R. = 0.055$, which is almost twice as low as the $C.R.$ threshold of 0.1.

### 2.2.7 Comparative Analysis of Passenger Transport Alternatives

For each of the ten criteria, the expert compared three alternatives in pairs. The three ranked alternatives (road transport (BT), rail transport (RT) and air transport (AT) and their pair-wise comparison matrixes are shown in Table 2.5.

The ranks of the three transport alternatives show that road transport is the best for passenger transport on one criterion (J), average on five criteria (B, C, E, G, I) and the worst on four criteria (A, D, F, H). Rail is the best alternative for passenger transport on seven criteria (A, C, D, E, F, G, I), it is average on two criteria (H, J) and the worst on one criterion (B). Air transport is the best in terms of passenger transport on two criteria (B, H), medium on three criteria (A, D, F) and worst on five criteria (C, E, G, I, J).

The same sequence of transport alternatives is shown by the normalised weights $\omega_{i,a}$ and $\omega'_{i,a}$ (Table 2.5), which were used to calculate the overall priorities in both the distributive mode $Q_a^{DM}$ and ideal mode $Q_a^{IM}$ (Tables 2.6 and 2.7).

After synthesising the priorities from the weights of the criteria $\omega_i$ and the weights of the alternatives $\omega_{i,a}$, the global weights $Q_a^{DM}$ of each passenger transport alternative have been calculated using the distributive AHP method according to formula (2.11) (Table 2.6). They are the lowest for road transport $Q_{BT}^{DM} = 0.1752$, medium for air transport $Q_{AT}^{DM} = 0.2971$ and the highest for rail transport $Q_{RT}^{DM} = 0.5297$.

Priority synthesis has been carried out in two steps in the AHP ideal method. First, the idealised weights of the alternatives have been calculated $\omega'_{i,a}$ (Table 2.5). The sum of the weights $\omega_i$ and $\omega'_{i,a}$ products of the criteria indicates the normalised

**Table 2.4** Criteria ranks and their second level hierarchy pairwise comparison matrix

| Factor (criterion) $i = 1, 2, \ldots, m$ | | Rank | Required characteristic of passenger transport | | Criterion $j = 1, 2, \ldots, m$ | | | | | | | | | | $\omega_i$ |
|---|---|---|---|---|---|---|---|---|---|---|---|---|---|---|---|
| | | | | | A | B | C | D | E | F | G | H | I | J | |
| A | Safety | 1 | Criterion $i = 1, 2, \ldots, m$ | A | 1 | 3 | 4 | 7 | 3 | 9 | 6 | 5 | 8 | 6 | 0.341 |
| B | Trip duration | 2 | | B | 1/3 | 1 | 3 | 6 | 2 | 8 | 6 | 4 | 7 | 5 | 0.171 |
| C | Expenses | 4 | | C | 1/4 | 1/3 | 1 | 5 | 1/2 | 6 | 4 | 2 | 5 | 3 | 0.085 |
| D | Environmental friendliness | 8 | | D | 1/7 | 1/6 | 1/5 | 1 | 1/5 | 3 | 1/3 | 1/4 | 2 | 1/3 | 0.043 |
| E | Comfort | 3 | | E | 1/3 | 1/2 | 2 | 5 | 1 | 7 | 5 | 3 | 6 | 4 | 0.114 |
| F | Vehicle capacity | 10 | | F | 1/9 | 1/8 | 1/6 | 1/3 | 1/7 | 1 | 1/4 | 1/5 | 1/2 | 1/4 | 0.034 |
| G | Wheatear conditions | 7 | | G | 1/6 | 1/6 | 1/4 | 3 | 1/5 | 4 | 1 | 1/3 | 3 | 1/3 | 0.049 |
| H | The quality of services | 5 | | H | 1/5 | 1/4 | 1/2 | 4 | 1/3 | 5 | 3 | 1 | 4 | 2 | 0.068 |
| I | Possibility of COVID-19 | 9 | | I | 1/8 | 1/7 | 1/5 | 1/2 | 1/6 | 2 | 1/3 | 1/4 | 1 | 1/4 | 0.038 |
| J | Door-to-door mobility | 6 | | J | 1/6 | 1/5 | 1/3 | 3 | 1/4 | 4 | 3 | 1/2 | 4 | 1 | 0.057 |

**Table 2.5** Comparison of passenger transport alternatives with respect to the ten characteristics (criteria)

| Safety(A) | | | $\omega_A = 0.341$ | | | $\omega_{A,a}$ | $\omega'_{A,a}$ | Trip duration(B) | | | $\omega_B = 0.171$ | | | $\omega_{B,a}$ | $\omega'_{B,a}$ |
|---|---|---|---|---|---|---|---|---|---|---|---|---|---|---|---|
| | | | BT | RT | AT | | | | | | BT | RT | AT | | |
| Rank | 3 | BT | 1 | 1/7 | 1/3 | 0.081 | 0.111 | Rank | 2 | BT | 1 | 3 | 1/7 | 0.153 | 0.197 |
| | 1 | RT | 7 | 1 | 5 | 0.731 | 1.000 | | 3 | RT | 1/3 | 1 | 1/8 | 0.070 | 0.090 |
| | 2 | AT | 3 | 1/5 | 1 | 0.188 | 0.257 | | 1 | AT | 7 | 8 | 1 | 0.777 | 1.000 |
| | | | | | | $C.R. = 0.056$ | | | | | | | | $C.R. = 0.090$ | |
| Expenses (efficient) (C) | | | $\omega_C = 0.085$ | | | $\omega_{C,a}$ | $\omega'_{C,a}$ | Environmental friendliness (D) | | | $\omega_D = 0.043$ | | | $\omega_{D,a}$ | $\omega'_{D,a}$ |
| | | | BT | RT | AT | | | | | | BT | RT | AT | | |
| Rank | 2 | BT | 1 | 1/3 | 3 | 0.268 | 0.436 | Rank | 3 | BT | 1 | 1/7 | 1/3 | 0.081 | 0.111 |
| | 1 | RT | 3 | 1 | 4 | 0.614 | 1.000 | | 1 | RT | 7 | 1 | 5 | 0.731 | 1.000 |
| | 3 | AT | 1/3 | 1/4 | 1 | 0.118 | 0.192 | | 2 | AT | 3 | 1/5 | 1 | 0.188 | 0.257 |
| | | | | | | $C.R. = 0.063$ | | | | | | | | $C.R. = 0.056$ | |
| Comfort (E) | | | $\omega_E = 0.114$ | | | $\omega_{E,a}$ | $\omega'_{E,a}$ | Vehicle capacity (F) | | | $\omega_F = 0.034$ | | | $\omega_{F,a}$ | $\omega'_{F,a}$ |
| | | | BT | RT | AT | | | | | | BT | RT | AT | | |
| Rank | 2 | BT | 1 | 1/4 | 3 | 0.218 | 0.315 | Rank | 3 | BT | 1 | 1/7 | 1/6 | 0.067 | 0.105 |
| | 1 | RT | 4 | 1 | 6 | 0.691 | 1.000 | | 1 | RT | 7 | 1 | 3 | 0.641 | 1.000 |
| | 3 | AT | 1/3 | 1/6 | 1 | 0.091 | 0.132 | | 2 | AT | 6 | 1/3 | 1 | 0.292 | 0.455 |
| | | | | | | $C.R. = 0.046$ | | | | | | | | $C.R. = 0.086$ | |
| Wheater conditions (G) | | | $\omega_G = 0.049$ | | | $\omega_{G,a}$ | $\omega'_{G,a}$ | The quality of services (H) | | | $\omega_H = 0.068$ | | | $\omega_{H,a}$ | $\omega'_{H,a}$ |
| | | | BT | RT | AT | | | | | | BT | RT | AT | | |
| Rank | 2 | BT | 1 | 1/4 | 3 | 0.211 | 0.299 | Rank | 3 | BT | 1 | 1/3 | 1/6 | 0.091 | 0.132 |
| | 1 | RT | 4 | 1 | 7 | 0.705 | 1.000 | | 2 | RT | 3 | 1 | 1/4 | 0.218 | 0.315 |

(continued)

**Table 2.5** (continued)

| Wheater conditions (G) | | | $\omega_G = 0.049$ | | | $\omega_{G,a}$ | $\omega'_{G,a}$ | The quality of services (H) | | | $\omega_H = 0.068$ | | | $\omega_{H,a}$ | $\omega'_{H,a}$ |
|---|---|---|---|---|---|---|---|---|---|---|---|---|---|---|---|
| | | | BT | RT | AT | | | | | | BT | RT | AT | | |
| | 3 | AT | 1/3 | 1/7 | 1 | 0.084 | 0.119 | | 1 | AT | 6 | 4 | 1 | 0.691 | 1.000 |
| | | | | | | $C.R. = 0.028$ | | | | | | | | $C.R. = 0.046$ | |
| Possibility of COVID-19(I) | | | $\omega_I = 0.038$ | | | $\omega_{I,a}$ | $\omega'_{I,a}$ | Door-to-door mobility (J) | | | $\omega_J = 0.057$ | | | $\omega_{J,a}$ | $\omega'_{J,a}$ |
| | | | BT | RT | AT | | | | | | BT | RT | AT | | |
| Rank | 2 | BT | 1 | 1/3 | 3 | 0.258 | 0.405 | Rank | 1 | BT | 1 | 5 | 7 | 0.731 | 1.000 |
| | 1 | RT | 3 | 1 | 5 | 0.637 | 1.000 | | 2 | RT | 1/5 | 1 | 3 | 0.188 | 0.257 |
| | 3 | AT | 1/3 | 1/5 | 1 | 0.105 | 0.165 | | 3 | AT | 1/7 | 1/3 | 1 | 0.081 | 0.111 |
| | | | | | | $C.R. = 0.033$ | | | | | | | | $C.R. = 0.056$ | |

**Table 2.6** Synthesis of global priorities for passenger transport alternatives by applying the distributive mode of the AHP method

| Alternative | Normalized weights of criteria | | | | | | | | | | Global weight |
|---|---|---|---|---|---|---|---|---|---|---|---|
| | A<br>0.341 | B<br>0.171 | C<br>0.085 | D<br>0.043 | E<br>0.114 | F<br>0.034 | G<br>0.049 | H<br>0.068 | I<br>0.038 | J<br>0.057 | $Q_a^{DM}$ |
| BT | 0.341 × 0.081 = 0.0276 | 0.171 × 0.153 = 0.0262 | 0.085 × 0.268 = 0.0228 | 0.043 × 0.081 = 0.0035 | 0.114 × 0.218 = 0.0023 | 0.034 × 0.067 = 0.0023 | 0.049 × 0.211 = 0.0062 | 0.068 × 0.091 = 0.0062 | 0.038 × 0.258 = 0.0098 | 0.057 × 0.731 = 0.0417 | 0.1752 |
| RT | 0.341 × 0.731 = 0.2493 | 0.171 × 0.070 = 0.0120 | 0.085 × 0.614 = 0.0522 | 0.043 × 0.731 = 0.0314 | 0.114 × 0.691 = 0.0788 | 0.034 × 0.641 = 0.0218 | 0.049 × 0.705 = 0.0345 | 0.068 × 0.218 = 0.0148 | 0.038 × 0.637 = 0.0242 | 0.057 × 0.188 = 0.0107 | 0.5297 |
| AT | 0.341 × 0.188 = 0.0641 | 0.171 × 0.777 = 0.1329 | 0.085 × 0.118 = 0.0100 | 0.043 × 0.188 = 0.0081 | 0.114 × 0.091 = 0.0104 | 0.034 × 0.292 = 0.0099 | 0.049 × 0.084 = 0.0041 | 0.068 × 0.691 = 0.0470 | 0.038 × 0.105 = 0.0040 | 0.057 × 0.081 = 0.0046 | 0.2951 |

**Table 2.7** Synthesis of global priorities for passenger transport alternatives by applying the ideal mode of the AHP method

| Alternative | Normalized weights of criteria | | | | | | | | | | $\sum_{i=1}^{m} \omega_i \cdot \omega'_{i,a}$ |
|---|---|---|---|---|---|---|---|---|---|---|---|
| | A<br>0.341 | B<br>0.171 | C<br>0.085 | D<br>0.043 | E<br>0.114 | F<br>0.034 | G<br>0.049 | H<br>0.068 | I<br>0.038 | J<br>0.057 | |
| BT | 0.341 ×<br>0.111 =<br>0.0378 | 0.171 ×<br>0.197 =<br>0.0337 | 0.085 ×<br>0.436 =<br>0.0371 | 0.043 ×<br>0.111 =<br>0.0048 | 0.114 ×<br>0.315 =<br>0.0359 | 0.034 ×<br>0.105 =<br>0.0036 | 0.049 ×<br>0.299 =<br>0.0147 | 0.068 ×<br>0.132 =<br>0.0090 | 0.038 ×<br>0.405 =<br>0.0154 | 0.057 ×<br>1.000 =<br>0.0057 | 0.2490 |
| RT | 0.341 ×<br>1.000 =<br>0.341 | 0.171 ×<br>0.090 =<br>0.0154 | 0.085 ×<br>1.000 =<br>0.085 | 0.043 ×<br>1.000 =<br>0.043 | 0.114 ×<br>1.000 =<br>0.114 | 0.034 ×<br>1.000 =<br>0.034 | 0.049 ×<br>1.000 =<br>0.049 | 0.068 ×<br>0.315 =<br>0.0214 | 0.038 ×<br>1.000 =<br>0.038 | 0.057 ×<br>0.257 =<br>0.0146 | 0.7554 |
| AT | 0.341 ×<br>0.257 =<br>0.0876 | 0.171 ×<br>1.000 =<br>0.171 | 0.085 ×<br>0.192 =<br>0.0163 | 0.043 ×<br>0.258 =<br>0.0111 | 0.114 ×<br>0.132 =<br>0.0150 | 0.034 ×<br>0.455 =<br>0.0155 | 0.049 ×<br>0.119 =<br>0.0058 | 0.068 ×<br>1.000 =<br>0.068 | 0.038 ×<br>0.165 =<br>0.0063 | 0.057 ×<br>0.111 =<br>0.0063 | 0.4029 |
| | | | | | | | Total: | $\sum_{a=1}^{N} \sum_{i=1}^{m} \omega_i \cdot \omega'_{i,a} = 1.4073$ | | | |

priorities of the alternatives. In a second step, these products are normalised for each transport alternative according to formula (2.13).

According to formula (2.13) the products of the weights of the criteria $\omega_i$ and the weights of the alternatives $\omega'_{i,a}$ are normalised, i.e., divided by their sum:

$$Q_{BT}^{IM} = \frac{0.2490}{1.4073} = 0.1769,$$
$$Q_{RT}^{IM} = \frac{0.7554}{1.4073} = 0.5368,$$
$$Q_{AT}^{IM} = \frac{0.4029}{1.4073} = 0.2863.$$

The findings of the study show that the global weights for passenger transport per mode calculated by the AHP method distributive and ideal modes are almost identical (Fig. 2.2).

A comparative analysis of the different modes of transport shows that rail transport is the best alternative for passenger transport, with a global weight $Q_{RT}^{DM} = 0.530$ and $Q_{RT}^{IM} = 0.534$. The rail priority is about 1.6 times higher than the mean of 0.333, which corresponds to equal importance of the alternatives. The global weight of this mode of transport (RT) is about 1.85 times higher than air transport and about 3.02 times higher than road transport. The findings of the study therefore show that rail transport is the best option for passenger transport.

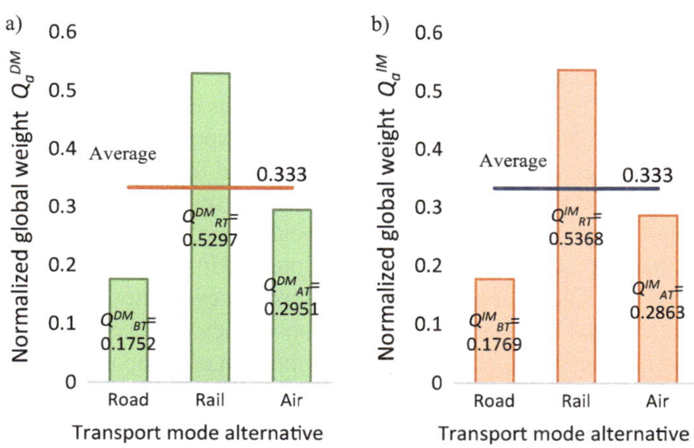


**Fig. 2.2** Best alternatives for the passenger transport according to the global weights of the alternatives calculated by the AHP method for: **a** distributive mode; **b** ideal mode

## 2.3 Methods to Establish the Significance of Hierarchically Unstructured and Structured Criteria

### *2.3.1 The Methods of the Average Rank Transformation into Weight*

The importance expressed by weights of the evaluation criteria (sub-criteria or key criteria) largely determine the evaluation result. In practice, the subjective weights assigned by experts or respondents to the considered criteria are commonly used. These weights present the judgements of highly qualified experts having long-term practical experience and theoretical knowledge in the considered field [51, 59]. Passengers themselves make decisions about the mode of transport they choose for travel and, therefore, are the experts themselves. However, because of their low competence, they should be referred to as respondents answering the survey questions rather than experts.

Most of the widely known and used methods for evaluating the weights of multiple criteria are based on experts' judgements. These methods embrace a thorough problem analysis by the experts, the organization of this process as well as quantitative evaluation of decisions and the arrangement of the obtained results. Therefore, the problem of practical determination of the accurate weights of the considered criteria arises. The subjective weights of the evaluation criteria can also be found from the ranks assigned to these criteria by the experts. The estimates (judgements) of various experts differ considerably, often being inconsistent, which implies that the obtained weights (significances) of the criteria as well as their order of preference may be different.

The result of the experts' evaluation largely depends on their qualification, experience in assessing the objectives as well as their responsibility for providing the appropriate estimates of the criterion significance and readiness to take part in the experimental study. The judgements of specialists and respondents about the relative significance of the criteria and their arrangement in the priority (preference) order often differ, therefore, the ranks and weights expressed in terms of the average values of the experts' estimates can be used in multi-criteria evaluation only if the consistency of the estimates has been proved. The consistency of the estimates given by a group of experts in terms of ranks is based on the idea of compactness.

In the case of expert evaluation, the average estimate obtained from a group of experts (respondents) is the problem solution (a result of decision making) only when the judgements of all the experts are consistent. If a decision is made based on the average estimate of the experts or respondents, the level of consistency of the experts' estimates is described by the concordance coefficient $W$. To determine the concordance coefficient $W$, the ranks of the evaluation criteria assigned by the experts or respondents are required. If their estimates are given in other units (for example, in percentage points), they should be ranked.

The consistency of the weights of the criteria describing an object and the estimates provided by experts is usually determined by using the AHP method approach [21, 35]. The consistency of group evaluation results is determined by using the method of rank correlation [19, 60].

Experts usually assign the ranks $R_{ij}$ to the criteria by arranging them according to their significance and giving them the appropriate numbers. This method of determining the criteria weights is logical, however, its accuracy is low. Therefore, it can be used only at the initial stage of analysis. Using more accurate and complicated methods still requires preliminary ranking of the criteria.

The weights $\omega_i$ of the ranked criteria can be determined by applying dif-ferent methods (algorithms) that do not have theoretical advantages over each other. However, the general principle of all algorithms is the same: the most important criterion is assigned the highest weight. The values of the weights $\omega_i$ must correspond to the criteria ranks (lower rank – higher weight). The sum of the weights $\omega_i$ of all the criteria describing the research object must be equal to one, i.e., weights must be normalized.

It is convenient to transform the ranks assigned to the criteria by a group of experts into weights by using the ARTIW-L and ARTIW-N methods [20], whose sequence of operations and calculations is given below (Fig. 2.3). The average rank $\overline{R}_i$, representing the ranks assigned by all $j$-th experts ($j = 1, 2, \ldots, n$) is calculated for each $i$-th criterion ($i = 1, 2, \ldots, m$) by the formula:

$$\overline{R}_i = \frac{\sum_{j=1}^{n} R_{ij}}{n}. \tag{2.16}$$

Significances (weights) of the evaluation criteria of an object can be determined in the process of their normalization (setting their sum equal to one) by transforming the average ranks into weights (the ARTIW method). This method had first been offered in 2011 [61], however, then it was not called ARTIW. A relative weight $\omega_i$ of the criterion is calculated as follows:

$$\omega_i = \frac{(m + 1) - \overline{R}_i}{\sum_{i=1}^{m} \overline{R}_i}, \tag{2.17}$$

where $m$ is the number of criteria describing the quality of the considered object, $\overline{R}_i$ is the average $i$-th criterion rank calculated by the formula (2.16).

The normalized weights $\omega_i$ of the $i$-th criteria calculated according to formula (2.17) have a linear inverse correlation (functional) relationship with the rank averages (average ranks) $\overline{R}_i$ of these criteria calculated according to formula (2.16). Therefore, this method is called the Average Rank Transformation into Weight-Linear (ARTIW-L).

The normalized weights $\omega_i'$ of the criteria can be calculated by using another method of transforming rank averages into weights. The criterion weights calculated according to formulas (2.18) and (2.19) are related to criteria ranks averages $\overline{R}_i$ by

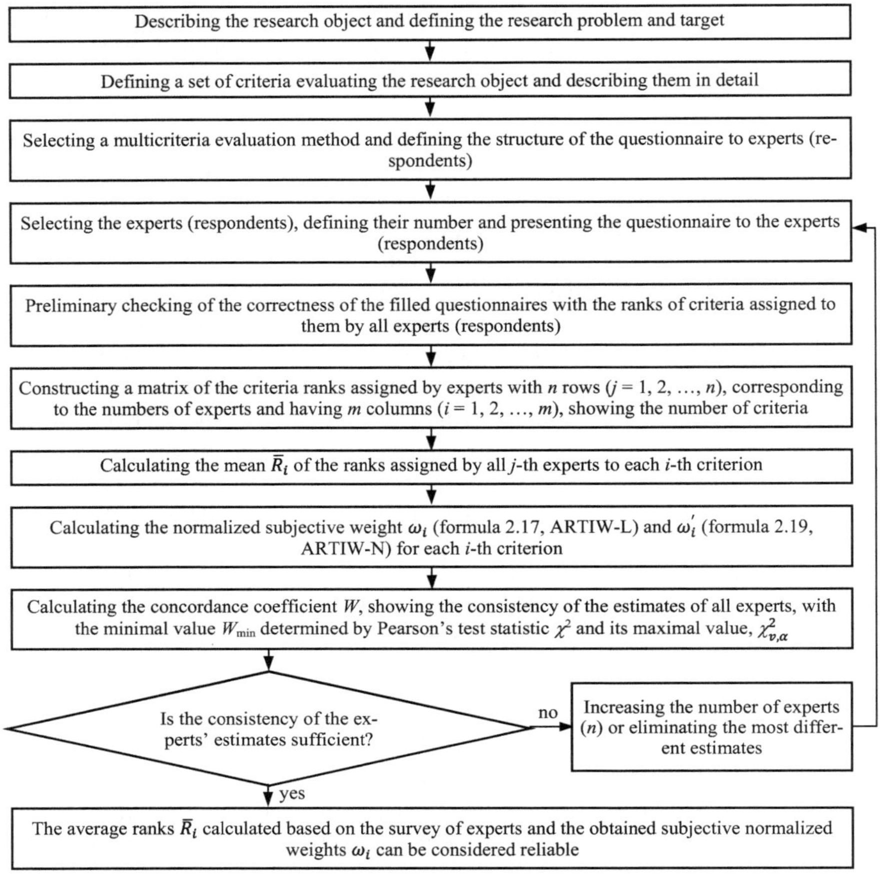


**Fig. 2.3** The algorithm of ARTIW-L and ARTIW-N methods of the average rank transformation into weight used for calculating normalized subjective weights of the criteria describing the research object

a non-linear inverse correlation dependence. Therefore, this method is called the Average Rank Transformation into Weight-Non-Linear (ARTIW-N).

Using the ARTIW-N method, the ratio of the min $\overline{R}_i$ of the most important criterion (with the lowest average ranks $\overline{R}_i$) to the average of the ranks $\overline{R}_i$ of all other $i$-th criteria is initially calculated:

$$u_i = \frac{\min_i \overline{R}_i}{\overline{R}_i}. \tag{2.18}$$

After normalizing the values $u_i$ for each criterion, their subjective significances $\omega_i'$ are calculated. The more important the criterion, the smaller its average rank $\overline{R}_i$. In practice, it is more convenient to use the estimates of the criteria significance, whose numerical values show higher importance. For this purpose, the normalized weights of criteria $i$, expressing relative importance, are used:

$$\omega_i' = \frac{u_i}{\sum_{i=1}^{m} u_i}. \tag{2.19}$$

Neither of these two methods (ARTIW-L and ARTIW-N) can be considered more accurate than the other and neither of them can be looked at as the reference method. The average of the weights $\overline{\omega}_i$ calculated for each criterion by these two methods can be considered as the result of the task calculation:

$$\overline{\omega}_i = \frac{\omega_i + \omega_i'}{2}. \tag{2.20}$$

The consistency of expert group estimates is determined by the concordance coefficient $W$. The concordance coefficient $W$ in the absence of tied ranks is expressed in terms of the relationship between the obtained sum $S$ and the respective largest sum $S_{max}$ [60]:

$$W = \frac{12S}{n^2\left(m^3 - m\right)}. \tag{2.21}$$

When the estimates provided by the experts or respondents agree, the Kendall coefficient of concordance, $W$, is about one. When the estimates differ considerably, the value of $W$ is close to zero.

The squares sum $S$ deviations of the ranks $R_{ij}$ of each $i$-th criterion from rank mean $\overline{R}$ can be calculated as follows:

$$S = \sum_{i=1}^{m}\left[\sum_{j=1}^{n} R_{ij} - \frac{1}{2}n(m+1)\right]^2, \tag{2.22}$$

where $m$ is the number of criteria ($i = 1, 2, \ldots, m$), $n$ is the number of experts (respondents) ($j = 1, 2, \ldots, n$).

The random value for $S$ was calculated by (formula 2.22) adding the squared values of all the criteria given in parentheses.

The concordance coefficient may be used in practice when its ultimate value, showing that expert estimates can be considered consistent, has been found. Kendall [19] has shown that if the number of criteria is $m > 7$, the significance of the concordance coefficient $W$ can be determined using Pearson's chi-squared test statistic $\chi^2$. The random value

$$\chi^2 = Wn(m-1), \tag{2.23}$$

is distributed according to $\chi^2$ distribution, with the degree of freedom $v = m - 1$.

When the number of the compared criteria $m$ ranges from 3 to 7, the distribution $\chi^2$ cannot be used in all cases because sometimes the critical value of $\chi^2_{v.\alpha}$ may be larger than calculated value (even though the consistency of the estimates is still sufficiently high). In this case, the probability tables of the concordance coefficient or the tables of critical values $S$ (with $3 \leq m \leq 7$) can be used [62].

The smallest value of the concordance coefficient $W_{min}$ allowing the authors to con-sider that the estimates of $n$ experts of the quality of the research object based on $m$ criteria, with the assigned (required) significance level $\alpha$ and degree of freedom $v = m - 1$, are consistent, can be calculated as follows [61]:

$$W_{min} = \frac{\chi^2_{v,\alpha}}{n(m-1)}, \tag{2.24}$$

where $\chi^2_{v,\alpha}$ is the critical Pearson's statistic found in the table [63], assuming the degree of freedom $v = m - 1$ and the significance level $\alpha$.

The quality of the research object is evaluated by the additive mathematical model used for calculating its comprehensive quality index (CQI), which allows for describing the quality of the object by a single number. It also allows for comparing it with the quality of other similar objects, and the $\overline{\omega}_i$ of the normalized criteria weights (rather than the average criteria ranks $\overline{R}_i$, which cannot show how one criterion is more important that another) are used.

The weights of the criteria describing the research object (the selection of rail transport rather than road transport and air transport by passengers) can be calculated by using a very popular but complicated approach referred to as the Analytic Hierarchy Process (AHP) offered by Saaty [21], Saaty and Vargas [30] and Saaty and Ozdemir [35]. Passengers are not highly qualified experts and, therefore, can hardly fill in a pairwise comparison matrix properly, particularly, if the number of the criteria compared is large. This number may be more than nine (e.g., fifteen) criteria. In the study (Maskeliūnaitė et al. 2009), passengers completed 32 pairwise comparison matrices, 22 of which were rejected because their *C.R.* (consistency ratio) was greater than 0.1. Only 10 matrices were applicable for the study on the quality of passenger transport by train. Therefore, it is not rational to apply the AHP method in passenger interviews. Not every passenger can complete the pairwise comparison matrix properly. The AHP method can only be applied to interview highly qualified experts.

The objective weights of the criteria and sub-criteria can be calculated by using the entropy method [64, 65] as well as the new IDOCRIW method [59], which combines (integrates) the entropy and the criterion impact loss (CILOS) methods.

The number of times the calculated concordance coefficient $W$ is greater than its minimum (critical) value $W_{min}$ and the number of times $\chi^2$ is greater than its critical value $\chi^2_{v,\alpha}$, is shown by the new consistency coefficient $k_c$, which is calculated from the formula:

$$k_c = \frac{W}{W_{min}} = \frac{\chi^2}{\chi^2_{v,\alpha}}. \quad (2.25)$$

When the opinions of the experts (respondents) are in agreement, $k_c > 1$. Otherwise (when $k_c < 1$), the opinions of the group of experts differ significantly, i.e., they are not harmonised, and the average of the weights of the criteria cannot be taken as a reliable result of the task.

### 2.3.2 *The Inverse Hierarchy for Assessment of Min Criteria Importance Method*

The quality of rail passenger transport is difficult to quantify and measure directly by one single factor (criterion). It can only be defined in terms of a set of criteria. The criteria showing the advantages of train travel as an alternative to road transport (bus) or train travel as an alternative to air travel (plane) can be presented in a hierarchically unstructured model (Fig. 2.4a).

Its first level is made up of the objective (the choice travelling by train) and the second level is made up of the criteria (C1, C2, ..., C16, the number of criteria depends on the complexity of the object of interest and can vary).

Frequently, the number of elements in the structure being analysed is so large that it exceeds the ability of the researcher/expert to take in the full information. In this case, the system is divided into subsystems. One such division is the hierarchy (Fig. 2.4b). A hierarchy is a particular design (form) of a system based on the assumption that its elements can be grouped into non-connected wholes (sets). We assume that the elements of each group in the hierarchy, called levels, are independent. A hierarchy is a multi-level organisational structure, with links from the top to the bottom of the hierarchy forming a strictly regulated ladder of dependence.

In the analysis of the functionality of a system, the first requirement is to build a hierarchy that reproduces the functional relationships (interfaces). To do this, it starts by listing all the elements belonging to the hierarchy. They are then divided into groups corresponding to the influence/impact between the groups. This is how hierarchy levels develop. The goals for which the task is being studied are identified and a hierarchy is established.

In the classical hierarchy model (Fig. 2.4b), the first level is made up of the goal (the choice travelling by train), the second level is made up of several key criteria (first key criterion, second key criterion, third key criterion) and the third level is made up of sub-criteria, the number of which can vary within each key criterion. In the model (Fig. 2.4b), the first key criterion consists of 5 sub-criteria (F1, F2, ..., F5), the second key criterion consists of 4 sub-criteria (S1, S2, ..., S4), the third key criterion consists of 7 sub-criteria (T1, T2, ..., T7). This model was applied in Chaps. 3–7 of the monograph, where the quality of the train journey was assessed by four main criteria made up of 16, 19, 6 and 8 sub-criteria.

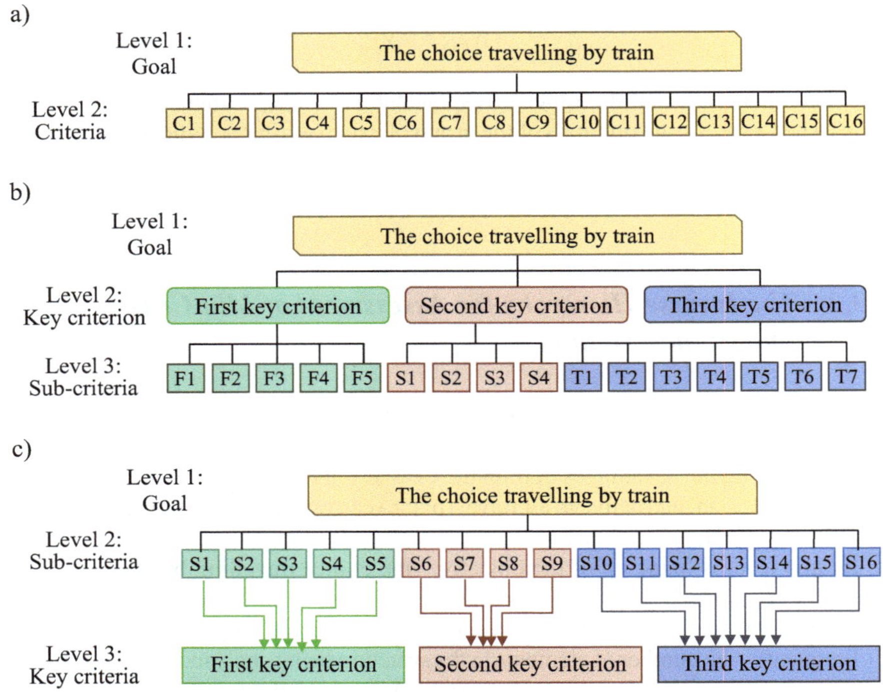


**Fig. 2.4** Models of the determinants for passengers' choice to travel by train: **a** hierarchically unstructured criteria; **b** hierarchically structured three-level key criteria and sub-criteria; **c** inverse hierarchy of structured sub-criteria and key criteria

Given the significance of the criteria C1, C2, …, C$m$ (where $m$ is the number of criteria), expressed as normalised subjective weights (Fig. 2.4a), it is possible to structure them (criteria) hierarchically, breaking them down into separate groups and calculating the weights of their groups (key criteria). The Inverse Hierarchy for Assessment of Main Criteria Importance (IHAMCI) method can be used for this purpose [15, 66].

In the IHAMCI model (Fig. 2.4c), the first level is made up of the choice travelling by train, the second level is made up of the sub-criteria S1, S2, …, S$m$, with weights corresponding to the weights of the criteria C1, C2, …, C$m$, calculated from the model of hierarchically unstructured variables (Fig. 2.4a). The third level includes the key criteria (first key criterion, second key criterion, third key criterion), which may be at least 2 in number, and the number of sub-criteria ($k$) for each key criterion depends on the grouping attributes.

When the global weights $\overline{\omega}_i$ of all criteria (Fig. 2.4a) were determined and criteria were divided into three groups (key criteria) as shown in Fig. 2.4c, the weights of the key criterion $\tilde{\omega}_g$ were calculated as follows using the IHAMCI method:

$$\tilde{\omega}_g = \frac{\sum_{i=1}^{k} \overline{\omega}_i / k}{\sum_{b=1}^{g} \sum_{i=1}^{k} \overline{\omega}_i / k} = \frac{\omega_g / k}{\sum_{b=1}^{g} \omega_g / k}, \tag{2.26}$$

where $\overline{\omega}_i$ is the global weight of $i$-th criterion; $k$ is the number of sub-criteria in the key criterion ($i = 1, 2, ..., k$); $g$ is the number of key criteria describing the research object ($b = 1, 2, ..., g$). In the present study, $g = 3$.

### 2.3.3 Characteristics and Number of Experts (Respondents) Required

Passengers are free to decide which mode of public transport they want to use based on the factors (criteria) that are most important to them. Passengers rank the criteria established by transport professionals. The outcome of the passenger group survey is taken as an indication of the general tendency of the public to travel by train as an alternative to other modes of transport. Passengers can therefore be considered as respondents doing the work of experts. An expert is a person who is an expert in science, art, or other fields and who is called upon to answer questions that require special knowledge. A respondent is a person who completes a questionnaire or responds to oral questions asked by a researcher in sociological research.

Given the importance/significance of the issue at hand, the expertise is organised by the decision-makers or a management team appointed by them. The composition and number of experts is based on the size of the problem, the required validity of the assessments, the characteristics of the experts and the cost of resources. The issue under investigation requires the involvement of specialists from different fields. The minimum number of experts is decided in accordance with the various aspects and areas that need to be assessed to solve the issue. In most cases, the average opinion of a group of experts, rather than a single expert, is relied upon. The reliability of the panel's assessments depends on the knowledge of each expert and the number of experts. Assuming that the experts are reasonably accurate assessors, increasing the number of experts increases the validity of the overall panel result.

The cost of research resources is proportional to the number of experts. As the number of experts increases, the time and costs involved in building the panel, conducting the survey, and processing/managing the results increase. The higher the reliability of the test, the higher the cost. Available financial and time resources usually reduce the maximum number of experts in a team. The performance of the panel of experts depends on the characteristics (qualities) of each expert: competence, creativity, attitude towards the study, conformism, ability to think logically, ability to work in a team (if needed in the study), self-criticism.

When applying expert research methods to assess the significance of criteria, there is a problem of determining the required (necessary) minimum number of experts or respondents. In practice, the mathematically unsound provision (principle) that the number of experts must be equal to or greater than the number of criteria is often observed. There is another common position that is often applied in practice which maintains that the amount of data required for studies $n \geq 30$ is also not substantiated, because in some cases the number $n$ is sufficient (if the group range is small), while in other cases it is too small. The credibility of expert group assessments depends on the level of knowledge of individual experts and their number. Having assumed that the experts are accurate assessors, it can be stated that as their number increases, the reliability of the expertise of the whole group of experts (average of the opinion estimate) also increases. The minimum number of experts or respondents to be interviewed $n_{min}$ can be calculated according to the sample size formula [67]:

$$n_{min} = \frac{t^2 \sigma_{R_i}^2}{\Delta_i^2}, \tag{2.27}$$

where $t$ is the value of $t$ distribution, which depends on the probability taken to assess the importance of the criterion in deciding to go by train as an alternative to the bus. When the probability $P = 95\%$ (significance level $\alpha = 0.05$ for one-sided test), $t = 1.96$; $\sigma_{R_i}$—standard deviation of the ranks $R_{ij}$ of the evaluated $i$-th criterion; $\Delta_i$ is the absolute error of passengers (respondents) rank values $i$-th criterion, indicating the accuracy of the survey results.

From formula (2.27), the absolute error of the experts' or respondents' estimates can be calculated, which is equal to:

$$\Delta_i = \sqrt{\frac{t^2 \cdot \sigma_{R_i}^2}{n}}. \tag{2.28}$$

The absolute error $\Delta_i$ of the survey shows how much the average of the ranks $\overline{R}_i$ calculated for the $i$-th criterion of $n$ surveyed passengers may differ from the average of the population set $\overline{R}_{ip}$ that would be determined by surveying all passengers. Due to the limited sample size $n$, $\overline{R}_i$ always differs from $\overline{R}_{ip}$ by no more than plus or minus $\Delta_i$. This difference is greater the smaller the $n$ and the larger $\sigma_{R_i}^2$.

By interviewing the $n$ passengers and calculating the standard deviation $\sigma_{R_i}^2$ of the ranks of $i$-th criterion with the 95% probability recommended in practice, the absolute error $\Delta_i$ of the $i$-th criterion value can be determined from formula (2.27) and compared with the permissible value (if any).

A questionnaire for ranking the criteria by using the method of correlation was prepared by the authors. An anonymous survey was carried out, 52 questionnaires were presented to passengers on the international passenger train No. 6 (IPT No. 6). About 48% of the trip that lasts for 14 h, and 05 min, took place during the night. Respondent characteristics are presented in Table 2.8.

The same passengers completed questionnaires and assessed the criteria that determine the choice to travel by train as an alternative to the aircraft. The results of this research were published in the article [15] and in the Chap. 2.

The number of the respondents (52) was tree times that of criteria (15 or 16). Therefore, it was sufficient because ($n \geq m$). A description of 15 or 16 criteria was

**Table 2.8** Details of 52 respondents who gave the judgement on ranks

| Items | Descriptions |
|---|---|
| Gender | 28 male respondents, 24 female respondents |
| Age | Average age is 46 years |
| Citizenship | Lithuanian (24), Ukraine (1), Russia (26), passenger with dual Russian–Lithuanian citizenship took part (1) |
| Education | Most of respondents (46) had higher education 6 respondents had secondary education |
| Aim (goal) of trip | Business (21), tourism (15), visiting relatives and friends (10), medical treatment (5), research (1) |

presented in the questionnaire and the respondents assigned different ranks to them (all the ranks had different assignable values).

## 2.4 Significance of the Criteria for Train Travel as an Alternative to Bus Travel

The famous American writer Mark Twain wrote that "Travel is fatal to prejudice, bigotry, and narrow-mindedness". On the other hand, people get tired when travelling and, therefore, the choice of an appropriate mode of transport is very important. Now, there is a wide choice of modes of travel, which include pedestrian travelling, cycling, and travelling by automobile, by bus as well as by rail, air or water transport. A passenger decides which mode of transport is most safe and comfortable for travel. The criteria determining the choice of a particular mode of transport can be identified when a set of the evaluation criteria is defined, and a certain number of passengers are surveyed. The passengers, who chose a particular mode of transport (e.g., rail transport) as an alternative to another mode of transport, assign the ranks to the considered criteria. All the criteria describing a particular mode of transport have some advantages over the criteria describing another means of transport.

The significance of hierarchically unstructured criteria is identified using a two-level model (Fig. 2.4a).

The study was based on the survey of passengers travelling from IPT No. 6. There is a regular rail and road service between the capital of Lithuania, one of the Baltic States (and a member-state of the EU) and the foreign country. Therefore, passengers can choose between the two modes of transport in covering a distance of 944 km between these cities.

A set of criteria was defined to determine their influence on passengers' choice to travel by train rather than by bus. For this purpose, passengers had to rank the considered criteria according to their importance for their choice of this mode of transport. The following criteria were included in the questionnaire presented to the passengers:

A. A trip by train is safer than a trip by bus (according to the statistical data),
B. Trains take preference over buses at road crossings, which helps them to cut the time of stops,
C. Trains pollute the environment less heavily,
D. A trip by train is cheaper than a trip by bus,
E. Passengers can lie down on berths, rest, and change clothes in the compartment,
F. The compartment of a passenger car is a closed space (ensuring fewer disturbances from other passengers),
G. There is a possibility for passengers to use WC on the train and smoke on the platform of a passenger car at any time,
H. Travelling by train does not depend on weather conditions,
I. There is less noticeable rocking and vibration in a passenger car,
J. A dining car is available for passengers,
K. There is freedom of movement on the train,
L. Trains have fewer stops than buses and are rarely overdue,
M. Trains are travelling day and night,
N. There is a possibility for passengers to order food or newspapers and magazines to the compartment,
O. There is a simpler border control for passengers on the train (they don't need to leave the cars and their luggage is with them).

The 15-criteria model, unstructured in a hierarchical way, is shown in Fig. 2.5a.

The ranks of the significance of criteria, determining the choice by passengers of a trip by train rather than travelling by bus were used for calculating the average values of the ranks $\overline{R}_i$, the concordance coefficient $W$, Pearson's chi-square statistic $\chi^2$ and criteria weights $\omega_i$ and $\omega_i'$ (Table 2.9). The following table also presents the

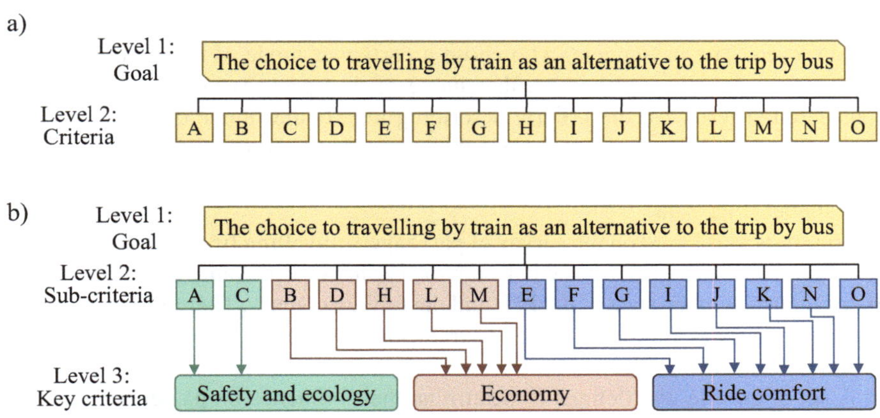


**Fig. 2.5** Models of the determinants (criteria) for passengers' choice to travel by train as an alternative to travel by bus: **a** hierarchically unstructured criteria; **b** inverse hierarchy of structured sub-criteria and key criteria

**Table 2.9** Averages of criteria ranks $\overline{R}_i$ of passenger choice to travel by train as an alternative to travel by bus and calculated weights $\omega_i$ and $\omega_i'$ with reference to linear ARTIW-L and non-linear ARTIW-N methods

| Formula | Criterion $i = 1, 2, \ldots, m$ | | | | | | | | | | | | | | | Sum |
|---|---|---|---|---|---|---|---|---|---|---|---|---|---|---|---|---|
| | A | B | C | D | E | F | G | H | I | J | K | L | M | N | O | |
| $\sum_{j=1}^{n} R_{ij}$ | 351 | 405 | 566 | 665 | 270 | 396 | 386 | 349 | 447 | 444 | 418 | 351 | 287 | 507 | 398 | 6240 |
| $\overline{R}_i = \frac{\sum_{j=1}^{n} R_{ij}}{n}$ (2.16) | 6.751 | 7.788 | 10.885 | 12.788 | 5.192 | 7.615 | 7.423 | 6.712 | 8.596 | 8.538 | 8.038 | 6.751 | 5.519 | 9.750 | 7.654 | 120.000 |
| $\sum_{j=1}^{n} R_{ij} - \frac{n(m+1)}{2}$ | − 65 | − 11 | 150 | 249 | − 146 | − 20 | − 30 | − 67 | 31 | 28 | 2 | − 65 | − 129 | 91 | − 18 | 0 |
| $\left[\sum_{j=1}^{n} R_{ij} - \frac{n(m+1)}{2}\right]^2$ (2.22) | 4225 | 121 | 22500 | 62001 | 21316 | 400 | 900 | 4489 | 961 | 784 | 4 | 4225 | 16641 | 8281 | 324 | 147172 |
| ARTIW-L $\omega_i = \frac{(m+1)-\overline{R}_i}{\sum_{i=1}^{m} \overline{R}_i}$ (2.17) | 0.077 | 0.0684 | 0.0426 | 0.0268 | 0.0901 | 0.0699 | 0.0715 | 0.0774 | 0.0617 | 0.0622 | 0.0664 | 0.077 | 0.0873 | 0.0521 | 0.0696 | 1.0000 |
| Priority ARTIW-L | 4–5 | 9 | 14 | 15 | 1 | 7 | 6 | 3 | 12 | 11 | 10 | 4–5 | 2 | 13 | 8 | 12 |
| ARTIW-N $u_i = \frac{\min_j \overline{R}_i}{\overline{R}_i}$ (2.18) | 0.7691 | 0.6667 | 0.4770 | 0.4060 | 1 | 0.6818 | 0.6994 | 0.7735 | 0.6040 | 0.6081 | 0.6459 | 0.7691 | 0.9408 | 0.5325 | 0.6783 | 10.2522 |
| $\omega_i' = \frac{u_i}{\sum_{i=1}^{m} u_i}$ (2.19) | 0.0750 | 0.0650 | 0.0465 | 0.0396 | 0.0976 | 0.0665 | 0.0682 | 0.0755 | 0.0589 | 0.0593 | 0.0630 | 0.0750 | 0.0918 | 0.0519 | 0.0662 | 1.0000 |
| Priority ARTIW-N | 4–5 | 9 | 14 | 15 | 1 | 7 | 6 | 3 | 12 | 11 | 10 | 4–5 | 2 | 13 | 8 | 120 |
| $\overline{\omega}_i = \frac{\omega_i + \omega_i'}{2}$ (2.20) | 0.0760 | 0.0667 | 0.0445 | 0.0332 | 0.0939 | 0.0682 | 0.0699 | 0.0764 | 0.0603 | 0.0607 | 0.0647 | 0.076 | 0.0896 | 0.052 | 0.0679 | 1.0000 |
| Standard deviation$\sigma_{Ri}$ | 4.414 | 4.267 | 4.081 | 3.392 | 4.039 | 4.303 | 3.957 | 3.972 | 3.604 | 4.377 | 3.242 | 3.725 | 3.654 | 3.602 | 3.875 | – |

$n$—the number of experts ($j = 1, 2, \ldots, n$), $n = 52$; $m$—the number of criteria ($i = 1, 2, \ldots, m$), $m = 15$

mean values $\overline{\omega}_i$ of the criteria weights $\omega_i$ and $\omega_i'$ and the standard deviations of the ranks $R_{ij}$.

The total of criteria ranks was $\sum_{i=1}^{15} R_i = 6240$, while the sum of average ranks $\overline{R}_i$ of all $i$ criteria was $\sum_{i=1}^{15} \overline{R}_i = 120.0$. The average value $\overline{R} = \frac{n(m+1)}{2} = \frac{52(15+1)}{2} = 416$ of criteria ranks or $\overline{R} = 6240/15 = 416$. The sum of squared deviations $S = 147172$ (formula 2.22). The concordance coefficient $W = 0.194$, showing the consistency of the estimates of respondents (52 passengers), was calculated by formula (2.21).

In this study, the choice of passengers to take the train as an alternative to the bus is assessed on 15 criteria, i.e., more than 7. Therefore, the significance of the concordance coefficient can be assessed by applying the $\chi^2$ statistic. Based on the data of passengers' survey and using formula (2.23), $\chi^2 = 141.5$ was obtained. The critical value $\chi^2_{v,\alpha}$ taken from the table of chi-squared distribution with $v = 15 - 1 = 14$ degrees of freedom and the significance level $\alpha = 0.01$ was equal to 29.14. The empirical value $\chi^2 = 141.5$ was 4.85 times that of the critical value $\chi^2_{v,\alpha} = 29.14$, which allowed researchers to assume that the respondents' estimates were consistent.

The smallest value of the concordance coefficient $W_{min}$, with the significance level $\alpha = 0.01$ and the degree of freedom $v = m - 1 = 15 - 1 = 14$, allowing the authors to assume that the respondents' estimates were consistent, was calculated by formula (2.24). The smallest value of the concordance coefficient $W_{min} = 0.040$ made only about one fifth of the calculated concordance coefficient $W = 0.194$.

The coefficient of concordance (formula 2.25), which indicates the non-contradictory nature of the opinions of passengers who evaluated the significance of the criteria determining their choice of using the train as an alternative to taking the bus, is equal:

$$k_c = \frac{W}{W_{min}} = \frac{\chi^2}{\chi^2_{v,\alpha}} = \frac{0.194}{0.040} = \frac{141.5}{29.14} = 4.85.$$

A diagram of the calculated average ranks $\overline{R}_i$ of the 15 criteria determining the choice of travelling by train rather than by bus by the passengers is given in Fig. 2.6.

By applying the ARTIW-L and ARTIW-N methods, the passengers' reasons for selecting a trip by train rather than a trip by bus, which are described by criteria and their weights $\omega_i$, $\omega_i'$, and $\overline{\omega}_i$ were determined. The calculation data for criteria beginning from the most important (E) to the least important (D) ones are shown in Fig. 2.7.

The calculated rank means $\overline{R}_i$ (Fig. 2.6) and global weights $\omega_i$ and $\omega_i'$ (Fig. 2.7) of criteria, determining the choice of the respondents of the trip by train rather than the travel by bus, show that criteria E, M and H are much more important than criteria N, C and D. This implies that their priority order should be as follows:

$$\mathrm{E} \succ \mathrm{M} \succ \mathrm{H} \succ \mathrm{L} \succ \mathrm{A} \succ \mathrm{G} \succ \mathrm{F} \succ \mathrm{O} \succ \mathrm{B} \succ \mathrm{K} \succ \mathrm{J} \succ \mathrm{I} \succ \mathrm{N} \succ \mathrm{C} \succ \mathrm{D}.$$

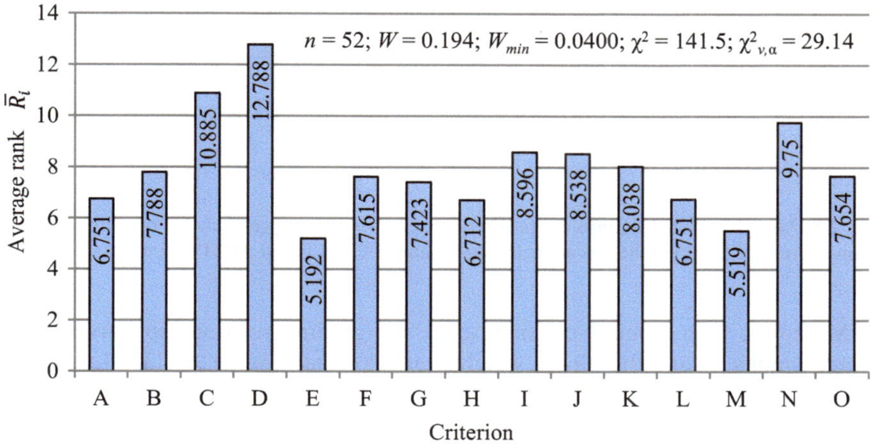


**Fig. 2.6** The average ranks of criteria $\overline{R}_i$

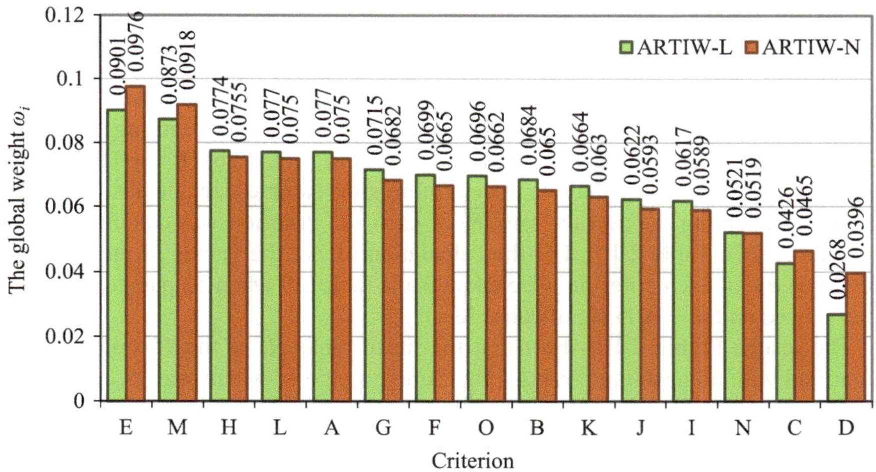


**Fig. 2.7** Relative weights of criteria in descending order

The arithmetic means of the weights of the criteria calculated by the two MCDM methods that determine the choice of passengers to take the train as an alternative to the bus are presented in Fig. 2.8. The means of the weights of the criteria were calculated without any preference of one method over the other for either of the two methods (ARTIW-L and ARTIW-N). For passengers, the most significant criterion E ($\overline{\omega}_E = 0.0939$) is more important (more relevant) than the least significant criterion D ($\overline{\omega}_D = 0.0332$) by about 2.8 times.

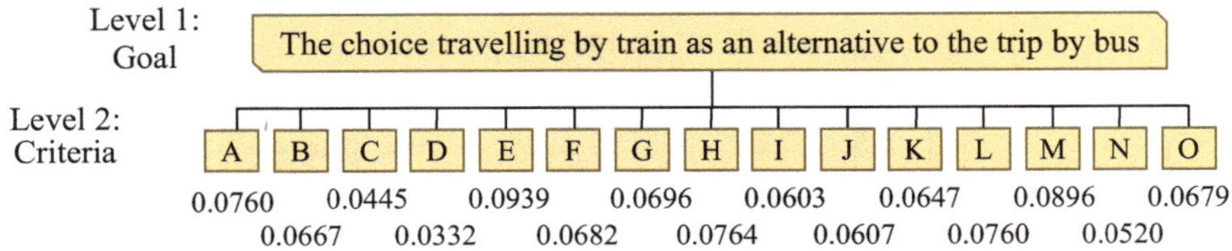


**Fig. 2.8** Means of global weights of the criteria determining passengers' choice of taking the train as an alternative to taking the bus are calculated using ARTIW-L and ARTIW-N methods

Moreover, there should be an inverse straight-line relationship between the average ranks $\overline{R}_i$ and the global weights $\omega_i$ calculated by the ARTIW-L method. The determination coefficient of the regression equation of 15 criteria, $R^2 = 1$, and coefficient of correlation $r = -1$, shows that this is a functional linear relationship $\omega_i = -0.083\overline{R}_i + 0.1333$. The correlation between $\overline{R}_i$ and the weights $\omega_i'$ calculated using the ARTIW-N method is non-linear (Fig. 2.9a). Ranks and weights are related by the quadratic regression equation $\omega_i' = 0.0009\,\overline{R}_i^2 - 0.0234\,\overline{R}_i + 0.1932$, coefficient of determination of which is $R^2 = 0.9955$.

The data obtained in the performed study show that the estimates of the significance (importance) of criteria, determining the passenger choice of travelling by train rather than by bus, are consistent (in agreement) and reflect their general opinion, shown by the averages $\overline{\omega}_i$ of criteria weights $\omega_i$ and $\omega_i'$ calculated according to formula (2.20) and their correlation with the rank averages $\overline{R}_i$ (Fig. 2.9b) ($R^2 = 0.9989$).

With reference to the principle of determining the sample size, the absolute error $\Delta_i$ for determining the average rank $\overline{R}_i$ of each $i$-th criterion was calculated from formula (2.28) (Table 2.10). For the calculation of $\Delta_i$, the values of the standard deviation $\sigma_{R_i}$ of each criterion were taken from Table 2.9 when 52 passengers were interviewed and $\alpha = 0.05$ significance level.

The results (Table 2.10) show that the mean ranks $\overline{R}_i$ of the 15 criteria were identified with an absolute error $\Delta_i$, the sample range of which was $\Delta_{imax} - \Delta_{imin}$

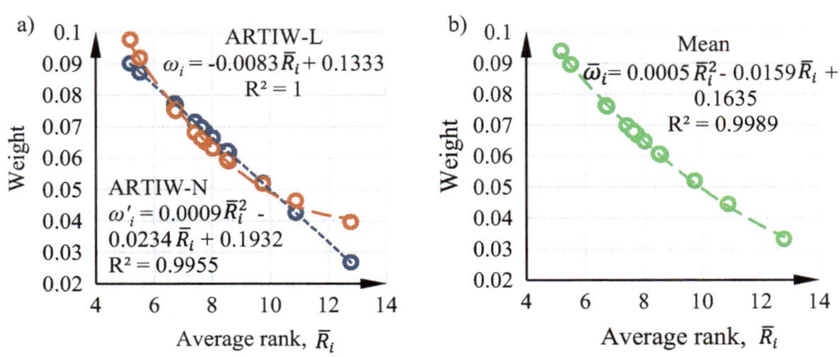


**Fig. 2.9** Correlation between the means of the criteria ranks and the normalized weights of these criteria calculated using: **a** ARTIW-L and ARTIW-N methods; **b** average values of two methods

**Table 2.10** The absolute error $\Delta_i$ in determining the averages of the 15 criteria ranks $\overline{R}_i$ of the group of 52 surveyed passengers of train

| Criterion $i = 1, 2, \ldots, m$ | | | | | | | | | | | | | | |
|---|---|---|---|---|---|---|---|---|---|---|---|---|---|---|
| A | B | C | D | E | F | G | H | I | J | K | L | M | N | O |
| 1.20 | 1.16 | 1.11 | 0.92 | 1.10 | 1.17 | 1.08 | 1.08 | 0.98 | 1.19 | 0.88 | 1.01 | 0.99 | 0.98 | 1.05 |

$= \Delta_{iA} - \Delta_{iK} = 1.20 - 0.88 = 0.32$, and the mean value $\overline{\Delta}_i = 1.06$. By taking $\Delta_i$ for each criterion or the mean value $\overline{\Delta}_i$ of 15 criteria, it is possible to calculate the confidence interval $\overline{R}_i \pm \Delta_i$ of ranks average $\overline{R}_i$, with the population mean rank $\overline{R}_{ip}$ of 95% probability. For example, the population mean rank $\overline{R}_A$ of the criterion A with the highest rank variation is in the range $6.751 \pm 1.20$, i.e., between 5.551 and 7.951. The range would reduce if more than 52 passengers were interviewed. We believe that $\overline{\Delta}_i = 1.06$ is close to one, so the number of passengers $n = 52$ who completed the survey is sufficient and allowed to reliably assess the factors determining the choice of passengers to travel by train as an alternative to the bus.

All 15 criteria presented in the questionnaire, which determined the choice of passengers to travel by train rather than by bus, were divided into three groups, and named key criteria (Fig. 2.5b). The group of safety and environmental protection included two criteria (A and C), the economy group embraced five criteria (B, D, H, L, M) and ride comfort group consisted of eight criteria (E, F, G, I, J, K, N, O) (Table 2.11).

The normalised global weights $\overline{\omega}_i$, of the 15 hierarchically unstructured criteria, obtained by the ARTIW-L and ARTIW-N methods, allowed to calculate the weights $\tilde{\omega}_b$ of the factors grouped under the key criteria using the IHAMCI method. When the weights of the ungrouped factors are calculated, they are referred to as criteria, and when they are grouped according to the hierarchy models (Fig. 2.4b, c), they are referred to as sub-criteria and key criteria.

The weight $\tilde{\omega}_{bSa}$ of two sub-criteria, A + C, making the travel safety and environmental protection group (key criterion), which was calculated by formula (2.26), is the smallest:

$$\tilde{\omega}_{bSa} = \frac{0.1205 : 2}{0.0602 + 0.0684 + 0.0672} = 0.3077.$$

The weight $\tilde{\omega}_{bEc}$ of five sub-criteria, B + D + H + L + M, included in the key criterion describing economy was calculated in the same way as follows:

$$\tilde{\omega}_{bEc} = \frac{0.3419 : 5}{0.0602 + 0.0684 + 0.0672} = 0.3492.$$

The weight $\tilde{\omega}_{bCo}$ of key criterion of eight sub-criteria, E + F + G + I + J + K + N + O, describing ride comfort was the largest:

**Table 2.11** Global weights of the criteria determining passengers' choice to take the train as an alternative to the bus and the weights of the key criteria calculated by the IHAMCI method

| Factor (criterion) | Mean global weight $\overline{\omega}_i$ | Priority | Size of key criterion $k$ | Factor (Criterion) | Global weight $\overline{\omega}_i$ | The sum of global weights $\omega_b = \sum_{i=1}^{k} \overline{\omega}_i$ | Mean weight of key criterion $\frac{\omega_b}{k}$ | The weight of key criterion $\tilde{\omega}_b$ |
|---|---|---|---|---|---|---|---|---|
| A | 0.0760 | 4–5 | Safety and ecology protection $k_{Sa} = 2$ | A | 0.076 | 0.1205 | 0.1205:2 = 0.0602 | 0.3077 |
| B | 0.0667 | 9 | | C | 0.0445 | | | |
| C | 0.0445 | 14 | Economy $k_{Ec} = 5$ | B | 0.0667 | 0.3419 | 0.3419:5 = 0.0684 | 0.3492 |
| D | 0.0332 | 15 | | D | 0.0332 | | | |
| E | 0.0939 | 1 | | H | 0.0764 | | | |
| F | 0.0682 | 7 | | L | 0.0760 | | | |
| G | 0.0696 | 6 | | M | 0.0896 | | | |
| H | 0.0764 | 3 | Ride comfort $k_{Co} = 8$ | E | 0.0939 | 0.5376 | 0.5376:8 = 0.0672 | 0.3431 |
| I | 0.0603 | 12 | | F | 0.0682 | | | |
| J | 0.0607 | 11 | | G | 0.0699 | | | |
| K | 0.0647 | 10 | | I | 0.0603 | | | |
| L | 0.0760 | 4–5 | | J | 0.0607 | | | |
| M | 0.0896 | 2 | | K | 0.0647 | | | |
| N | 0.0520 | 13 | | N | 0.0520 | | | |
| O | 0.0679 | 8 | | O | 0.0679 | | | |
| Total | 1.0000 | – | – | – | 1.0000 | 1.0000 | 0.1958 | 1.0000 |

$$\tilde{\omega}_{bCo} = \frac{0.5376 : 8}{0.0602 + 0.0684 + 0.0672} = 0.3431.$$

The results of calculation show that the choice of passengers to travel by rail transport rather than by road transport (a bus) was determined by the key criteria describing ride comfort and economy (about 35%) as well as safety and environmental protection (only about 30%).

The obtained global weights $\overline{\omega}_i$ of criteria and the weights $\tilde{\omega}_g$ of key criteria divided into three groups, given in Fig. 2.10 allowed the authors to identify the criteria determining the choice of passengers of travelling by train (as an alternative to travel by bus). The obtained data can be used by the companies engaged in passenger transportation by rail to enhance the quality of services provided by this more environmentally friendly mode of land transport so that it will have a competitive edge over the rival modes of transportation.

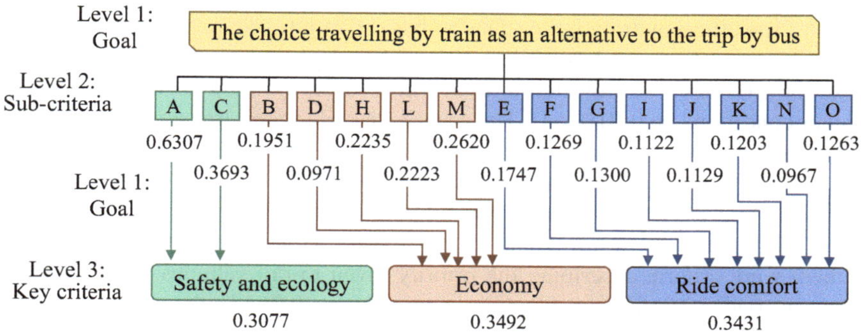


**Fig. 2.10** Normalized subjective weights of the sub-criteria and the key criteria determining passengers' choice to take the train as an alternative to taking the bus have been calculated using the inverse hierarchy model

The sub-criteria and key criteria weights calculated in this study, which show why passengers choose train travel as an alternative to the bus, are not absolutely accurate and constant. When interviewing citizens of other countries traveling on international trains, the significance of the sub-criteria and key criteria may differ. Their values can be influenced by the economic development of the country, passenger habits, the reliability factors relating to different modes of transport and risks.

The most important part of the study consists of the original criteria system and their weighting methodology, applying the new ARTIW-L and ARTIW-N methods, as well as the method of inverse hierarchy for assessment main criteria importance (IHAMCI). These methods can be used by other researchers to calculate normalized subjective weights when ranking the results of an expert or respondent survey.

Passengers, usually decide to travel by a particular means of transport after they evaluate travelling criteria. The selection of a particular (alternative) mode of transport is based on the significances (subjective weights) of the considered criteria, which can be determined by using expert evaluation methods. The average value of the estimates given by a considerable number of passengers (respondents) in ranking the criteria can be used as public opinion about a particular transport mode chosen for a particular route.

In the present work, the reasons behind the passengers' choice of travel by train rather than by bus made by passengers are identified by considering fifteen criteria. The significances of these criteria for choosing to travel by train were evaluated by 52 respondents (passengers of the international train No. 6) against a fifteen ranks' scale. The subjective total normalized weights of criteria by a new ARTIW-L and ARTIW-N methods allowed the authors to rank them in the priority (preference) order. A functional or close strong correlation between the means of the criteria ranks and the normalized weights of the criteria calculated from them indicates that the ARTIW-L and ARTIW-N methods are satisfactory to assess the significance of the criteria. The mean of the criteria weights calculated by these two methods was taken as the final significance of the criteria. The criteria determining the passengers' choice of travelling by train rather than by bus (as an alternative mode of transport) include ride comfort (the availability of berths in passenger cars for sleeping and relaxation) (0.0939), the selected time of travel (0.0896) and a negligible effect of weather conditions on it (0.0764). The criteria describing rail transport as safer than road transport, as well as such advantages as a smaller number of stops and delays on the way (0.0760), the availability of WCs and places for smoking (0.0696), a lower probability that passengers would disturb each other (0.0682) and simpler border control (0.0679) are less important for passengers. However, even less important for them were criteria describing the priority given to rail vehicles when crossing the motor roads (0.0667) and freedom of movement (0.0647), the availability of a dining car (0.0607) and unpleasant slight rocking and vibration (0.0603). The least important criteria for passengers include a possibility to order food or newspapers and magazines to the compartment (0.0520), lower environmental pollution of the rail transport (0.0445) and sometimes cheaper railways tickets (0.0332). The ratio of the largest total weight of criteria (0.0939) to their smallest total weight (0.0332), which is equal to 3.36, shows that the significance of particular criteria is different for passengers choosing a particular mode of transport.

The fifteen considered criteria were divided into three groups, using the IHAMCI method suggested by the second author. The normalized overall weights for these groups (key criteria) were calculated. The overall weight of two sub-criteria describing safety and environmental protection was equal to 0.3077, while the overall weight of five sub-criteria forming the "economy" group was 0.3492, and the overall weight of eight sub-criteria referring to ride comfort was the largest and made 0.3431. The overall weights of the key criteria of any sub-criteria group were calculated using a new IHAMCI method, which allowed to assessment of different numbers of sub-criteria in key criterion.

The minimum number of experts or respondents to be consulted to obtain reliable results has been calculated using the sample size principle. After interviewing 52 passengers, the values of the standard deviations of the criteria ranks were identified, which were used to determine the absolute error of the mean of the ranks of each criterion. The results showed that with 95% probability, the criteria rank averages differ from the population averages by no more than 1.20–0.88 rank (on average 1.06 rank). This difference is close to the unit and indicates that the significance of the criteria was determined with sufficient accuracy.

The decision-makers in the countries engaged in passenger transportation by the considered international train, should primarily improve the services described by the criteria, most strongly influencing a decision of passengers to choose a trip by train rather than a trip by bus. A company providing passenger transportation by any means of transport can win the competition in this area only if the provided services are of the highest quality and satisfy the ever-growing demands of passengers.

## 2.5 Significance of the Criteria for Train Travel as an Alternative to Air Travel

The advantages and disadvantages of rail transport in competition with air transport are important and valued by passengers. There is no fair competition between air transport and railways, as there is a large supply of low-cost flights on the market. A level playing field for all modes of transport is crucial. For example, fuel tax may be introduced, or State support for night trains should be the exception and not the rule. Also, if a public service commitment is a requirement, the contracting process should be as transparent as possible [3].

A set of criteria determining the choice of passengers to travel by train was defined when studying the reasons for their decision to use this mode of transport. The passengers who chose rail transport were asked to answer the question: why am I riding in the train, but not in the plane? They were also asked to assign the ranks to the evaluation criteria, depending on their importance. The criteria given below were included in the questionnaire provided to the passengers of the train (Fig. 2.11).

A. In rail transport, the check-in is not as long as it is in air transport,
B. The luggage can be taken by the passengers of the train exactly at the time of its arrival at the station (they need not wait for it long),
C. I prefer travelling by train because I am afraid of travelling by plane (because of air pockets, air toss, storms, etc.),
D. The difference in the time of travelling to a particular place by a fast train and by plane is not considerable,
E. Electronic devices of passengers should not be switched off in the train as it should be done in the plane (during its take-off or landing),

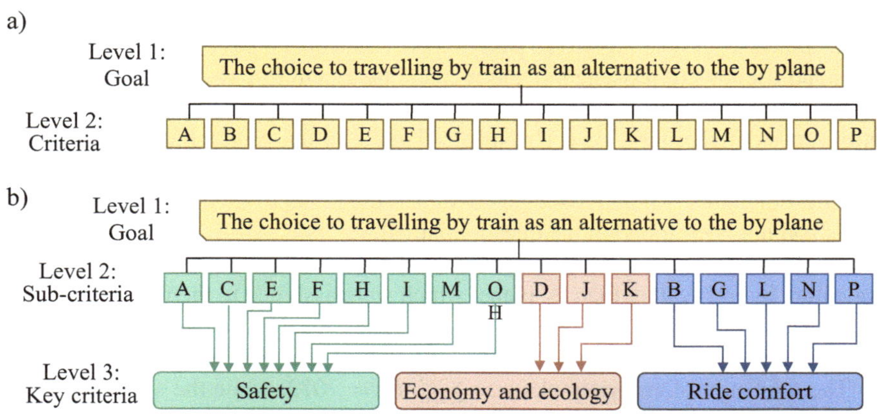


**Fig. 2.11** Models of the factors (criteria) ithat influence passengers' choice to travel by train as an alternative to aircraft: **a** hierarchically unstructured criteria; **b** inverse hierarchy of structured sub-criteria and key criteria

F. A trip by train does not strongly depend on weather and cannot be delayed because of bad weather conditions,
G. Railway stations are close to the departure and arrival places of the train, while airports are far from the areas, where planes take off and land,
H. Safety belts should be fastened by the passengers of the plane when it is taking off or landing, while they are not used in the train at all,
I. A danger of terrorists' attacks is greater in air transport than it is in rail transport,
J. An airplane emits more burnt gases into the air, polluting it more heavily than a rail vehicle, and consumes more fuel,
K. A trip by train is cheaper than a trip by plane,
L. The occupancy of the space by passengers in the train is lower than that in the plane,
M. A person, buying a railway ticket or getting into the railway car, is not required to show a document (e.g., a passport or an identification card), which is the case in air transport,
N. The acceleration or deceleration of the train does not affect the passengers as strongly as does the plane's take-off or landing,
O. The luggage of all passengers in the train is not examined, and they can carry liquids,
P. The persons accompanying the passengers of the train can stay in the railway car until the train departs.

A questionnaire was specially devised for a survey of passengers and the application of the method of ranks' correlation. The anonymous survey of the passengers of the train (IPT No. 6) was carried out in June. The passengers had to fill in 52 questionnaires. There were 24 women and 28 men among them. Most of the respondents had higher education (88.5%) and were 46 year-old on the average. This indicates that most passengers were sufficiently qualified and able to make experience-based decisions (in selecting travel by train). The aims of their journey were associated with work (21 passengers), tourism (15 passengers), education (1 passenger), visiting relatives and friends (10 passengers) and medical treatment (5 passengers) (Table 2.8). The number of the respondents was 52, which was more than 3 times the number of the considered criteria (16). This means that the sufficient number of respondents took part in the survey ($n \geq m$). A description of these 16 criteria was provided in the questionnaires, where passengers assigned the ranks to each of them (the same ranks had not been found in the questionnaires).

The ranks of the importance (significance) of criteria, determining the choice of travelling by train rather than by plane, which were assigned to them by the respondents, were used for calculating the average ranks $\overline{R}_i$, the concordance coefficient $W$, Pearson's chi-squared test $\chi^2$ and the criteria weights $\omega_i$ (Table 2.12).

The total sum of criteria ranks was $\sum_{i=1}^{16} R_i = 7072$, while the sum of average ranks of 16 criteria was $\sum_{i=1}^{16} \overline{R}_i = 136.0$. The calculation of the average criteria rank $\overline{R} = \frac{n(m+1)}{2} = \frac{52(16+1)}{2} = 442$ which was performed by using formula (2.16), or some other mathematical operation, yielded $\overline{R} = \frac{7072}{16} = 442$. The sum of squared deviations $S = 247124$ (formula 2.22). The concordance coefficient $W$, showing the

**Table 2.12** Criteria rank means $\overline{R}_i$ of passenger choice to travel by train as an alternative to travel by plane and calculated weights $\omega_i$ and $\omega_i'$ with reference to linear ARTIW-L and non-linear ARTIW-N methods

| Formula | Criterion $i = 1, 2, \ldots, m$ | | | | | | | | | | | | | | | | Suma |
|---|---|---|---|---|---|---|---|---|---|---|---|---|---|---|---|---|---|
| | A | B | C | D | E | F | G | H | I | J | K | L | M | N | O | P | |
| $\sum_{j=1}^{n} R_{ij}$ | 265 | 319 | 409 | 504 | 449 | 231 | 277 | 603 | 382 | 663 | 613 | 421 | 532 | 481 | 409 | 514 | 7072 |
| $\overline{R}_i = \frac{\sum_{j=1}^{n} R_{ij}}{n}$ (2.16) | 5.096 | 6.135 | 7.865 | 9.692 | 8.635 | 4.442 | 5.327 | 11.596 | 7.346 | 12.750 | 11.789 | 8.096 | 10.231 | 9.250 | 7.865 | 9.885 | 136.000 |
| $\sum_{j=1}^{n} R_{ij} - \frac{n(m+1)}{2}$ | − 177 | − 123 | − 33 | 62 | 7 | − 211 | − 165 | 161 | − 60 | 221 | 171 | − 21 | 90 | 39 | − 33 | 72 | 0 |
| $\left[\sum_{j=1}^{n} R_{ij} - \frac{n(m+1)}{2}\right]^2$ (2.22) | 31329 | 15129 | 1089 | 3844 | 49 | 44521 | 27225 | 25921 | 3600 | 48841 | 29241 | 441 | 8100 | 1521 | 1089 | 5184 | 247124 |
| ARTIW-L<br>$\omega_i = \frac{(m+1)-\overline{R}_i}{\sum_{i=1}^{m} \overline{R}_i}$ (2.17) | 0.0875 | 0.0799 | 0.0672 | 0.0537 | 0.0615 | 0.0923 | 0.0858 | 0.0397 | 0.0710 | 0.0313 | 0.0383 | 0.0655 | 0.0498 | 0.0570 | 0.0672 | 0.0523 | 1.0000 |
| Priority ARTIW-L | 2 | 4 | 6–7 | 11 | 9 | 1 | 3 | 14 | 5 | 16 | 15 | 8 | 13 | 10 | 6–7 | 12 | 136 |
| ARTIW-N<br>$u_i = \frac{\min_i \overline{R}_i}{\overline{R}_i}$ (2.18) | 0.8717 | 0.7240 | 0.5648 | 0.4583 | 0.5144 | 1 | 0.8339 | 0.3831 | 0.6047 | 0.3484 | 0.3768 | 0.5487 | 0.4342 | 0.4802 | 0.5648 | 0.4494 | 9.1574 |
| $\omega_i' = \frac{u_i}{\sum_{i=1}^{m} u_i}$ (2.19) | 0.0952 | 0.0791 | 0.0617 | 0.0500 | 0.0562 | 0.1092 | 0.0911 | 0.0418 | 0.0660 | 0.0381 | 0.0411 | 0.0599 | 0.0474 | 0.0524 | 0.0617 | 0.0491 | 1.0000 |
| Priority ARTIW-N | 2 | 4 | 6–7 | 11 | 9 | 1 | 3 | 14 | 5 | 16 | 15 | 8 | 13 | 10 | 6–7 | 12 | 136 |
| $\overline{\omega}_i = \frac{\omega_i + \omega_i'}{2}$ (2.20) | 0.0914 | 0.0795 | 0.0645 | 0.0518 | 0.0589 | 0.1008 | 0.0884 | 0.0407 | 0.0685 | 0.0347 | 0.0397 | 0.0627 | 0.0486 | 0.0547 | 0.0644 | 0.0507 | 1.0000 |
| Standard deviation $\sigma_{Ri}$ | 3.5494 | 3.5975 | 5.5379 | 4.8250 | 3.8452 | 3.4380 | 4.1609 | 3.3152 | 4.1011 | 3.1924 | 4.8196 | 3.7743 | 3.9883 | 4.0092 | 3.2239 | 3.9090 | |

*n*—the number of experts ($j = 1, 2, \ldots, n$); *m*—the number of criteria ($i = 1, 2, \ldots, m$)

consistence level of the estimates provided by a group of respondents (52 passengers of the train), which was calculated by formula (2.21), was $W = 0.269$.

Based on the data of the survey of passengers and using formula (2.23), $\chi^2 = 209.7$ was obtained. The critical $\chi^2_{v,\alpha}$ value, taken from chi-squared test table, with the degree of freedom $v = 16 - 1 = 15$ and the confidence level $\alpha = 0.01$, was 30.5779. The empirical value $\chi^2 = 209.7$ was larger than the critical $\chi^2_{v,\alpha}$; therefore, it could be concluded that the respondents' estimates were consistent.

The smallest value of the concordance coefficient $W_{min}$, with the confidence level $\alpha = 0.01$ and the degree of freedom $v = m - 1 = 16 - 1 = 15$, allowing us to assume that the respondents' estimates were consistent, was calculated by formula (2.24). The smallest value of the concordance coefficient $W_{min} = 0.0392$ was much smaller that the calculated concordance coefficient $W = 0.269$.

The coefficient of concordance (formula 2.25) indicating the non-contradictory nature of the opinions of passengers who rated the significance of the criteria determining their choice of train travel as an alternative to air travel is equal to:

$$k_c = \frac{W}{W_{min}} = \frac{\chi^2}{\chi^2_{v,\alpha}} = \frac{0.269}{0.0392} = \frac{209.7}{30.58} = 6.86.$$

A diagram of the calculated average ranks $\overline{R}_i$ of the criteria determining the choice of travelling by train rather than by plane by the passengers is given in Fig. 2.12.

Using formula (2.17), the ARTIW-L method was applied to calculate the normalised weights $\omega_i$, for the 16 criteria, the values of which are given in Table 2.12. Table 2.12 also provides the weights $\omega_i'$, of these criteria, which were calculated using formulae (2.18) and (2.19) when the ARTIW-N method was used.

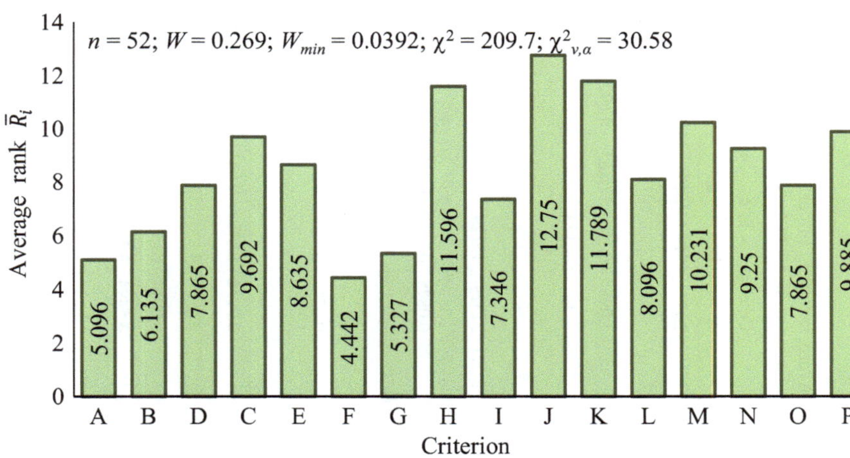


**Fig. 2.12** Rank means of the criteria determining passenger choice for taking the train as an alternative to taking the plane

The weights of the criteria calculated by the either MCDM methods are different, although the differences are not significant (Fig. 2.13). The weights demonstrate that the priority of the criteria determining passenger choice to take the train as an alternative to air travel is on the same level.

From formula (2.20), the arithmetic means $\overline{\omega}_i$ of the weights of each criterion were calculated, assuming that none of the methods used (ARTIW-L and ARTIW-N) has any advantage, i.e. preference has not been applied. For passengers, the most significant criterion F ($\overline{\omega}_F = 0.1008$) is more important (significant) than the least important criterion J ($\overline{\omega}_J = 0.0347$) by approximately 2.9 times (Fig. 2.14).

Moreover, there should be the inverse straight-line relationship between the average ranks $\overline{R}_i$ and the global weights $\omega_i$ calculated by the ARTIW method. The determination coefficient of the regression equation of 16 criteria, $R^2 = 1$, shows that this is a functional relationship $\omega_i = -0.0073\overline{R}_i + 0.125$ (Fig. 2.15a).

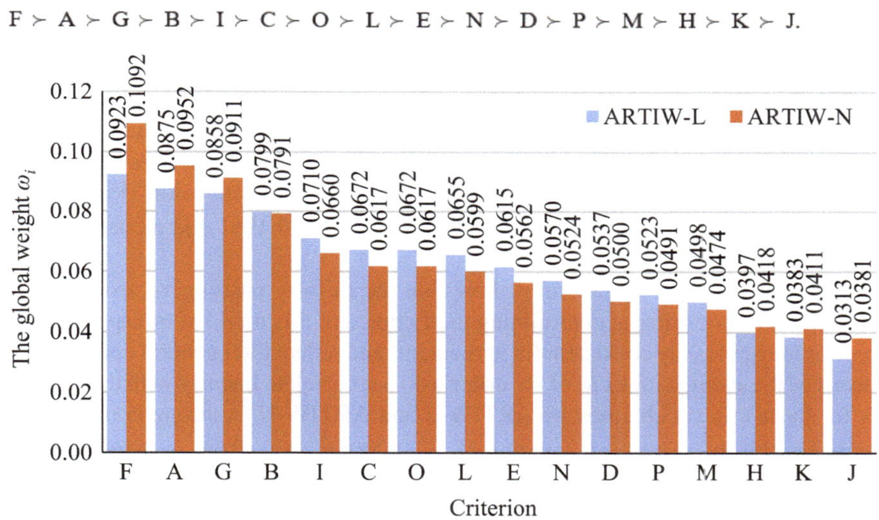


**Fig. 2.13** Normalised weights of the criteria determining passenger preference for train travel as an alternative to air travel, calculated by the ARTIW-L and ARTIW-N methods

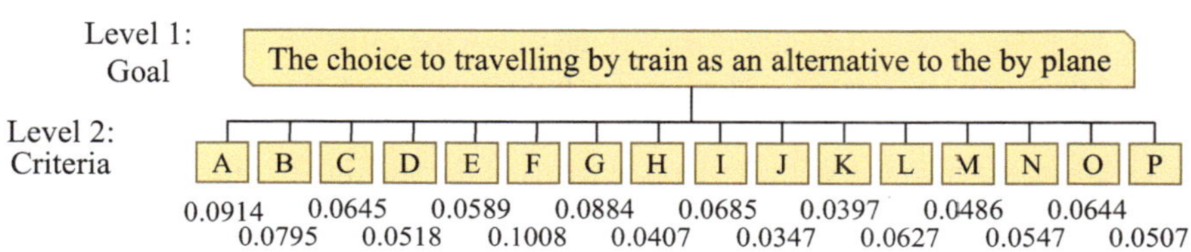


**Fig. 2.14** Means of the global weights for the criteria that determine passenger preference for train travel as an alternative to air travel calculated by the ARTIW-L and ARTIW-N methods

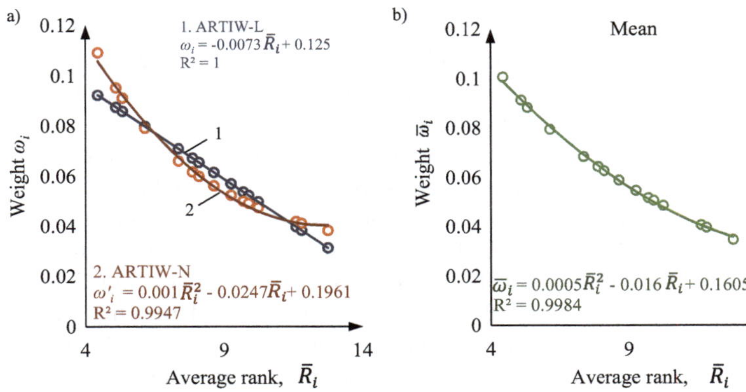


**Fig. 2.15** Correlation between the rank means and weights of the criteria determining the passenger choice to take the train as an alternative to air travel, calculated by: **a** ARTIW-L and ARTIW-N methods; **b** the average of the weights of both methods

The correlation between the criteria rank means $\overline{R}_i$ and the normalised weights $\omega_i'$, calculated applying the ARTIW-N method is non-linear. The arithmetic means and weights of the criteria that determine the passenger preference for train travel as an alternative to air travel are linked by a quadratic relationship (Fig. 2.15a), with a regression equation $\omega_i' = 0.001\overline{R}_i^2 - 0.024\overline{R}_i + 0.1961$ and a coefficient of determination $R^2 = 0.9947$.

The data obtained in the study show that the railway passengers' estimates of the importance of criteria, determining their choice of this mode of transport (rather than air transport), are consistent (in agreement) and reflect their general opinion. The means of the average ranks $\overline{R}_i$ of the criteria do not correlate linearly with the weights $\overline{\omega}_i$ for these criteria obtained by ARTIW-L and ARTIW-N. A strong correlation is evidenced by the regression equation $\overline{\omega}_i = 0.0005\overline{R}_i^2 - 0.016\overline{R}_i + 0.1605$ and its coefficient of determination $R^2 = 0.9984$.

By applying the principle of minimum sample size (formula 2.27), the deviation of the rank means for each criterion $\overline{R}_i$ from the population mean $\overline{R}_{ip}$ (formula 2.28), was calculated, i.e. the absolute error $\Delta_i$ (Table 2.13). The 95% probability (significance level $\alpha = 0.05$) was assumed, resulting in a Student's distribution statistics $t = 1.96$. The values of the standard deviations of the criterion ranks $\sigma_{R_i}$ are taken from Table 2.12. The standard deviation $\sigma_{R_i}$, which shows the variation in the criteria ranks $R_{ij}$ that determine passenger preference for train travel as an alternative to air travel, varied from 3.1924 (criterion J) to 5.5379 (criterion C). This suggests that passengers assessed the significance of the criteria in terms of the stability of different opinions. Whether the standard deviations $\sigma_{R_i}$ of the criteria ranks are significantly different from each other can be calculated using mathematical statistical methods for testing the homogeneity of variances, such as Bartlett's test.

The absolute error $\Delta_i$ of the criteria rank means $\overline{R}_i$, calculated from the questionnaires completed by 52 passengers on the train, varied from rank 0.87 (criterion J)

**Table 2.13** The absolute error $\Delta_i$ in determining the averages of the 16 criterion ranks $\overline{R}_i$ of the group of 52 surveyed passengers of train

| Criterion $i = 1, 2, \ldots, m$ | | | | | | | | | | | | | | | |
|---|---|---|---|---|---|---|---|---|---|---|---|---|---|---|---|
| A | B | C | D | E | F | G | H | I | J | K | L | M | N | O | P |
| 0.96 | 0.98 | 1.50 | 1.31 | 1.04 | 0.93 | 1.13 | 0.90 | 1.11 | 0.87 | 1.20 | 1.03 | 1.08 | 1.09 | 0.88 | 1.06 |

to rank 1.50 (criterion C). The average absolute error $\overline{\Delta}_i$ of the rank means $\overline{R}_i$ for all 16 criteria is $\overline{\Delta}_i = 1.07$, i.e. slightly greater than one. These results imply that the number of respondents to the survey is sufficient, as reliable significance levels for the criteria in terms of rank means have been obtained.

The 16 criteria that determine passenger preference for train travel as an alternative to air travel were grouped into 3 main criteria by applying the reverse hierarchy model (Fig. 2.11b). The safety key criterion consists of eight sub-criteria including A, C, E, F, H, I, M, O; the economy and ecology key criterion consists of three sub-criteria including D, J, K; and the ride comfort key criterion consists of five sub-criteria including B, G, L, N, P (Table 2.14).

The sum $\omega_b$ of the global weights $\overline{\omega}_i$ for each key criterion consisting of sub-criteria divided by the number $k$ of key criteria consisting of sub-criteria is used to calculate the mean non-normalised weight of key criterion. The normalisation is used to calculate the weight for each criterion $\tilde{\omega}_b$.

The weight $\tilde{\omega}_{bSa}$ of eight sub-criteria, A + C + E + F + H + I + M + O, included in the travel safety key criterion, which was calculated by formula (2.26) as follows:

$$\tilde{\omega}_{bSa} = \frac{0.5378 : 8}{0.06723 + 0.04207 + 0.0672} = \frac{0.067225}{0.1765} = 0.3809.$$

The weight $\tilde{\omega}_{bEc}$ of three sub-criteria, D + J + K, included in the economy and ecology key criterion, which was calculated by formula (2.26) in a similar way, was the smallest:

$$\tilde{\omega}_{bEc} = \frac{0.1262 : 3}{0.06723 + 0.04207 + 0.0672} = \frac{0.04207}{0.1765} = 0.2384.$$

The weight $\tilde{\omega}_{bCo}$ of five sub-criteria, B + G + L + N + P, included in the ride comfort key criterion, which was calculated in the same way, is to train passengers of the same significance as the safety key criterion:

$$\tilde{\omega}_{bCo} = \frac{0.3360 : 5}{0.06723 + 0.04207 + 0.0672} = \frac{0.0672}{0.1765} = 0.3807.$$

Passenger preference for taking the train instead of air travel is based on safety ($\tilde{\omega}_g = 0.3809$), ride comfort ($\tilde{\omega}_g = 0.3807$) and economic and environmental reasons ($\tilde{\omega}_g = 0.2384$). The most relevant reasons which are characterized by equal

**Table 2.14** Global weights of the criteria that influence passenger choice to travel by train as an alternative to air travel and the weights of the key criteria as derived through the IHAMCI method

| Factor (criterion) | Mean global weight $\overline{\omega}_i$ | Priority | Size of key criterion $k$ | Factor (criterion) | Global weight $\overline{\omega}_i$ | The sum of global weights $\omega_b = \sum_{i=1}^{k} \overline{\omega}_i$ | Mean weight of key criterion $\frac{\omega_b}{k}$ | The weight of key criterion $\tilde{\omega}_b$ |
|---|---|---|---|---|---|---|---|---|
| A | 0.0914 | 2 | Safety $k_{Sa} = 8$ | A | 0.0914 | 0.5378 | 0.5378:8 = 0.06723 | 0.3809 |
| B | 0.0795 | 4 | | C | 0.0645 | | | |
| C | 0.0645 | 6 | | E | 0.0589 | | | |
| D | 0.0518 | 11 | | F | 0.1008 | | | |
| E | 0.0589 | 9 | | H | 0.0407 | | | |
| F | 0.1008 | 1 | | I | 0.0685 | | | |
| G | 0.0884 | 3 | | M | 0.0486 | | | |
| H | 0.0407 | 14 | | O | 0.0644 | | | |
| I | 0.0685 | 5 | Economy and ecology $k_{Ec} = 3$ | D | 0.0518 | 0.1262 | 0.1262:3 = 0.04207 | 0.2384 |
| J | 0.0347 | 16 | | J | 0.0347 | | | |
| K | 0.0397 | 15 | | K | 0.0397 | | | |
| L | 0.0627 | 8 | Ride comfort $k_{Co} = 5$ | B | 0.0795 | 0.3360 | 0.3360:5 = 0.0672 | 0.3807 |
| M | 0.0486 | 13 | | G | 0.0884 | | | |
| N | 0.0547 | 10 | | L | 0.0627 | | | |
| O | 0.0644 | 7 | | N | 0.0547 | | | |
| P | 0.0507 | 12 | | P | 0.0507 | | | |
| Total | 1.0000 | 136 | – | – | 1.0000 | 1.0000 | 0.1765 | 1.0000 |

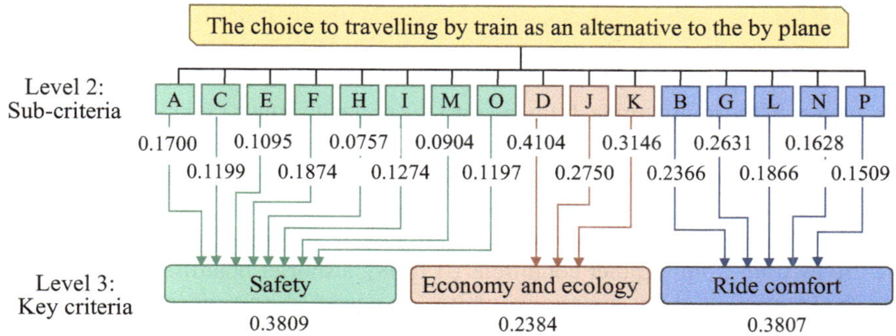


**Fig. 2.16** The inverse hierarchy model was used to calculate the normalised subjective weights of the sub-criteria and the key criteria that influence passenger preference for train travel as an alternative to air travel

significance include safety and comfort, while economics and the environment are of less importance.

The obtained calculation results show that railway passengers have chosen a trip by train as an alternative to a trip by plane because of the comfort (about 38%), safety (about 38%) and economy and ecology (only about 24%) of travelling by rail transport. The "economy and ecology" key criterion (sub-criteria D, J and K) is of least importance to passengers who selected travel by train as an alternative to plane, since ticket price and reducing environmental pollution were not as relevant as comfort and safety determining factors: passengers who travelled by train were wealthy or travelled to professional traineeships.

The global weights $\overline{\omega}_i$ of the criteria used in the study and the weights of key criteria divided into three groups $\tilde{\omega}_b$, which are given in Fig. 2.16, allowed the authors to define the key criteria, determining the choice of the trip by train rather than the travel by plane by the considered passengers. These data can help to determine the ways of increasing the quality of passenger transportation by rail so that this mode of land transport could effectively compete with air transport.

The study has led to the following findings and suggestions.

Passengers, having a possibility to travel from one place to another by various means of transport, choose a mode of transport which seems to be more suitable (or attractive) to them according to their personal experience and attitudes, largely depending on advertising. Usually, passengers make a choice of a particular mode of transport by considering the factors (evaluation criteria), whose significances (weights) are different from their point of view. The choice of the alternative mode of transport is determined by the significances of the criteria describing it (i.e., the subjective weights). These weights can be obtained by using the expert method. The average estimate of the criteria ranks or weights assigned by a great number of passengers (respondents) can be used as a result, showing the public opinion in choosing a means of transport for a particular route.

In the present study, the reasons for choosing a trip by train rather than a trip by plane by passengers are described by sixteen criteria. The importance of these criteria for choosing a particular mode of transport was evaluated by 52 respondents (the passengers of the international train) against the scale of sixteen ranks. The subjective normalized global weights of criteria calculated by the novel ARTIW-L and ARTIW-N methods allowed the authors to arrange them in the order of preference, showing the main reasons (expressed by criteria), why passengers preferred the trip by train to the trip by plane. This choice was primarily determined by the independence of the trip by train from weather conditions (0.1008), a shorter boarding time (0.0914) and the location of railway stations (which are nearer to passengers than airports) (0.0884). Less important criteria included the arrival of the passengers' luggage together with them (0.0795), a higher threat of terrorism on the plane (0.0685), fear of people to travel by plane (0.0645) and the absence of luggage examination on the train (0.0644). Other groups of less important criteria included larger spaces on the train (0.0627), the requirement to passengers to switch off their electronic devices on the plane (0.0589), lower acceleration of the train (0.0547), a shorter trip by a fast train (0.0518) and the possibility of the people accompanying a passenger to stay in the train until it departs (0.0507). The least important criteria included the identity check of people buying tickets for a trip by plane and during boarding (0.0486), the requirement to passengers to fasten safety belts on the plane (0.0407), the availability of cheaper tickets for a trip by train in some periods of time (0.0397), as well as lower pollution of the environment and lower fuel consumption by rail vehicles (0.0347). The ratio of the largest values (0.1008) to the smallest values (0.0347) of the global weights of criteria which is equal to 2.90 shows that criteria determining the choice of the mode of transport by passengers are of different importance.

The estimates of the criteria importance, provided by all 52 passengers taking part in the survey, are consistent because the calculated concordance coefficient is equal to 0.269, and Pearson's chi-squared test value (209.7) is much higher than the critical value, which is equal to 30.57, with the degree of freedom of 15 and the confidence level of 0.01. The concordance coefficient value of 0.269 is 6.86 times as large as its smallest value (0.0392), allowing us to assume that the estimates of all respondents are consistent. A very high consistence level of the respondents' estimates (with the concordance coefficient about one) could hardly be expected because the experience, habits, wishes and possibilities of passengers were completely different.

Sixteen criteria were divided into three groups (key criteria) based on using the inverse hierarchy model, and their normalized weights were calculated. The weight of eight sub-criteria making the "safety" criterion was found to be equal to 0.3809, while the weight of three sub-criteria forming the "economy and ecology" criterion was 0.2384, and the weight of five sub-criteria making the "ride comfort" criterion was equal to 0.3807. These results indicate that on a personal level, a passenger perceives comfort and safety of travelling by train as more important than economic and ecologic factors. Most often international train is selected by opulent people who are on profession-related business trip to another country. The weights of the criteria of any sub-criteria group (key criteria) were calculated using a new IHAMCI method, allowing for ignoring the influence of the number of sub-criteria in each key criterion.

The decision makers of company, engaged in passenger transportation by the considered international train, should primarily improve the services described by the criteria, most strongly influencing a decision of passengers to choose a trip by train rather than a trip by plane. A company providing passenger transportation by any particular means of transport can win the competition in this area only if the provided services are of the highest quality and satisfy the ever-growing demands of passengers.

## References

1. Sivilevičius H (2011) Modelling the interaction of transport system elements. Transport 26(1):20–34. https://doi.org/10.3846/16484142.2011.560366
2. Maskeliūnaitė L (2024) Funiculars in Lithuania: from the fruit basket to Gediminas Hill. Transport 39(3):250–263. https://doi.org/10.3846/transport.2024.22861
3. Cerny P (2021) 10 questions about night trains that you always wanted, but never dared, to ask. Heinrich Böll Stiftung, Brussels, European Union. https://eu.boell.org/en/10-questions-night-trains
4. Profillidis V (2016) Railway Management and Engineering, 4th Edition. Routledge, Taylor & Francis Group, London, p 552. https://doi.org/10.4324/9781315245362
5. Sakib N, Appiotti F, Magni F, Maragno D, Innocenti A, Gissi E, Musco F (2018) Addressing the passenger transport and accessibility enablers for sustainable development. Sustainability 10(4):903, 1–21. https://doi.org/10.3390/su10040903
6. Pijoan A, Kamara-Esteban O, Alonso-Vicario A, Borges CE (2018) Transport choice modeling for the evaluation of new transport policies. Sustainability 10(4):1230, 1–22. https://doi.org/10.3390/su10041230
7. Li X-H, Huang L, Li Q, Liu H-C (2020) Passenger satisfaction evaluation of public transportation using Pythagorean fuzzy MULTIMOORA method under-large group environment. Sustainability 12:4996. https://doi.org/10.3390/su12124996
8. White Paper on Transport (2011) Roadmap to a Single European Transport Area—towards a competitive and resource-efficient transport system. European Commission, Directorate-General for Mobility and Transport, Publications Office of the European Union, Luxembourg. https://doi.org/10.2832/30955
9. Al-Atawi A, Kumar R, Saleh W (2016) Transportation sustainability index for Tabuk sity in Saudi Arabia: an analytic hierarchy process. Transport 31(1):47–55. https://doi.org/10.3846/16484142.2015.1058857
10. Bauk S, Šekularac-Ivošević S, Jolič N (2015) Seaport positioning supported by the combination of some quantitative and qualitative approaches. Transport 30(4):385–396. https://doi.org/10.3846/16484142.2013.815657
11. Buliček J, Drdla P (2015) Quality of urban public transport websites according to extent of provided information. Transport 30(2):202–216. https://doi.org/10.3846/16484142.2015.1048721
12. Földes D, Csiszár C (2015) Route plan evaluation method for personalised passenger information service. Transport 30(3):273–285. https://doi.org/10.3846/16484142.2015.1086889
13. Paha J, Rompf D, Warnecke C (2013) Customer choise patterns in passenger rail competition. Transp Res Part A Policy Pract 50:209–227. https://doi.org/10.1016/j.tra.2013.01.037

14. Skačkauskienė I, Vilkaitė-Vaitonė N, Raudeliūnienė J, Davidavičienė V (2016) A model for measuring passenger loyalty. Transport 31(1):100–107. https://doi.org/10.3846/16484142.2015.1094748
15. Sivilevičius H, Maskeliūnaitė L (2018) Multiple criteria evaluation and the inverse hierarchy model for justifying the choice of rail transport mode. Promet-Traffic Transp 30(1):57–69. https://doi.org/10.7307/ptt.v30i1.2417
16. Kumar S, Dogiparthi KC, Vanajakshi L, Subramanian SC (2017) Integration of exponential smoothing with state space formulation for bus travel time and arrival time prediction. Transport 32(4):358–367. https://doi.org/10.3846/16484142.2015.1100676
17. Stenström C, Parida A, Kumar U (2016) Measuring and monitoring operational availability of rail infrastructure. Proc Inst Mech Eng Part F J Rail Rapid Transit 230(5):1457–1468. https://doi.org/10.1177/0954409715592189
18. Matuška J (2017) Railway system accessibility evaluation for wheelchair users: case study in the Czech Republic. Transport 32(1):32–43. https://doi.org/10.3846/16484142.2014.941396
19. Kendall ME (1970) Rank correlation methods, 4th edn. Griffin and Co., London
20. Maskeliūnaitė L, Meilus L, Sivilevičius H (2023) Application of expert evaluation methods in determining the significance of criteria for usability of railway traction rolling stock. Transport 38(2):77–86. https://doi.org/10.3846/transport.2023.19407
21. Saaty TL (1980) The analytic hierarchy process. McGraw-Hill, New York, NY
22. Saaty TL (2008) Decision making with the analytic hierarchy process. Int J Serv Sci 1(1):83–98
23. Belton V, Gear T (1983) On a short-coming of Saaty's method of analytic hierarchies. OMEGA Int J Manag Sci 11(3):228–230
24. Saaty TL, Kearns KP (1985) Analytical planning: the organization of systems. Pergamon Press, Oxford
25. Sahoo SK, Goswami SS (2023) A comprehensive review of Multiple Criteria Decision-Making (MCDM) methods: advancements, applications, and future directions. Decis Mak Adv 1:25–48. https://doi.org/10.31181/dma1120237
26. Barzilai J (2001) Notes on the analytic hierarchy process. In: Proceedings of the NSF Design and Manufacturing Research Conference, Tampa, Florida, January 2001, pp 1–6
27. Barzilai J, Cook WD, Golany B (1987) Consistent weights for judgements matrices of the relative importance of alternatives. Oper Res Lett 6(3):131–134
28. Brugha CM (2000) Theory and methodology. Relative measurement and the power function. Eur J Oper Res 121:627–640
29. Brugha CM (2003) Phased multicriteria preference finding. Eur J Oper Res 158(2004):308–316
30. Saaty TL, Ozdemir MS (2003) Why the magic number seven plus or minus two. Math Comput Model 38(2003):233–244. https://doi.org/10.1016/S0895-7177(03)00216-4
31. Byun D-H (2001) The AHP approach for selecting an automobile purchase model. Inf Manage 38(5):289–297
32. Chang S-C, Tsai P-H (2016) A hybrid financial performance evaluation model for wealth management banks following the global financial crisis. Technol Econ Dev Econ 22(1):21–46. https://doi.org/10.3846/20294913.2014.986771
33. Gudienė N, Banaitis A, Podvezko V, Banaitienė N (2014) Identification and evaluation of the critical success factors for construction projects in Lithuania: AHP approach. J Civ Eng Manag 20(3):350–359. https://doi.org/10.3846/13923730.2014.914082
34. Sivilevičius H, Žuraulis V, Bražiūnas J (2024) Expert evaluation of the significance of criteria for electric vehicle deployment: a case study of Lithuania. Smart Cities 7(4):2208–2231. https://doi.org/10.3390/smartcities7040087
35. Saaty L, Vargas LG (2012) Models, methods, concepts and applications of the analytic hierarchy process, 2nd Edition. Springer, New York, NY, 346 p. https://doi.org/10.1007/978-1-4614-3597-6
36. Kou G, Ergu D, Lin C, Chen Y (2016) Pairwise comparison matrix in multiple criteria decision making. Technol Econ Dev Econ 22(5):738–765. https://doi.org/10.3846/20294913.2016.1210694

37. Chen S, Leng Y, Mao B, Liu S (2014) Integrated weight-based multi-criteria evaluation on transfer in large transport terminals: a case study of the Beijing South Railway Station. Transp Res Part A 66(2014):13–26. https://doi.org/10.1016/j.tra.2014.04.015
38. Taherdoost H, Madanchian M (2023) Multi-criteria decision making (MCDM) methods and concepts. Encyclopedia 2023(3):77–87. https://doi.org/10.3390/encyclopedia3010006
39. Dotoli M, Epicoco N, Falagario M (2020) Multi-criteria decision making techniques for the management of public procurement tenders: a case study. Appl Soft Comput 88:106064. https://doi.org/10.1016/j.asoc.2020.106064
40. Šikšnelytė-Butkienė I, Štreimikienė D, Baležentis T, Agnusdei L (2024) Multi-criteria decision-making for solving transport sustainability issues: an overview. Transport 39(3):264–286. https://doi.org/10.3846/transport.2024.22915
41. Zavadskas EK, Cavallaro F, Podvezko V, Ubarte I, Kaklauskas A (2017) MCDM assessment of a healthy and safe built environment according to sustainable development principles: a practical neighborhood approach in Vilnius. Sustainability 9(5):702. https://doi.org/10.3390/su9050702
42. Liao H, Liu Z, Banaitis A, Zavadskas EK, Zhou X (2022) Battery supplier development for new energy vehicles by a probabilistic linguistic UTASTAR method. Transport 37(2):121–136. https://doi.org/10.3846/transport.2021.14710
43. Zavadskas EK, Vinogradova-Zinkevic I, Juodagalviene B, Lescauskiene I, Keizikas A (2024) Comparison of safety and sustainability of u-shaped internal staircase projects via a combined MCDM approach CORST. Appl Sci 14(1):158, 1–22. https://doi.org/10.3390/app14010158
44. Đalić I, Stević Ž, Ateljević J, Turskis Z, Zavadskas EK, Mardani A (2021) A novel integrated MCDM-SWOT-TOWS model for the strategic decision analysis in transportation company. Facta Universitatis. Mech Eng 19(3):401–422. https://doi.org/10.22190/FUME201125032D
45. Keshavarz-Ghorabaee M, Amiri M, Zavadskas EK, Turskis Z, Antuchevičienė J (2022) MCDM approaches for evaluating urban and public transportation systems: a short review of recent studies. Transport 37(6):411–425. https://doi.org/10.3846/transport.2022.18376
46. Zavadskas EK, Bausys R, Mazonaviciute I (2019) Safety evaluation methodology of urban public parks by multi-criteria decision making. Landsc Urban Plan 189(2019):372–381. https://doi.org/10.1016/j.landurbplan.2019.05.014
47. Saaty TL (1990) How to make a decision: The analytic hierarchy process. Eur J Oper Res 48(1):9–26
48. Podvezko V, Sivilevičius H (2013) The use of AHP and rank correlation methods for determining the significance of the interaction between the elements of a transport system having a strong influence on traffic safety. Transport 28(4):389–403. https://doi.org/10.3846/16484142.2013.866980
49. Lenze W, Nieβen N (2021) Quality in rail timetabling—a detailed review of stakeholders' interests compared with current practices in Germany. J Rail Transp Plan Manage 20(2021):1–14. https://doi.org/10.1016/j.jrtpm.2021.100282
50. Brunelli M (2015) Introduction to the analytic hierarchy process. In: SpringerBriefs in Operations Research. 83 p. https://doi.org/10.1007/978-3-319-12502-2
51. Stanujkic D (2016) An extension of the ratio system approach of MOORA method for group decision-making based on interval-valued triangular fuzzy numbers. Technol Econ Dev Econ 22(1):122–141. https://doi.org/10.3846/20294913.2015.1070771
52. Dong Q, Cooper O (2016) A peer-to-peer dynamic adaptive consensus reaching model for the group AHP decision making. Eur J Oper Res 250(2):521–530. https://doi.org/10.1016/j.ejor.2015.09.016
53. Saaty TL (1994) Fundamentals of decision making and priority theory with the analytic hierarchy process. Vol. 6 of the AHP series. RWS Publications Pittsburgh, 527 p
54. Taha H (1997) Urban climates and heat islands: albedo, evapotranspiration, and anthropogenic heat. Energy Build 25(2):99–103. https://doi.org/10.1016/S0378-7788(96)00999-1
55. Saaty TL (1996) Decision making with dependence and feedback: the analytic network process : the organization and prioritization of complexity. RWS Publications Pittsburgh, 370 p

56. Aguarón J, Moreno-Jiménez JM (2003) The geometric consistency index: approximated thresholds. Eur J Oper Res 147(1):137–145
57. Farhan J, Fwa TF (2009) Pavement maintenance prioritization using analytic hierarchy process. Transp Res Rec 2093(1):12–24. https://doi.org/10.3141/2093-02
58. Šakalys R, Sivilevičius H, Miliauskaitė L, Šakalys A (2019) Investigation and evaluation of main indicators impacting synchromodality using ARTIW and AHP methods. Transport 34(3):300–311. https://doi.org/10.3846/transport.2019.9718
59. Zavadskas EK, Podvezko V (2016) Integrated determination of objective criteria weights in MCDM. Int J Inf Technol Decis Mak 15(2):267–283. https://doi.org/10.1142/S0219622016500036
60. Kendall M, Gibbons JD (1990) Rank correlation methods, 5th edn. Oxford University Press, New York
61. Sivilevičius H (2011) Application of expert evaluation method to determine the importance of operating asphalt mixing plant quality criteria and rank correlation. Baltic J Road Bridge Eng 6(1):48–58. https://doi.org/10.3846/bjrbe.2011.07
62. Podvezko V (2005) Determining the level of agreement of expert estimates. Technol Econ Dev Econ 11(2):101–107. https://doi.org/10.1080/13928619.2005.9637688
63. Montgomery DC (2008) Introduction to statistical quality control, 6th edn. Wiley, New York, NY, p 734
64. Chen S, Hung H-F (2016) Formative measurement test of supplier evaluation indicators and weight models. J Test Eval 44(3):1350–1363. https://doi.org/10.1520/JTE20140426
65. Sun C-C (2016) A novel approach for evaluating TFT-LCD manufacturer operational performance. J Test Eval 44(3):1421–1429. https://doi.org/10.1520/JTE20140040
66. Maskeliūnaitė L, Sivilevičius H (2021) Identifying the importance of criteria for passenger choice of sustainable travel by train using ARTIW and IHAMCI methods. Appl Sci 11(23):11503. https://doi.org/10.3390/app112311503
67. Navikas D, Sivilevičius H, Bulevičius M (2018) Investigation and evaluation of railway ballast properties variation during technological processes. Constr Build Mater 185:325–337. https://doi.org/10.1016/j.conbuildmat.2018.07.090

# Chapter 3
# Significance of Train Travel Quality Criteria

*The train is a small world moving through a larger world.*
Elisha Cooper "Train"

## 3.1 Model and Quality Criteria for International Train Travel

Transport is an integral part of the country economic and social infrastructure designed for the purpose of meeting public needs. Five modes of transport are currently in use: railway, road, sea, air, and inland water transport. Railway transport represents a promising and competitive area of activity which plays an important role in the reduction of the increasingly growing traffic of motor vehicles and minimizing environmental pollution, enhancing the safety of road traffic, and solving other problems encountered by the transport system. Many countries have reformed their transportation policies for the sake of environmental and economic considerations [1].

Quality, like efficiency, is one of the most complex philosophical, geopolitical, economic, social, and industrial-technical concepts. Philosophers who attempt to explain the concept of quality, do not only relate it to quantity but also to structure [2]. Quality is the set of product or service attributes and characteristics that enable satisfaction of the consumer's stated or implied values. The quality of a product or service can be described as a characteristic that meets the needs of the consumer. It is obvious that consumer satisfaction with a service or product depends on more than one parameter of that product or service. Satisfaction with the service provided by rail transport for passengers is therefore a function of many parameters.

The study of the development of railways in different countries of the European Union is essential for transport policy at both national and international level. The development of passenger transport by rail is dependent on the economic, technical,



H. Sivilevičius and L. Maskeliūnaitė, *Assessment of the Quality of Passenger Transportation by Train Using Multiple Criteria Decision Making Methods*, Lecture Notes in Intelligent Transportation and Infrastructure,
https://doi.org/10.1007/978-3-031-85877-2_3

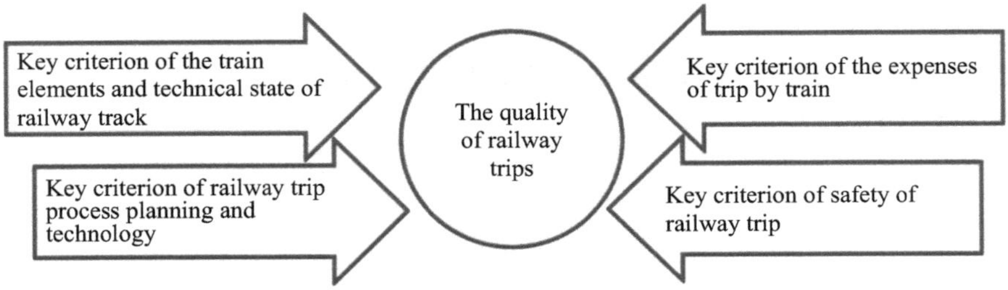


**Fig. 3.1** The model of key criteria describing the railway trip quality

technological, infrastructural, and financial state of transport companies and the countries. The classification of rail passenger transport can be done in terms of one or more indicators, such as the number of passengers, passenger kilometers, train kilometers, transport costs and other indicators. Each classification could serve different purposes and it is important for the development of railway passenger market as an attractive mode for transportation. The classification can contribute to strategic planning, regulation measures, guidance for investment and management, and the quality of transportation [3].

For performing research on the quality of passenger transportation by railway, the most widely recognized and accepted Analytic Hierarchy Process (AHP) method of pairwise comparison developed by the Saaty [4] was selected.

The carried-out work included the identification and ranking of criteria defining the quality of a railway trip, drafting survey questionnaires suitable for application in this method and their distribution to the respective respondents (passengers) and experts (train staff and administration staff). The affinity diagram of railway trip quality factors (corresponding to the structure of the questionnaire) is presented in Fig. 3.1 [5].

Passenger transportation is a specific area of social and economic importance associated with the main problem of the state—the provision of the freedom of travel (movement). Social expenses on railway transport consist of the maintenance costs of railways (infrastructure), the costs of renewing and maintaining rolling stock (repairing old and purchasing new locomotives) (Fig. 3.2) and passenger transportation costs. The costs of passenger transportation services may differ depending on their quality [6].

An increase in the quality of rolling stock and railways track decreases the costs of travelling. Due to the increase in speed, the travel time is reduced and chances to be involved in a railway traffic accident decrease (medical and insurance costs are spared).

The costs of railway maintenance and development are related to the railway construction, repair, and maintenance. Expenses on stock renewal, purchasing, repair and maintenance, including the costs of fuel, electric power, spare parts, oil, workers' payment, etc. make a major part of these costs.

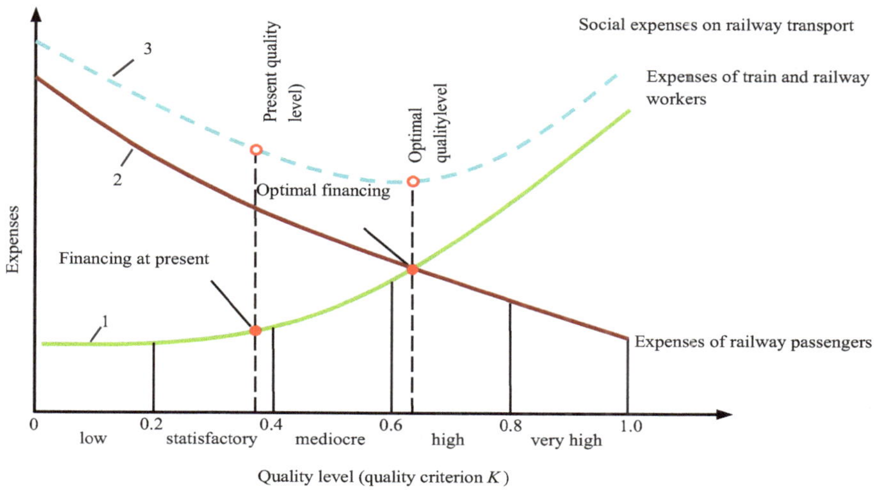


**Fig. 3.2** Model of the relationship between rail passenger costs and the travel quality level: 1—train staff and railway staff costs; 2—rail passenger costs; 3—(total) social costs

The expenses of rail transport users are related to fare losses experienced due to delays of trips, traffic accidents, environmental pollution, and a lack of comfort during the trip.

Some increase in railway staff costs can decrease the total expenses of passengers and the society. When the variation curves of railway staff and passengers cross each other, social expenses are the lowest, i.e., optimal quality level is achieved. Lack of investment in rail transport infrastructure and rolling stock results in poor quality rail travel or makes this mode of transport unattractive for passengers.

To attract more passengers, the quality of transportation should be improved: in addition to high-quality rolling stock, highly qualified staff should be trained. Therefore, for several decades, the problems of theoretical and practical knowledge of employees have been in the focus of researchers from various countries. The suggested model of a potential evaluation of knowledge has been adapted to the transport sector. Considering a specific character of criteria describing it, the model is based on education, professional experience, position, dutifulness, the scope of decision making and responsibility as well as on self-dependence at work and work culture. Some researchers also emphasize the use of technologies and the level of difficulty at work, motivation, and employee contribution to achievements in the goals of the organization [7]. An important point is identifying the problems of passenger transportation using rail transport in a particular country, i.e., a decrease in passenger flows, the growth of transportation cost and insufficient financing of unprofitable means of transportation.

Factors describing the quality of passenger transportation by railway (railway trip) were determined and grouped as key (main) criteria and sub-criteria (Fig. 3.3). The conducted analysis was based on the application of the AHP method. Criteria

**Fig. 3.3** A hierarchical structure of railway trip quality (RTQ) key criteria A, B, C, D and its sub-criteria

describing the quality of the railway trip were collected and appropriate questionnaires on a survey were prepared and later distributed among the respondents (passengers) and experts (train staff and administrative staff (representatives of the Passenger Transportation Board of the Joint-Stock Company "Lithuanian Railways"). The hierarchical structure of all criteria for establishing the quality of railway trips was divided into criteria A, B, C and D (structure of the questionnaire) and given in Fig. 3.3.

The quality of travel by train (QTT) is described by employing both qualitative and quantitative criteria. It would be convenient to quantitatively evaluate the significance of all the criteria for QTT by a single number. The significance of particular criteria differs to some extent. In the work [5], 49 QTT sub-criteria belonging to four various groups (A, B, C, D) called key criteria are considered. Their average weights are determined by using the expert evaluation method.

Trains made up for various routes have railway carriages of different types and technical state and are serviced by the staff members who have different work experience and education. The quality of passenger transportation by any train can be determined only roughly, subjectively, and intuitively. Therefore, to evaluate it more accurately, a quantitative method and the quality level indicator (QLI) $K$, allowing the quality of travel on a particular route to be expressed by a single number (varies from 0 to 1), were developed. The QLI $K$ is calculated by the formula:

$$K = K_A + K_B + K_C + K_D, \tag{3.1}$$

where $K_A$, $K_B$, $K_C$, $K_D$ denote CQI proportion of the key criteria A, B, C and D, describing the quality of travel by an international train.

It is convenient to measure the quality of a subject or object by a single number obtained by quantifying all or the most significant characteristics (criteria). Numerical estimates, indices, rankings, scores, sum of scores, are used in many areas of life to compare subjects or objects with each other. Popular sociological indices include the Quality of Life Index, the Gender Equality Index, the Corruption Perception Index, the Happiness Index, the Wealth of Nations Index and others. In sport, rankings are given to players in some sports. Their sporting performance is often measured collectively by a single number, for example in the athletics decathlon. Scores are awarded in figure skating, ice dancing, springboard diving, gymnastics synchronised swimming and other competitions. In the technical field, a single number (on a seven-point scale) measures the quality of roads in individual countries. The quality of an asphalt mixing plant is assessed comprehensively in terms of a single number [8, 9]. Universities and the output of individual researchers are also ranked. The idea and methodology of comprehensively assessing the quality of a train journey with a single number is novel.

## 3.2 Structure of Questionnaires, Respondents and Experts

The analysis was based on the application of the Analytic Hierarchy Process (AHP) method. The sub-criteria describing railway trip quality were collected, the appropriate survey questionnaires were prepared and later distributed among the respondents (passengers) and experts (train staff and administrative staff). The diagram of the railway trip quality key criterion A, B, C, D (questionnaire's structure) is given in Fig. 3.3. Thirty-two questionnaires were distributed among passengers including eighteen Lithuanian citizens, nine passengers having the citizenship of Russia, one passenger with the US citizenship, one having Spanish, one having Italian, one German and one British citizenship. However, only the data on 10 questionnaires completed by passengers (whereof 3 came from Lithuania, 4 from Russia, 1 from the US, 1 from Germany and 1 from Italy) were used in the survey as the remaining 22 questionnaires were found to be inconsistent (consistency ratio $C.R. > 0.1$) and, therefore, rejected.

17 questionnaires were given to the experts (i.e., service staff) too. Only 11 questionnaires were used in the survey.

Six questionnaires were handed over to the managers of Passenger Transportation Board of the JSC "Lithuanian Railways". Only five of them were used in the survey. Meanwhile, the evaluation of one questionnaire was rejected for the reason described above.

The inertia of the service staff of the train (e.g., train manager, car attendant) and passengers (preferring to read a book or visit a dining car) did not allow us to conduct surveys in the trains operating on other routes and thereby to increase the number of respondents.

The number of questionnaires handed out and completed by respondents and experts, then evaluated by the researchers (authors of the monograph), and used in the study is shown in Table 3.1. The number of completed and counted pairwise comparison matrices $m$ is shown without brackets. For all of them (32 + 17 + 6), Eigen-vector or priority vector $\omega_i$ and consistency ratio ($C.R.$) were calculated. Matrices with $C.R. > 0.1$ were discarded as non-consistent. After discarding them, only those matrices with $C.R. < 0.1$ were used in the study and their number is given in brackets (Table 3.1).

**Table 3.1** The number of questionnaires, distributed among the respondents and experts

| Category of respondents or experts* | Submitted (used) questionnaires | | | |
|---|---|---|---|---|
| | The key criterion | | | |
| | A | B | C | D |
| P | 32 (10) | 32 (10) | 32 (10) | 32 (10) |
| TS | 17 (11) | 17 (11) | 17 (11) | 17 (11) |
| AS | 6 (5) | 6 (5) | 6 (5) | 6 (5) |

* Categories of respondents and experts: P—passengers, TS—train staff, AS—administration staff

The questionnaires filled in by all respondents and experts were collected, processed, and analyzed.

## 3.3 Methodology for Identifying the Significance of Sub-criteria Through AHP and Rank Correlation Methods

### *3.3.1 Determining the Estimates of the Sub-criteria Weight Elicited from One Expert by AHP Method*

The AHP is one of the most popular MCDM methods proposed by Saaty [4]. AHP is a structured technique for organizing and solving complex decision-making problems based on mathematics and psychology. AHP provides a comprehensive and logical framework to quantify each structural decision-making element within a hierarchical structure. The AHP begins with choosing the decision criteria. The alternatives are then evaluated based on the selected criteria [10].

It is important to determine the significance of the obtained sub-criteria by applying a decision support system. A method of pair wise comparison of sub-criteria was chosen for indicating the railway trip quality.

Methods for determining the weights of the sub-criteria to describe the railway trip quality are considered to be subjective if they are evaluated by respondents or experts. In that case, experts' qualification should be high because the agreement of their estimates depends on it. For this purpose, the method of pairwise comparison of sub-criteria suggested by Saaty [4] and widely known as AHP is well suited. This approach allows the researchers to determine the weights of the sub-criteria of the same hierarchical level with respect to higher level sub-criteria or to determine hierarchically unstructured criteria weights. Experts compare all the evaluated criteria $R_i$ and $R_j$ $(i, j = 1, \ldots, m)$, where $m$ is the number of the compared sub-criteria.

The application of AHP requires highly developed logical thinking, the estimate of one highly qualified expert may be more important than the estimates made by several inexperienced (not logically thinking) specialists. Therefore, researchers usually interview several highly qualified experts, for example, 5 experts [11, 12].

The method described above is easy to use because it is easier to compare pairs of criteria than all of them at a time. Pairwise comparisons are made between the criteria on the left-hand side of the matrix and the criteria at the top of the matrix, giving them T. Saaty's scale inter-criterion intensity scores.

The matrix of the comparison of evaluation criteria $(a_{ij} = 1/a_{ji})$ is as follows (2.2):

Let us find the eigenvector which may (2.2) be calculated in four ways [13]. We will use the 4-th steps:

Step 1. The elements of each row are multiplied together, and the results obtained are written as follows:

$$\omega_i'' = \prod_{j=1}^{m} a_{ij}. \tag{3.2}$$

Step 2. $m$-th root is extracted from the element of each row (the $m$-th root is extracted). The results obtained are written as follows:

$$\omega_i' = \sqrt[m]{\prod_{j=1}^{m} a_{ij}}. \tag{3.3}$$

Step 3. Let us add together the elements of this row:

$$\sum_{i=1}^{m} \omega_i' = \sum_{i=1}^{m} \sqrt[m]{\prod_{j=1}^{m} a_{ij}}. \tag{3.4}$$

Step 4. Let us divide each element of this row by the sum obtained, i.e. the evaluations normalization (formula 2.6):

Thus, the eigenvector (subjective weight) $\omega_i$ is found (step 4). The sum of its elements is equal to unity: $\left(\sum_{i=1}^{m} \omega_i = 1\right)$.

### 3.3.2 *Example of Matrix Calculation for the AHP Method*

Let us calculate a pairwise comparison questionnaire completed by a respondent to illustrate the used technique [14].

To evaluate the quality of a railway trip, the weights of 6 sub-criteria, describing the trip expenses (the sub-criteria of key criterion C) (Fig. 3.3) were determined. The respondents (passengers) were given questionnaires of pairwise comparison of sub-criteria. The completed comparison matrix is given in Table 3.2 as an illustration.

**Table 3.2** An example of the 4th respondent's pairwise comparison of the sub-criteria, describing ticket price

| Rank | Sub-criterion | C1 | C2 | C3 | C4 | C5 | C6 | Eigenvector |
|---|---|---|---|---|---|---|---|---|
| 2 | C1 | 1 | 7 | 7 | 3 | 1 | 3 | 0.3178 |
| 4 | C2 | 1/7 | 1 | 3 | 1 | 1/5 | 1 | 0.0765 |
| 6 | C3 | 1/7 | 1/3 | 1 | 1/3 | 1/9 | 1/5 | 0.0306 |
| 5 | C4 | 1/3 | 1 | 3 | 1 | 1/5 | 1/3 | 0.0733 |
| 1 | C5 | 1 | 5 | 9 | 5 | 1 | 7 | 0.3929 |
| 3 | C6 | 1/3 | 1 | 5 | 3 | 1/7 | 1 | 0.1089 |

Priorities C5 $\succ$ C1 $\succ$ C6 $\succ$ C2 $\succ$ C4 $\succ$ C3

Now, the eigenvector should be found.

The 4-th step of calculation is used. It is implemented in the following order: the elements of each row of the questionnaire are multiplied together and the data obtained are written in a row; $m$-th root is extracted from each element of the obtained row; its elements are added together; each element is divided by the sum obtained. Using any of the four possible techniques in the case of an ideal matrix, we will get the same accurate result.

Based on the data presented in Table 3.2, we can find eigenvector by using the 4-th steps of calculation.

The steps of eigenvector's calculation and their results are given in Table 3.3.

Now, the largest eigenvalue $\lambda_{max}$ should be found in the considered matrix (formula 2.7).

The 1st row elements $a_{ij}$ of Table 3.2 are multiplied by eigenvector $\omega_i$:

$$\sum_{j=1}^{m} a_{ij} \cdot \omega_j = 1 \cdot 0.3178 + 7 \cdot 0.0765 + 7 \cdot 0.0306$$
$$+ 3 \cdot 0.0733 + 1 \cdot 0.3929 + 3 \cdot 0.1089$$
$$= 2.0071.$$

Dividing the 1st element from the obtained row by the respective (first) element of the initial multiplication row (step 4), we get:

$$\frac{\sum_{j=1}^{m} a_{ij} \cdot \omega_j}{\omega_i} = \frac{2.0071}{0.3178} = 6.3160.$$

The largest eigenvalue are calculated by formula (2.7). Their closest average value is found. Then, we get:

$$\lambda_{max} = \frac{1}{6}(6.3160 + 6.2047 + 6.2512 + 6.3056 + 6.3572 + 6.6166) = 6.3419.$$

It is known that the largest eigenvalue of the inverse symmetrical $m$—row matrix is $\lambda_{max} \geq m$. In the case considered, the matrix row or size is $m = 6$. Therefore, 6.34 > 6. The condition is satisfied.

Now, it is easy to calculate the consistency index $C.I.$, which is expressed as follows (2.8):

$$C.I. = \frac{{}_{max} - m}{m - 1} = \frac{6.3419 - 6}{6 - 1} = 0.0684.$$

The smaller the $C.I.$ value, the higher consistency of the matrix.

The relationship between the consistency index of the matrix $C.I.$ and the average value of the random index $R.I.$, found from the table [4], is referred to as consistency ratio, $C.R.$, showing the degree of matrix consistency:

**Table 3.3** Steps of eigenvector's calculation and the results obtained in each step

| Steps of eigenvector's calculation | Evaluation sub-criterion No | | | | | |
|---|---|---|---|---|---|---|
| | C1 | C2 | C3 | C4 | C5 | C6 |
| | The data obtained in eigenvector's calculation | | | | | |
| **Step 1** | $\omega_1''$ | $\omega_2''$ | $\omega_3''$ | $\omega_4''$ | $\omega_5''$ | $\omega_6''$ |
| The elements of each row are multiplied together $\omega_1' = \prod_{j=1}^{6} a_{ij}$ and the data obtained are written in a row | | | | | | |
| The elements of the 1st row of Table 3.2 are multiplied together | $\omega_1'' = 1 \cdot 7 \cdot 7 \cdot 3 \cdot 1 \cdot 3 = 441$ | | | | | |
| The results of multiplying together the elements of the 1st, 2nd, 3rd …, 6-th rows of Table 3.2 are as follows: | 441 | 0.0857 | 0.0004 | 0.0667 | 1575 | 0.7143 |
| **Step 2** | $\omega_1'$ | $\omega_2'$ | $\omega_3'$ | $\omega_4'$ | $\omega_5'$ | $\omega_6'$ |
| $m$-th root is extracted from each element of the obtained row (since the number of the criteria compared is $m = 6$, 6th root is extracted) $\omega_1' = \sqrt[6]{\prod_{j=1}^{6} a_{ij}}$ | | | | | | |
| 6-th root is extracted from the 1st element of the obtained row: | $\omega_1' = \sqrt[6]{441} \approx 2.7589$ | | | | | |
| The following values are obtained by extracting 6th root from the 1st, 2nd, 3rd … 6th element of the obtained row: | 2.7589 | 0.6640 | 0.2658 | 0.6368 | 3.4110 | 0.9455 |
| **Step 3** The elements of this row are added together $\sum_{i=1}^{m} \omega_i' = \sum_{i=1}^{6} \sqrt[6]{\prod_{j=1}^{6} a_{ij}}$: | $2.7589 + 0.6640 + 0.2658 + 0.6368 + 3.4110 + 0.9455 = 8.6820$ | | | | | |
| **Step 4** | $\omega_1$ | $\omega_2$ | $\omega_3$ | $\omega_4$ | $\omega_5$ | $\omega_6$ |
| Each element of this row is divided by the sum obtained $\omega_i = \frac{\sqrt[6]{\prod_{j=1}^{6} a_{ij}}}{\sum_{i=1}^{6} \sqrt[6]{\prod_{j=1}^{6} a_{ij}}}$ | | | | | | |
| The 1st element of this row is divided by the sum obtained | $\omega_1 = \frac{2.7589}{8.6820} = 0.3178 \ldots, \omega_6 = \frac{0.9455}{8.6820} = 0.1089$ | | | | | |
| The following values are obtained by dividing the 1st, 2nd, 3rd … 6th element of this row by the sum obtained: | 0.3178 | 0.0765 | 0.0306 | 0.0733 | 0.3929 | 0.1089 |

$$C.R. = \frac{C.I.}{R.I.} = \frac{0.0684}{1.24} = 0.0551.$$

The $C.R.$ needs to be kept "small" for example, less than 10%, indicating deviations from non-random entries (informed judgments) of less than an order of magnitude [15].

The matrix is consistent if the value of $C.R.$ is smaller than or equal to 0.1. Since $0.0551 < 0.1$, the considered matrix is consistent. The calculated $C.R.$ indicates that the matrix is correct and the eigenvector calculated from it can reasonably be taken as the normalised subjective weight of criterion $i$.

### 3.3.3 Consistency of Views Across Experts or Respondents

The reliability of expert studies is enhanced when the study considers the mean significance value of a criterion for the opinions of a group of experts/respondents, rather than for the opinion of a single expert or respondent. The mean of the ranks given by a group of experts ($j = 1, 2, \ldots, n$) for each main criterion or sub-criterion ($i = 1, 2, \ldots, m$) can only be taken as the result of the problem when all experts have similar, i.e. non-significantly different, opinions. The degree of agreement between the experts' opinions is indicated by the concordance coefficient $W$.

The idea of the concordance coefficient is associated with the sum of ranks of each criterion (key criterion or sub-criterion) $R_i$ ($i = 1, 2, \ldots, m$) with respect to the estimates of experts and respondents:

$$R_i = \sum_{j=1}^{m} R_{ij}, \tag{3.5}$$

where $n$ is the number of respondents or experts ($j = 1, \ldots, n$). More exactly, to the deviation of values $R_i$ from the sum of squares $S$ of the average value $\overline{R}$ (the analogue of variance):

$$S = \sum_{i=1}^{m} \left(R_i - \overline{R}\right)^2. \tag{3.6}$$

The average value $\overline{R}$ is calculated by the formula:

$$\overline{R} = \frac{n(m+1)}{2}. \tag{3.7}$$

When there is absolute agreement (uniformity) between the experts (respondents) in the ranking of the significance of the criteria, the maximum possible value $S_{\max}$ is calculated from the formula:

$$S_{max} = \frac{n^2 \cdot m(m^2 - 1)}{12}. \tag{3.8}$$

The level of consistency of respondents' and experts' estimates is shown by the concordance coefficient $W$:

$$W = \frac{S}{S_{max}} = \frac{12S}{n^2 m(m^2 - 1)}, \tag{3.9}$$

where: $S$ calculated by the Eq. (3.6).

If the estimates of the respondents or experts are consistent, the concordance coefficient $W$ is about one (unity). When the estimates differ to a great extent, $W$ is about zero.

The smallest value of the concordance coefficient $W_{min}$, not allowing us to state that the estimates of the quality of the research object described by $m$ criteria provided by all $n$ respondents or experts with the prescribed significance level $\alpha$ and degree of freedom $\nu = m - 1$, are consistent, may be calculated by the suggested formula (2.24).

The concordance $W$ coefficient can be calculated based on the ranks of the criteria assigned by experts. If their estimates were presented in any other form, they should be ranked. Ranking is a procedure, when the most significant criterion is given the rank of one, the second is ranked by two, etc.

### 3.3.4 A Numerical Illustration of Consistency from Passengers' Estimates

The ranking of the sub-criteria, describing expenses of a trip by train (key criterion C) according to their significance, carried out by ten respondents (passengers) is demonstrated in Table 3.4.

**Table 3.4** Ranking of expenses of a trip by train related sub-criteria in passengers questionnaires

| Sub-criterion $i = 1, 2, \ldots, m$ | Respondent (passenger) $j = 1, 2, \ldots, n$ | | | | | | | | | | Sum of ranks$R_i = \sum_{j=1}^{n} R_{ij}$ |
|---|---|---|---|---|---|---|---|---|---|---|---|
| | E1 | E2 | E3 | E4 | E5 | E6 | E7 | E8 | E9 | E10 | |
| C1 | 1 | 1 | 1 | 2 | 1 | 1 | 1 | 1 | 3 | 4 | 16 |
| C2 | 3 | 5 | 2 | 3 | 4 | 6 | 3 | 2 | 4 | 3 | 35 |
| C3 | 4 | 6 | 5 | 6 | 6 | 5 | 6 | 4 | 6 | 6 | 54 |
| C4 | 5 | 3 | 4 | 5 | 4 | 3 | 4 | 5 | 2 | 2 | 37 |
| C5 | 2 | 2 | 6 | 1 | 2 | 2 | 2 | 3 | 1 | 1 | 22 |
| C6 | 6 | 4 | 3 | 5 | 4 | 4 | 5 | 6 | 5 | 5 | 46 |
| Total | 21 | 21 | 21 | 21 | 21 | 21 | 21 | 21 | 21 | 21 | 210 |

**Table 3.5** The ranks assigned by passenger No 4 to the sub-criteria describing expenses of a trip by train according to their significance

| Sub-criterion | Weight $\omega_i^{ARTIW-L}$ | Weight $\omega_i^{AHP}$ | Rank $R_{ij}$ |
|---|---|---|---|
| C1 | 0.2381 | 0.3178 | 2 |
| C2 | 0.1429 | 0.0765 | 4 |
| C3 | 0.0476 | 0.0306 | 6 |
| C4 | 0.0952 | 0.0733 | 5 |
| C5 | 0.2857 | 0.3929 | 1 |
| C6 | 0.1905 | 0.1089 | 3 |
| Total | 1.0000 | 1.0000 | 21 |

The example of sub-criteria ranking which describes expenses of a trip by train according to their significance performed by the passenger No 4 is given in Table 3.5.

The sum of ranks of all the sub-criteria (of the last elements of Table 3.4) which is $\sum_{j=1}^{n} R_i - 210$. The average value of the criteria ranks, calculated by formula (3.7) is $\overline{R} = 35$, or, calculating by another method, it is $\overline{R} = \frac{210}{6} = 35$. The sum of deviations of the squares is $S = 1016$. The maximum possible value $S_{max} = 1750$.

The concordance coefficient of the criteria associated with the ticket price obtained from a group of respondents (ten passengers) by formula (3.9) is expressed as $W = 0.581$.

A random value:

$$\chi^2 = \frac{12S}{nm(m+1)}. \tag{3.10}$$

is distributed according to $\chi^2$ distribution with the degree of freedom $v = m - 1$. Based on the selected significance level $\alpha$ (which is usually 0.05 or 0.01), the critical value $\chi^2_{v,\alpha}$ may be found from the distribution table of $\chi^2$ with the degree of freedom $v = m - 1$. If the value of $\chi^2$ calculated by formula (3.10) is larger than $\chi^2_{v,\alpha}$, the estimates of respondents are consistent.

Using the data given in Table 3.4 and calculating by formula (3.10), we get $\chi^2 = 29.03$. The value of $\chi^2$ was calculated to be equal to 29.03, while the value of $\chi^2_{v,\alpha}$ taken from the distribution table with the degree of freedom $v = 6 - 1 = 5$ and the significance level $\alpha = 0.05$ is equal to 11.07. The empirical value $\chi^2$ is larger than the critical value $\chi^2_{5;0.05}$, therefore, the estimates of respondents are consistent ($k_c = 2.62$).

Now, let us calculate the smallest value of the concordance coefficient $W_{min}$. The minimum value of the concordance coefficient $W_{min}$ is calculated by formula (2.24), with the significance level $\alpha = 0.05$ and the degree of freedom $v = m - 1 = 6 - 1 = 5$, allowing us to state that the respondents' estimates are still consistent: $W_{min} = 0.221 < 0.581$.

## 3.4 Calculated Results for the Significance of the Criteria and Sub-criteria

### 3.4.1 Significance of Sub-criteria Derived from Passenger Ratings

The subjective normalised weights $\omega_{iPj}$ obtained from the pairwise comparison matrices of the ten ($n_P = 10$) passengers completing the $i$-sub-criteria ($m_A = 16$, $m_B = 19$, $m_C = 6$, $m_D = 8$) for each of the key criteria A, B, C, D, were used to derive the mean local weights for the sub-criteria. The local mean weights of the sub-criteria for each key criterion are calculated according to formulae (3.11)–(3.14):

$$Q_{AiP} = \frac{\sum_{j=1}^{n_P} \omega_{AiPj}}{n_P}, \tag{3.11}$$

$$Q_{BiP} = \frac{\sum_{j=1}^{n_P} \omega_{BiPj}}{n_P}, \tag{3.12}$$

$$Q_{CiP} = \frac{\sum_{j=1}^{n_P} \omega_{CiPj}}{n_P}, \tag{3.13}$$

$$Q_{DiP} = \frac{\sum_{j=1}^{n_P} \omega_{DiPj}}{n_P}, \tag{3.14}$$

where $\omega_{AiPj}$, $\omega_{BiPj}$, $\omega_{CiPj}$, $\omega_{DiPj}$ is the normalised subjective weight (eigenvector) for the $i$-th sub-criterion of the key criterion A, B, C, or D, respectively, of the $j$-th respondent (passenger (P), as determined by the AHP method; $n_P$ is the number of respondents ($j = 1, 2, \ldots n_P$), $n_P = 10$.

The mean values of the local weights for the sub-criteria calculated according to formulae (3.11)–(3.14) are shown in Fig. 3.4.

The survey revealed that passenger opinions were consistent. Figure 3.4a presents a diagram showing the importance of sub-criteria related to train elements and the technical state of the railway track (key criterion A) determined based on respondents' opinion. The diagram shows that the respondents attach high importance to the speed of train travel (the duration of the trip) (A2), the required level of temperature within a passenger carriage (A7) and the operation of ventilation, air conditioning, cooling, and lighting systems and their timely switch on/off (A6). No importance is attached to such sub-criteria as the radio broadcasting unit and its centralized switch on/off (A13), the unevenness of the railway track (A1) and the possibility of using a hairdryer, iron etc. (A16).

We understand that the speed of train travel is important to passengers because it is the factor, upon which the duration of the trip (the time of arrival at the destination) depends. The required temperature inside the passenger coach and the operation of

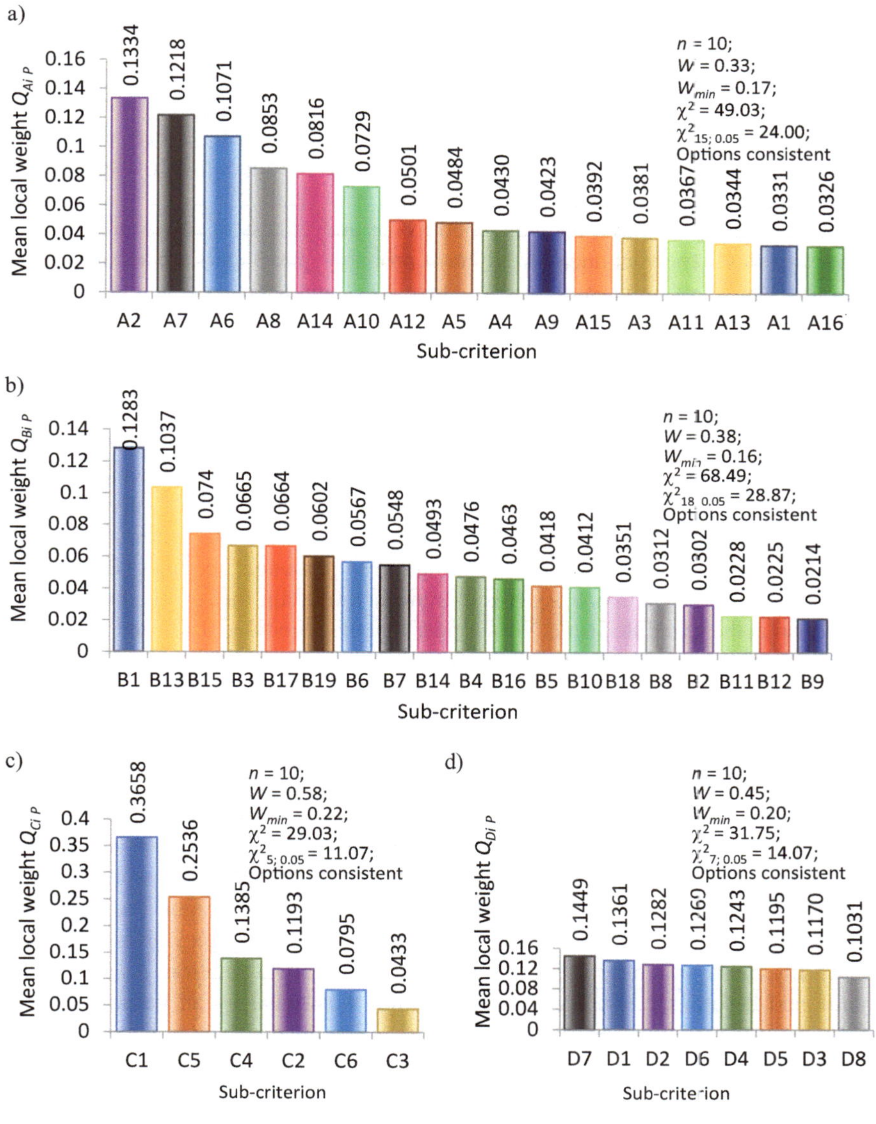


**Fig. 3.4** The data on the evaluation of the significance of the sub-criteria describing the quality of passenger transportation by rail elicited from the passengers of the train (respondents category P): **a** the technical state of the train parts and the railway (key criterion A); **b** denotes the organization and technology of railway trip (key criterion B); **c** the expenses of a trip by train (key criterion C); **d** railway trip safety (key criterion D)

ventilation, air conditioning, cooling, and lighting systems and their timely switch on/off naturally determine the level of comfort enjoyed during the trip.

We can also see that passengers consider the availability of the radio broadcasting unit and its centralized switch on/off absolutely unimportant as some people may find listening to music undesirable or even annoying, especially if they do not like the music broadcast or it is played too loud. Some passengers prefer taking rest or reading a book while traveling.

The low rating of importance attached to the unevenness of rails (railway track geometry) was somewhat unexpected; most probably, passengers are accustomed to the uneven travel of trains, and therefore take it as a matter-of-course and inevitable reality.

Figure 3.4b presents a diagram showing the importance of sub-criteria related to railway transportation process planning and technology (key criterion B) based on respondents' opinion.

The above diagram indicates that the departure and arrival of trains at the scheduled time (B1), the safekeeping of passenger luggage (B13) and personal items as well as the possibility of acquiring a visa at the cross-border office (B15) are very important to passengers. No importance seems to be attached to the onboard sales of souvenirs (B11), the conformity of broadcast information to passenger requests (B12) or the possibility of calling a taxi (B9). It is extremely important that trains should arrive and depart at the scheduled time. We suppose that the scheduled departure and arrival of trains is important to passengers as any deviations from the schedule may upset their plans (they may miss another train, flight etc.). The safekeeping of passenger luggage and personal items are important issues as the risk of losing documents or personal items exists for passengers and when it happens, such situations always entail extra troubles and the mood of passengers.

If passengers had the possibility of acquiring visas at the cross-border station, this would solve a problem of passenger detraining. The passenger is forced to discontinue the trip and change his/her plans which, naturally, is upsetting. Thus, acquiring visas at the cross-border office would be very convenient. Passengers do not consider the onboard sales of souvenirs being important as they always have the possibility of buying these items in a town or at the railway station. As it has been mentioned before, the sounds of music are not welcomed by all passengers who may always get information related to their trip from the coach attendant or the train manager that may explain why they do not deem the sub-criterion on the conformity of broadcast music and information to their requests being important. We believe that the possibility of calling a taxi does not enjoy popularity among passengers since many of them are met at the station of arrival or call taxi services themselves. Passengers who need to go a long way cannot afford going by taxi, and therefore they prefer taking the other means of transport.

Figure 3.4c presents a diagram showing the importance the respondents have attached to the criterion related to the expenses of a trip by train (key criterion C).

We can notice that the price of the trip ticket (C1) as well as the cost of a visa (C5) and health insurance card valid abroad (C4) is important to passengers. However, such sub-criteria as the price of meals served in the dining-car (C2), the delivery of meals (included into the cost of the ticket) to passengers going in the first-class double compartment (C6) and the price of newspapers and magazines (C3) are rated by passengers as unimportant. Passengers consider the price of the ticket, visa, and health insurance card valid abroad important because such situation constitutes the major share of trip expenses. The price of the ticket has relevance to those who need to take frequent trips (for e.g., on business purposes).

The price of meals served in the dining-car is not important because visiting this place is not common practice among passengers. The delivery of meals (included into the cost of the ticket) to passengers going in the first-class double compartment is not important, either, as the price of the first-class tickets is high. Most passengers prefer going in cheaper lower class passenger coaches.

The price of newspapers and magazines available on trains and at press kiosks is similar. Moreover, the daily *Respublika* is delivered free of charge. Besides, many passengers take books with them, thus, they can do very well without any press.

Figure 3.4d presents a diagram showing the importance of sub-criteria related to the safety of the railway trip (key criterion D) determined based on respondents' opinion.

We can see that the operation of a hand brake (D7) and the availability of fire protection equipment (D1) and first medical aid means (stretchers, first aid kit) (D2) are important to passengers. Less important seem to be such indicators as the availability of emergency exits (D5), the possibility of calling an ambulance (D3) or the police (D8).

The hand brake may be used for stopping the train in emergency cases, for instance, when fire breaks out or the axle-box overheat alarm system goes off etc. Therefore, we believe that such high passenger rating of this sub-criterion was prompted namely by the inherent instinct of self-preservation as escaping from the train in a standstill position is obviously less dangerous than attempting to jump off the train in motion. The availability of fire protection equipment ensures fire safety, and since the cases of burns and injuries on trains are rather frequent, the availability of first medical aid means and pharmaceuticals must be important to passengers. Many people do not take pharmaceuticals with them when setting off on a trip. Rather low rating in terms of importance attached by passengers to such criteria as the availability of emergency exits and the possibility of calling an ambulance or the police is somewhat unexpected. The availability of emergency exits, in fact, ensures the safe evacuation of passengers in the case of fi re or in other emergency situations posing risk to human health and life. Maybe passengers believe that in emergency situations, it is possible to escape through the window or doors, and therefore they do not rate the availability of emergence exits as important.

The diagram (Fig. 3.4d) shows that passengers are self-confident in that respect that they do not anticipate being in any serious need during the trip for medical aid or the interference of the police officers and that they will be able to handle any problems that may arise during the trip on their own or with the help of service staff. It should be noted that of those 10 respondents, the questionnaires of whose completed correctly, included 3 females and 7 males. This could have determined the results of rating indicating that males usually avoid seeing a doctor or dealing with the police officers.

### 3.4.2 *The Evaluation of the Sub-criteria Significance by Train Service Staff*

The questionnaires of the same type were used in the research (Fig. 3.3). The analysis has shown that the estimates given by the service staff are quite consistent (Fig. 3.5).

The subjective local normalised weights $\omega_{iTSj}$ calculated from the 11 ($n_{TS} = 11$) pairwise comparison matrices completed by the train staff for the $i$-th sub-criteria ($m_A = 16, m_B = 19, m_C = 6, m_D = 8$) of each of the key criteria A, B, C and D, were used to derive the mean weights of the sub-criteria. The local mean weights of the sub-criteria for each key criterion are calculated according to formulae (3.15)–(3.18):

$$Q_{AiTS} = \frac{\sum_{j=1}^{n_{TS}} \omega_{AiTSj}}{n_{TS}}, \tag{3.15}$$

$$Q_{BiTS} = \frac{\sum_{j=1}^{n_{TS}} \omega_{BiTSj}}{n_{TS}}, \tag{3.16}$$

$$Q_{CiTS} = \frac{\sum_{j=1}^{n_{TS}} \omega_{CiTSj}}{n_{TS}}, \tag{3.17}$$

$$Q_{DiTS} = \frac{\sum_{j=1}^{n_{TS}} \omega_{DiTSj}}{n_{TS}}, \tag{3.18}$$

where $\omega_{AiTSj}$, $\omega_{BiTSj}$, $\omega_{CiTSj}$, $\omega_{DiTSj}$ is the normalised subjective weight (eigenvector) of the $i$-th sub-criterion for the key criterion A, B, C, or D, respectively, identified by the $j$-th expert (train staff (TS) through the AHP method; $n_{TS}$ is the number of experts (TS) ($j = 1, 2, \ldots, n_{TS}$), $n_{TS} = 11$.

The diagram significance of the sub-criteria of the key criterion A (sub-criteria A1–A16) to TS is given in Fig. 3.5a. As shown by the diagram, the highest local weights are assigned by service people to the sub-criteria A6, A8 and A7, while the lowest local weights are given to the sub-criteria A1, A12 and A3. We think that service people of the train could not avoid the influence of their work and its specific nature and therefore were not able to assess the sub-criteria from passengers' perspective. The comfort of the trip described by the sub-criteria A6, A8 and A7 is important for them, while the sub-criterion A1, influencing the speed of the train (trip duration) does not seem to be significant because the train is their workplace.

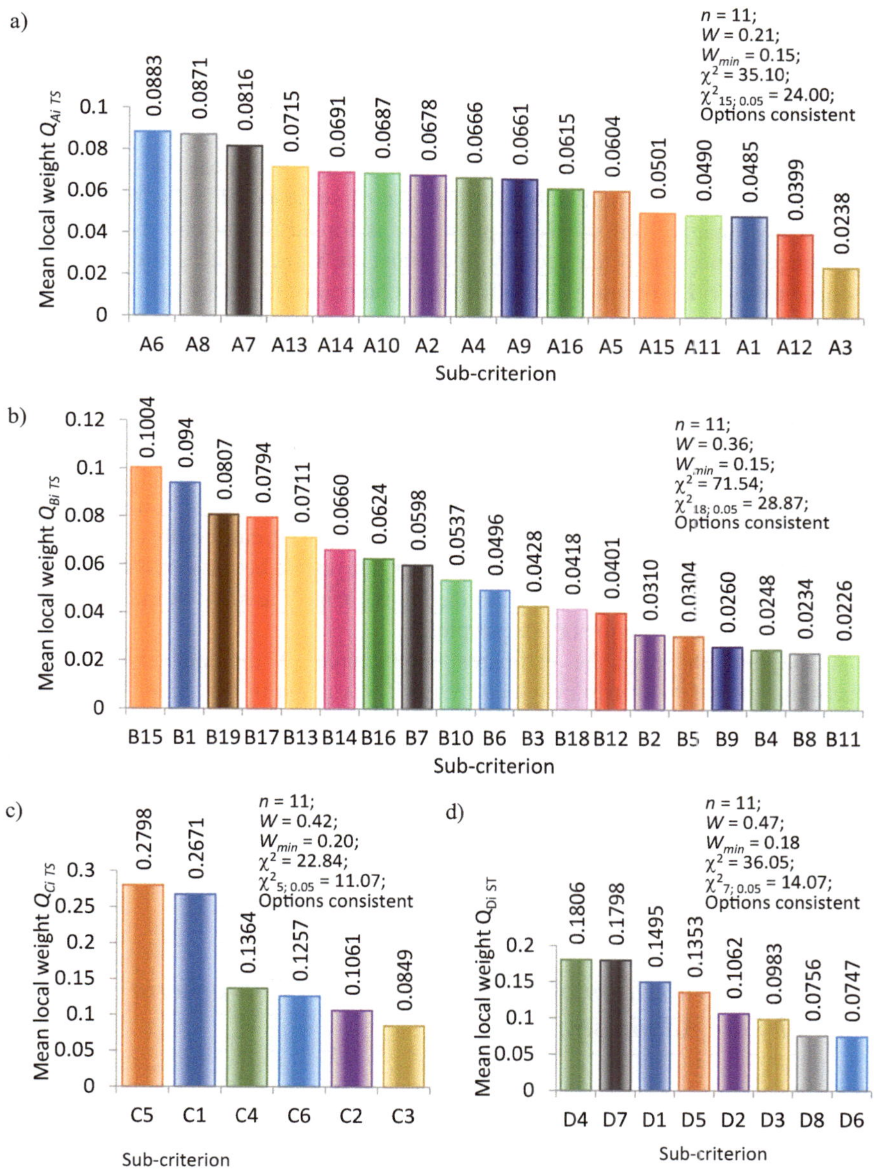


**Fig. 3.5** The data on the evaluation of the significance of the grouped criteria describing the quality passenger transportation by rail elicited from the train staff (category of experts TS): **a** the technical state of the train parts and the railway (key criterion A); **b** denotes the organization and technology of railway trip (key criterion B); **c** the expenses of a trip by train (key criterion C); **d** railway trip safety (key criterion D)

A diagram showing the significance of the key criterion B (sub-criteria B1–B19) to TS is given in Fig. 3.5b. As shown in the diagram (Figs. 3.3 and 3.5b), the sub-criteria B15, B1 and B19 are significant to service people, while the sub-criteria B4, B8 and B11 seem to be insignificant to them. We think that the sub-criteria B15 and B1 are important for service people because they are perfectly aware of the troubles caused to passengers by delayed trains. They also often witness the detraining of passengers at the cross-border station because of some problems with their visas or other documents. In this case, a passenger is forced to suspend the trip. Then, a passenger's mood is often spoiled, and plans are ruined. The sub-criterion B19 is important for TS because the behaviour, competence, objectivity and contacts with people such as custom's officers and frontier guards are perceived to be a sort of visiting card of the country they represent. The sub-criteria B4, B8, B11 are insignificant to the TS. This attitude may be explained by the specific character of their work because car attendants (as service people) provide these services. Sometimes, the sub-criterion does not have any influence on the comfort of a service person, e.g., the sub-criterion B8.

A diagram evaluating the significance of the sub-criteria C1–C6 describing the expenses of a trip by train to TS is given in Fig. 3.5c. It can be observed that the sub-criteria C5, C1 and C4 are significant to service people because passengers spend most of their money on these services during the trip. The sub-criteria C6, C2 and C3 are of no importance to service people.

A diagram showing the estimates significance of the sub-criteria D1–D8 describing the safety of trip provided by the TS is given in Fig. 3.5d. It is thought that the sub-criteria D4, D7 and D1 are important for service staff because as specialists they realize the significance of these sub-criteria for ensuring trip safety. The sub-criteria D3, D8 and D6 are assessed as being not important because all failures during the trip are fixed by the train electrician. Besides, they do not affect the safety of the trip. No problems arise if it is necessary to call the emergency service or the police from the train.

### *3.4.3 The Evaluation of the Significance of Sub-criteria by the Administration Staff*

The significance all sub-criteria of the key criteria A, B, C, D (Figs. 3.3 and 3.6a–d) was determined by the administration staff (AS) of the Passenger Transportation Board of the Joint-Stock Company "Lithuanian Railways".

The subjective normalised weights $\omega_{iASj}$ calculated from pairwise comparison matrices completed by the AS (5 people) ($n_{AS} = 5$) for the $i$-th sub-criteria ($m_A = 16$, $m_B = 19$, $m_C = 6$, $m_D = 8$) of each of the key criteria A, B, C and D, were used to derive the mean local weights of the sub-criteria. The local mean weights of the sub-criteria for each key criterion are calculated according to formulae (3.19)–(3.22):

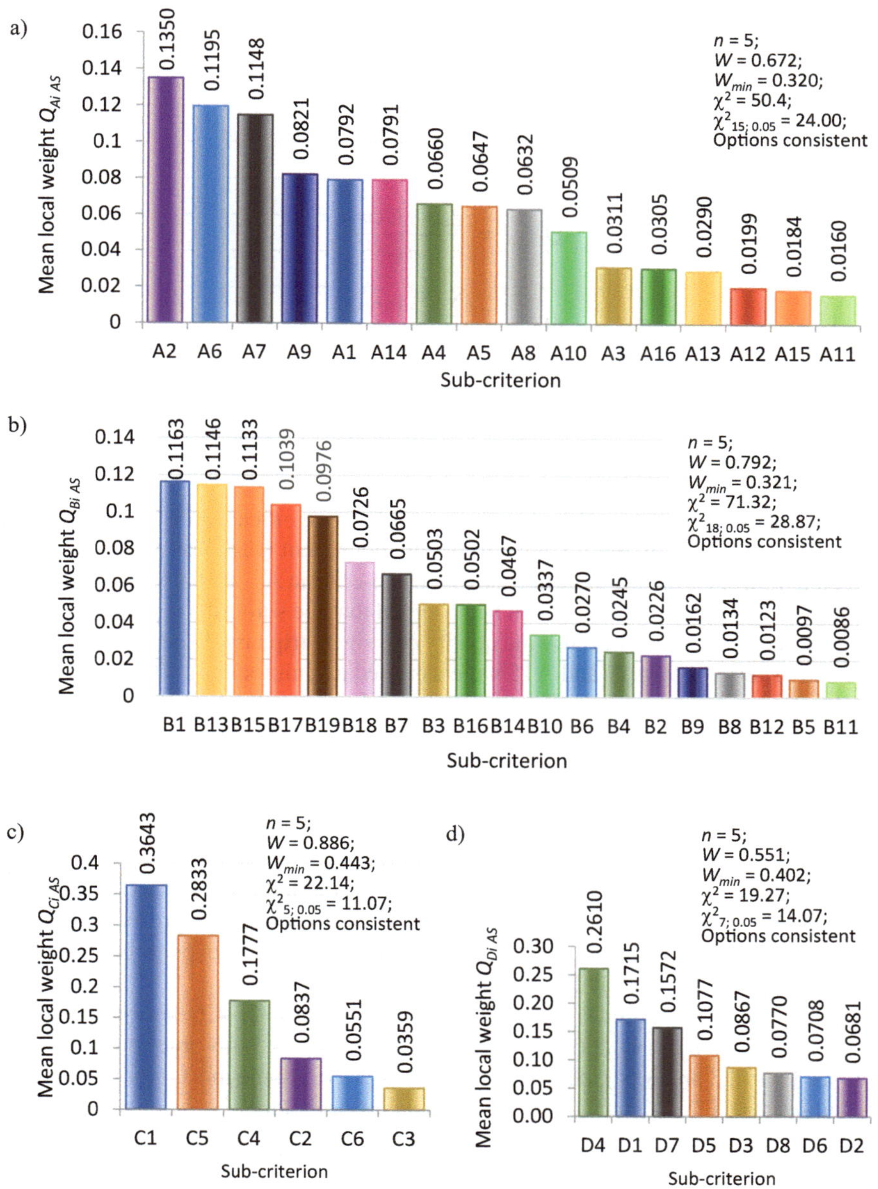


**Fig. 3.6** The data on the evaluation of the significance of the grouped criteria describing the quality passenger transportation by rail elicited from the administration staff (category of experts AS): **a** the technical state of the train parts and the railway (key criterion A); **b** denotes the organization and technology of railway trip (key criterion B); **c** the expenses a trip by train (key criterion C); **d** railway trip safety (key criterion D)

$$Q_{AiAS} = \frac{\sum_{j=1}^{n_{AS}} \omega_{AiASj}}{n_{AS}}, \tag{3.19}$$

$$Q_{BiAS} = \frac{\sum_{j=1}^{n_{AS}} \omega_{BiASj}}{n_{AS}}, \tag{3.20}$$

$$Q_{CiAS} = \frac{\sum_{j=1}^{n_{AS}} \omega_{CiASj}}{n_{AS}}, \tag{3.21}$$

$$Q_{DiAS} = \frac{\sum_{j=1}^{n_{AS}} \omega_{DiASj}}{n_{AS}}, \tag{3.22}$$

where $\omega_{AiASj}$, $\omega_{BiASj}$, $\omega_{CiASj}$, $\omega_{DiASj}$ is the normalised subjective weight (eigenvector) of the $i$-th sub-criterion for the key criterion A, B, C, or D, respectively, identified by the $j$-th expert (AS) through the AHP method; $n_{AS}$ is the number of AS ($j = 1, 2, \ldots, n_{AS}$), $n_{AS} = 5$.

The analysis has shown that the estimates of the significance of the key criteria A, B, C and D made by the AS are completely consistent (Fig. 3.6a–d).

A diagram showing the significance of the sub-criteria A1–A16 of key criterion A to the AS is given in Fig. 3.6a. It can be observed that the sub-criteria A2, A6 and A7 are important for the AS because the quality of railway trip largely depends on them. The sub-criteria A12, A11 and A15 were assessed as unimportant. The members of the AS do not attach high significance to these sub-criteria because there are special areas for smoking in the trains on international routes. Moreover, a passenger is allowed to carry a bicycle with him/her if it is taken apart (the wheels are removed) and packed, while two-birth compartments of the first-class cars are provided with the buttons for emergency calling or for car attendant. It seems that service people do not attach great importance to the sub-criteria given above for these reasons.

The service people of the train consider the sub-criteria B1, B13 and B15 (Figs. 3.3 and 3.6b) to be significant because they, as experienced specialists, are aware of their important role in ensuring high quality of railway trip. The experts consider the sub-criteria B11, B5 and B12 to be unimportant.

A diagram, showing the significance of the sub-criteria C1–C6 of key criterion C to the AS, is given in Fig. 3.6c. As shown in this figure, the sub-criteria C1, C5 and C4 (see Figs. 3.3 and 3.6c), describing the ticket price, are the most important, while the sub-criteria C2, C6 and C3 are assessed as insignificant.

A diagram showing the estimates significance of the sub-criteria D1–D8 and provided by the AS is presented in Figs. 3.3 and 3.6d. It can be observed that the sub-criteria D4, D1 and D7 determining trip safety are important for service people, while the sub-criteria D8, D6 and D2 are insignificant to them.

The critical value $\chi^2_{\alpha,v}$ (taken from the distribution table with the respective degree of freedom and the significance level $\alpha = 0.05$) and the smallest concordance coefficient $W_{min}$ values found in the questionnaires aimed to evaluate the significance of the key criteria A, B, C, D, which were elicited from respondents and experts of

**Table 3.6** The value of the concordance coefficient $W$ the calculated value $\chi^2$ and the lowest value of the concordance coefficient $W_{min}$ for the key criteria A, B, C, D, obtained in the questionnaires elicited from respondents (P) and experts (TS, AS)

| The category of respondents or experts* | Key criterion | | | | | | | | | | | |
|---|---|---|---|---|---|---|---|---|---|---|---|---|
| | A | | | B | | | C | | | D | | |
| | Calculated value | | | | | | | | | | | |
| | $W$ | $\chi^2$ | $W_{min}$ | $W$ | $\chi^2$ | $W_{min}$ | $W$ | $\chi^2$ | $W_{min}$ | $W$ | $\chi^2$ | $W_{min}$ |
| P | 0.33 | 49.03 | 0.17 | 0.38 | 68.49 | 0.16 | 0.58 | 29.03 | 0.22 | 0.45 | 31.75 | 0.20 |
| TS | 0.21 | 35.10 | 0.15 | 0.36 | 71.54 | 0.15 | 0.42 | 22.84 | 0.20 | 0.47 | 36.05 | 0.18 |
| AS | 0.67 | 71.32 | 0.32 | 0.79 | 71.32 | 0.32 | 0.89 | 22.14 | 0.44 | 0.55 | 19.27 | 0.40 |
| $\chi^2_{v;0.05}$ | 2400 | | | 2887 | | | 1107 | | | 1407 | | |

* The categories of respondents and experts: *P* passengers, *TS* train staff, *AS* administration staff

various categories (10 passengers, 11 TS and 5 representatives of the AS) are given in Table 3.6.

## 3.5 The Comparative Analysis of the Respondents and Experts Estimates

The study involved passengers, the staff of the train and the administration staff of the joint-stock company. The most reliable results for measuring the significance of the sub-criteria are obtained by calculating the mean numerical estimates of the opinions expressed by respondents and experts in these three categories. The following assumptions have been used to calculate the total (local) weights $\overline{Q}_i$ of the $i$-th sub-criteria ($i = 1, 2, \ldots, m$) for each key criterion (A, B, C or D):

1. The opinions of all three categories of respondents and experts are equally important (no preference for opinions);
2. The number $n$ of respondents and experts completing the provided questionnaires varies ($n_P = 10$, $n_{TS} = 11$, $n_{AS} = 5$). However, this difference is not factored into the calculation of $\overline{Q}_i$ (any increase in the number $n$ of respondents or experts in a category does not increase its contribution).

The significant estimates of the $i$-th sub-criterion in the each key criterion A, B, C or D elicited from the respondents (P) and experts (TS, AS) of all three categories are expressed by the local weight $\overline{Q}_{Ai}$, $\overline{Q}_{Bi}$, $\overline{Q}_{Ci}$ and $\overline{Q}_{Di}$ calculated by the formulas:

$$\overline{Q}_{Ai} = \frac{\sum_{e=1}^{E} Q_{Aie}}{E} = \frac{Q_{AiP} + Q_{AiTS} + Q_{AiAS}}{3}, \tag{3.23}$$

$$\overline{Q}_{Bi} = \frac{\sum_{e=1}^{E} Q_{Bie}}{E} = \frac{Q_{BiP} + Q_{BiTS} + Q_{BiAS}}{3}, \tag{3.24}$$

$$\overline{Q}_{Ci} = \frac{\sum_{e=1}^{E} Q_{Cie}}{E} = \frac{Q_{CiP} + Q_{CiTS} + Q_{CiAS}}{3}, \tag{3.25}$$

$$\overline{Q}_{Di} = \frac{\sum_{e=1}^{E} Q_{Die}}{E} = \frac{Q_{DiP} + Q_{DiTS} + Q_{DiAS}}{3}, \tag{3.26}$$

where $Q_{Aie}$ is mean weight assigned to $i$-th sub-criterion ($i = 1, 2, \ldots, m$) of the key criterion A (in the questionnaires) by respondents' or experts (representing the category $e$; $e = 1, 2, \ldots, E$); $Q_{AiP}$ is mean weight assigned to $i$-th sub-criterion of the key criteria A by passengers; $Q_{AiTS}$ is mean weight assigned to $i$-th sub-criterion of the key criterion A by service staff; $Q_{AiAS}$ is mean weight assigned by administration staff to $i$-th sub-criterion of the key criterion A (Table 3.7).

The local weight $\overline{Q}_{Bi}$, $\overline{Q}_{Ci}$, $\overline{Q}_{Di}$ of the $i$-th sub-criteria in other key criteria (B, C, D) were calculated using similar formulas (3.24)–(3.26) (Tables 3.8, 3.9 and 3.10).

**Table 3.7** Mean weights and priorities for the key criterion A of the sub-criteria of respondents (P) and experts (TS, AS) as established through the AHP method

| Sub-criterion $i = 1, 2, \ldots, m$ | Respondents (P) and experts (TS, AS); $e = 1, 2, \ldots, E$; $j = 1, 2, \ldots, n$ | | | | | | Average values of sub-criterion | |
|---|---|---|---|---|---|---|---|---|
| | P $n_P = 10$ | | TS $n_{TS} = 11$ | | AS $n_{AS} = 5$ | | | |
| | Mean weight $Q_{AiP}$ | Rank $R_{AiP}$ | Mean weight $Q_{AiTS}$ | Rank $R_{AiTS}$ | Mean weight $Q_{AiAS}$ | Rank $R_{AiAS}$ | Local weight $\overline{Q}_{Ai}$ | Priority (rank) $\overline{R}_{Ai}$ |
| A1 | 0.0331 | 15 | 0.0485 | 14 | 0.0792 | 5 | 0.0536 | 10 |
| A2 | 0.1334 | 1 | 0.0678 | 7 | 0.1350 | 1 | 0.1121 | 1 |
| A3 | 0.0381 | 12 | 0.0238 | 16 | 0.0311 | 11 | 0.0310 | 16 |
| A4 | 0.0430 | 9 | 0.0666 | 8 | 0.0660 | 7 | 0.0585 | 8 |
| A5 | 0.0484 | 8 | 0.0604 | 11 | 0.0647 | 8 | 0.0578 | 9 |
| A6 | 0.1071 | 3 | 0.0883 | 1 | 0.1195 | 2 | 0.1050 | 3 |
| A7 | 0.1218 | 2 | 0.0816 | 3 | 0.1148 | 3 | 0.1061 | 2 |
| A8 | 0.0853 | 4 | 0.0871 | 2 | 0.0638 | 9 | 0.0787 | 4 |
| A9 | 0.0423 | 10 | 0.0661 | 9 | 0.0821 | 4 | 0.0635 | 7 |
| A10 | 0.0729 | 6 | 0.0687 | 6 | 0.0509 | 10 | 0.0642 | 6 |
| A11 | 0.0367 | 13 | 0.0490 | 13 | 0.0160 | 16 | 0.0339 | 15 |
| A12 | 0.0501 | 7 | 0.0399 | 15 | 0.0199 | 14 | 0.0366 | 13 |
| A13 | 0.0344 | 14 | 0.0715 | 4 | 0.0290 | 13 | 0.0450 | 11 |
| A14 | 0.0816 | 5 | 0.0691 | 5 | 0.0791 | 6 | 0.0766 | 5 |
| A15 | 0.0392 | 11 | 0.0501 | 12 | 0.0184 | 15 | 0.0359 | 14 |
| A16 | 0.0326 | 16 | 0.0615 | 10 | 0.0305 | 12 | 0.0415 | 12 |
| Total | 1.0000 | 136 | 1.0000 | 136 | 1.0000 | 136 | 1.0000 | 136 |

**Table 3.8** Mean weights and priorities for the key criterion B of the sub-criteria of respondents (P) and experts (TS, AS) as established through the AHP method

| Sub-criterion $i = 1, 2, \ldots, m$ | Respondents (P) and experts (TS, AS); $e = 1, 2, \ldots, E; j = 1, 2, \ldots, n$ | | | | | | Average values of sub-criterion | |
|---|---|---|---|---|---|---|---|---|
| | P $n_P = 10$ | | TS $n_{TS} = 11$ | | AS $n_{AS} = 5$ | | | |
| | Mean weight $Q_{BiP}$ | Rank $R_{BiP}$ | Mean weight $Q_{BiTS}$ | Rank $R_{BiTS}$ | Mean weight $Q_{BiAS}$ | Rank $R_{BiAS}$ | Local weight $\overline{Q}_{Bi}$ | Priority (rank) $\overline{R}_{Bi}$ |
| B1 | 0.1283 | 1 | 0.0940 | 2 | 0.1163 | 1 | 0.1128 | 1 |
| B2 | 0.0302 | 16 | 0.0310 | 14 | 0.0226 | 14 | 0.0279 | 14 |
| B3 | 0.0665 | 4 | 0.0428 | 11 | 0.0503 | 8 | 0.0532 | 8 |
| B4 | 0.0476 | 10 | 0.0248 | 17 | 0.0245 | 13 | 0.0323 | 13 |
| B5 | 0.0418 | 12 | 0.0304 | 15 | 0.0097 | 18 | 0.0273 | 15 |
| B6 | 0.0567 | 7 | 0.0496 | 10 | 0.0270 | 12 | 0.0444 | 11 |
| B7 | 0.0548 | 8 | 0.0598 | 8 | 0.0665 | 7 | 0.0604 | 6 |
| B8 | 0.0312 | 15 | 0.0234 | 18 | 0.0134 | 16 | 0.0227 | 17 |
| B9 | 0.0214 | 19 | 0.0260 | 16 | 0.0162 | 15 | 0.0212 | 18 |
| B10 | 0.0412 | 13 | 0.0537 | 9 | 0.0337 | 11 | 0.0429 | 12 |
| B11 | 0.0228 | 17 | 0.0226 | 19 | 0.0086 | 19 | 0.0180 | 19 |
| B12 | 0.0225 | 18 | 0.0401 | 13 | 0.0123 | 17 | 0.0250 | 16 |
| B13 | 0.1037 | 2 | 0.0711 | 5 | 0.1146 | 2 | 0.0965 | 2 |
| B14 | 0.0493 | 9 | 0.0660 | 6 | 0.0467 | 10 | 0.0540 | 7 |
| B15 | 0.0740 | 3 | 0.1004 | 1 | 0.1133 | 3 | 0.0959 | 3 |
| B16 | 0.0463 | 11 | 0.0624 | 7 | 0.0502 | 9 | 0.0530 | 9 |
| B17 | 0.0664 | 5 | 0.0794 | 4 | 0.1039 | 4 | 0.0832 | 4 |
| B18 | 0.0351 | 14 | 0.0418 | 12 | 0.0726 | 6 | 0.0498 | 10 |
| B19 | 0.0602 | 6 | 0.0807 | 3 | 0.0976 | 5 | 0.0795 | 5 |
| Total | 1.0000 | 190 | 1.0000 | 190 | 1.0000 | 190 | 1.0000 | 190 |

The comparative analysis of railway trip evaluation by three categories of P and AS has shown that the criteria A2, A7 and A6 are significant for all of them. It can be considered that the estimates provided by the AS match those of passengers because they do not work in the train and therefore can easier assess the criteria significance from passengers' perspective. However, different estimates were given to insignificant criteria (Figs. 3.3, 3.5a and 3.6a). The similarity could be only observed in the estimates of the criterion A1 provided by P [5] and TS and in the evaluation of the criterion A12 by the service and AS.

The significance of the sub-criteria in the key criterion B was determined in a similar way (Figs. 3.3, 3.5b and 3.6b) by the TS and AS. They considered the sub-criteria B15, B19 and B1 to be significant. The experts, being highly experienced in

**Table 3.9** Mean weights and priorities for the key criterion C of the sub-criteria of respondents (P) and experts (TS, AS) as established through the AHP method

| Sub-criterion | Respondents (P) and experts (TS, AS); $e = 1, 2, \ldots, E$; $j = 1, 2, \ldots, n$ | | | | | | Average values of sub-criterion | |
|---|---|---|---|---|---|---|---|---|
| | P $n_P = 10$ | | TS $n_{TS} = 11$ | | AS $n_{AS} = 5$ | | | |
| | Mean weight $Q_{CiP}$ | Rank $R_{CiP}$ | Mean weight $Q_{CiTS}$ | Rank $Q_{RiTS}$ | Mean weight $Q_{CiAS}$ | Rank $R_{CiAS}$ | Local weight $\overline{Q}_{Ci}$ | Priority (rank) $\overline{R}_{Ci}$ |
| C1 | 0.3658 | 1 | 0.2671 | 2 | 0.3643 | 1 | 0.3324 | 1 |
| C2 | 0.1193 | 4 | 0.1061 | 5 | 0.0837 | 4 | 0.1030 | 4 |
| C3 | 0.0433 | 6 | 0.0849 | 6 | 0.0359 | 6 | 0.0547 | 6 |
| C4 | 0.1385 | 3 | 0.1364 | 3 | 0.1777 | 3 | 0.1509 | 3 |
| C5 | 0.2536 | 2 | 0.2798 | 1 | 0.2833 | 2 | 0.2722 | 2 |
| C6 | 0.0795 | 5 | 0.1257 | 4 | 0.0551 | 5 | 0.0868 | 5 |
| Total | 1.0000 | 21 | 1.0000 | 21 | 1.0000 | 21 | 1.0000 | 21 |

the discussed issues are aware of the significance of these sub-criteria to passenger transportation quality. The estimates which were given for insignificant sub-criteria B8 and B11 were not much different as well.

The estimates given by the P and AS to the significance of all sub-criteria in the key criterion C are actually the same, and the judgments of the TS are similar to them (Figs. 3.3, 3.5c and 3.6c).

The sub-criterion D4 was assessed to be equally significant to all experts (Figs. 3.3, 3.5d and 3.6d). As specialists, they are aware of its importance for traffic safety. The sub-criterion D1 was similarly assessed by P and AS (Figs. 3.3 and 3.6d). Moreover, the estimates given by 3 groups of respondents to the sub-criteria D3, D8 and D6 (Figs. 3.3, 3.5d and 3.6d) as insignificant ones did not differ much.

Thus, the profound analysis of the significance estimates of passenger transportation (trips by rail) quality sub-criteria has shown that the sub-criteria A7, A2 and A6 (Table 3.7), B1, B15 and B13 (Table 3.8), C1, C5 and C4 (Table 3.9), D4, D1 and D7 (Table 3.10) are the most significant to respondents and experts.

The estimates provided by respondents (P) and experts (TS, AS) show the highest similarity (Table 3.11).

Ranking the local weights $\overline{Q}_{Ai}$, $\overline{Q}_{Bi}$, $\overline{Q}_{Ci}$, $\overline{Q}_{Di}$ (Tables 3.7, 3.8, 3.9 and 3.10) of the sub-criteria for the main criteria A, B, C, D in descending order from highest to lowest values, their priority sequence has been established (Fig. 3.7). The local weights of the sub-criteria shown in these graphs are the outcome of the study, which was used to calculate the global weights $\overline{\omega}_i$ of the sub-criteria (Table 3.17). These local weights of the sub-criteria have been used in mathematical models of the key criteria to calculate and quantify the impact of the studied factors on the quality of the train passenger transportation (Chaps. 4, 5, 6 and 7).

**Table 3.10** Mean weights and priorities for the key criterion D of the sub-criteria of respondents (P) and experts (TS, AS) as established through the AHP method

| Sub-criterion | Respondents (P) and experts (TS, AS); $e = 1, 2, \ldots, E; j = 1, 2, \ldots, n$ | | | | | | Average values of sub-criterion | |
|---|---|---|---|---|---|---|---|---|
| | P $n_P = 10$ | | TS $n_{TS} = 11$ | | AS $n_{AS} = 5$ | | | |
| | Mean weight $Q_{DiP}$ | Rank $R_{DiP}$ | Mean weight $Q_{DiTS}$ | Rank $R_{DiTS}$ | Mean weight $Q_{DiAS}$ | Rank $R_{DiAS}$ | Local weight $\overline{Q}_{Di}$ | Priority (rank) $\overline{R}_{Di}$ |
| D1 | 0.1361 | 2 | 0.1495 | 3 | 0.1715 | 2 | 0.1524 | 3 |
| D2 | 0.1282 | 3 | 0.1062 | 5 | 0.0681 | 8 | 0.1008 | 5 |
| D3 | 0.1170 | 7 | 0.0983 | 6 | 0.0867 | 5 | 0.1007 | 6 |
| D4 | 0.1243 | 5 | 0.1806 | 1 | 0.2610 | 1 | 0.1887 | 1 |
| D5 | 0.1195 | 6 | 0.1353 | 4 | 0.1077 | 4 | 0.1208 | 4 |
| D6 | 0.1269 | 4 | 0.0747 | 8 | 0.0708 | 7 | 0.0908 | 7 |
| D7 | 0.1449 | 1 | 0.1798 | 2 | 0.1572 | 3 | 0.1606 | 2 |
| D8 | 0.1031 | 8 | 0.0756 | 7 | 0.0770 | 6 | 0.0852 | 8 |
| Total | 1.0000 | 36 | 1.0000 | 36 | 1.0000 | 36 | 1.0000 | 36 |

**Table 3.11** The number of matching estimates (ranks) provided by respondents and experts

| Key criterion | P = TS | P = AS | TS = AS |
|---|---|---|---|
| A | 2 | 3 | 2 |
| B | 1 | 1 | 2 |
| C | 2 | 6 | 2 |
| D | 0 | 1 | 1 |
| Total | 5 | 11 | 7 |

## 3.6 Difference in Opinions Between the Three Categories of Respondents and Experts

Given the ranks (priority's) assigned to the railway trip quality (RTQ) sub-criteria by the respondents and experts of the particular category and experts (Tables 3.7, 3.8, 3.9 and 3.10), the consistency of their estimates may be determined, based on the sum of the ranks' differences. The smaller the sum of the moduli of the differences in the ranks, assigned to the sub-criteria by two categories of respondents and experts or experts and experts, the more uniform are the estimates. The average ranks' differences are calculated by the formula:

$$\overline{\overline{R}}_{1-2} = \frac{\sum_{i=1}^{m} \left|\overline{R}_{1-2}\right|_i}{m}, \quad (3.27)$$

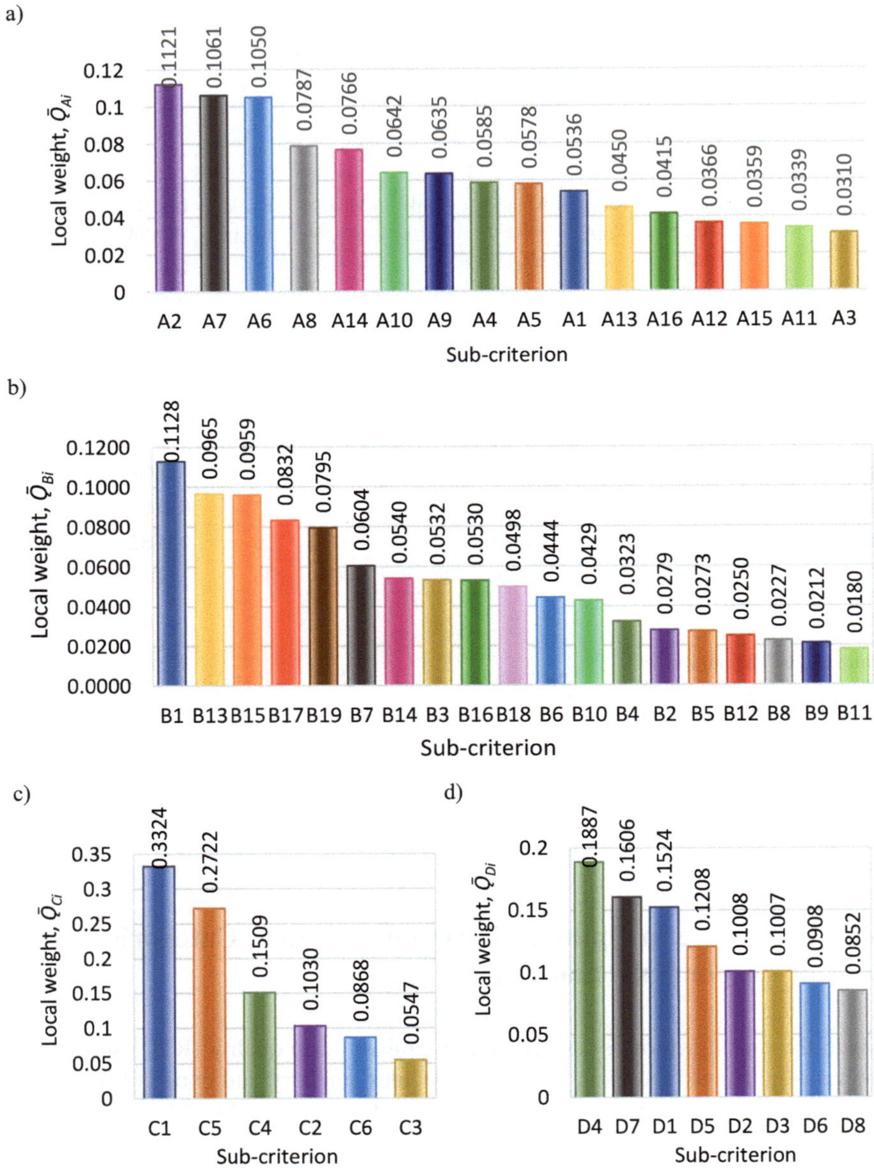


**Fig. 3.7** Mean subjective normalized local weights of sub-criteria for key criteria (**a** A, **b** B, **c** C, **d** D) as established through the AHP method and in relation to three categories of respondents (P) and experts (TS, AS)

where $\left|\overline{R}_{1-2}\right|_i$ is the modulus of differences (or the absolute value) of the ranks (priority), assigned to the key criteria (A, B, C or D) or to $i$-th sub-criterion by the respondents or by experts and experts; $m$ is the number of sub-criteria in the key criterion; the indices by $\overline{R}\,1$ and $\overline{R}\,2$ denote the respondents of the 1-st and 2-nd categories and experts, whose sub-criteria ranks are compared. When the ranks given to all sub-criteria of the key criterion by the respondents and experts or by experts and experts match each other, then, $\overline{\overline{R}}_{1-2} = 0$, and the estimates are the same. When the value $\overline{\overline{R}}_{1-2}$ is growing, the differences in the estimates are also increasing. The difference in the opinion between the three categories of survey participants (respondents and experts) was analysed and evaluated according to the algorithm presented in Fig. 3.8.

The calculated $\overline{R}_{1-2}$ values (Table 3.12) show that, in the key criterion A, the estimates of all the sub-criteria, elicited from P and the TS, are more consistent (close to each other) ($\overline{\overline{R}}_{P-TS} = 2.88$). In the sub-criteria key criterion B the estimates elicited from the TS and AS are more consistent ($\overline{\overline{R}}_{P-AS} = 2.21$). In the key criterion C the estimates of the sub-criteria significance, elicited from P and AS, are absolutely the same ($\overline{\overline{R}}_{P-AS} = 0$). In the key criterion D more consistent estimates of the sub-criteria significance were elicited from the TS and AS ($\overline{\overline{R}}_{TS-AS} = 1.00$).

## 3.7 Results of the Survey into Respondent and Expert Opinions on the Significance of the Key Criteria Describing the Quality of a Railway Trip

When grouping factors describing passenger transportation by railway and establishing their weight, the preference order or ranks of key criterion A, B, C and D were determined.

Twenty one passengers of the international train and 29 experts (including 20 TS and 9 AS of the JSC competent in describing the structure and constituent parts of the train, the technical state of the railway, the planning and technology of a railway trip, provisions for trip safety and requirements for the quality of transportation were given questionnaires (Table 3.13) and asked to assign ranks according to their importance and considering key criterion A, B, C and D (Fig. 3.3) describing the quality of the railway trip. Based on the above-described methods, the consistency of the estimates of each respondent/expert and the judgements of the whole key criterions were determined.

The diagrams of calculated average ranks $\overline{R}_f$ assigned by all estimators (respondents and experts) to key criterion A, B, C and D describing the quality of the railway trip are presented in Fig. 3.9.

Table 3.14 provides general data on questionnaires, including estimates $R_i$ showing the significance of key criterion A, B, C and D describing the quality of a railway trip, interpreted by the surveyed respondents (21 passengers) and experts

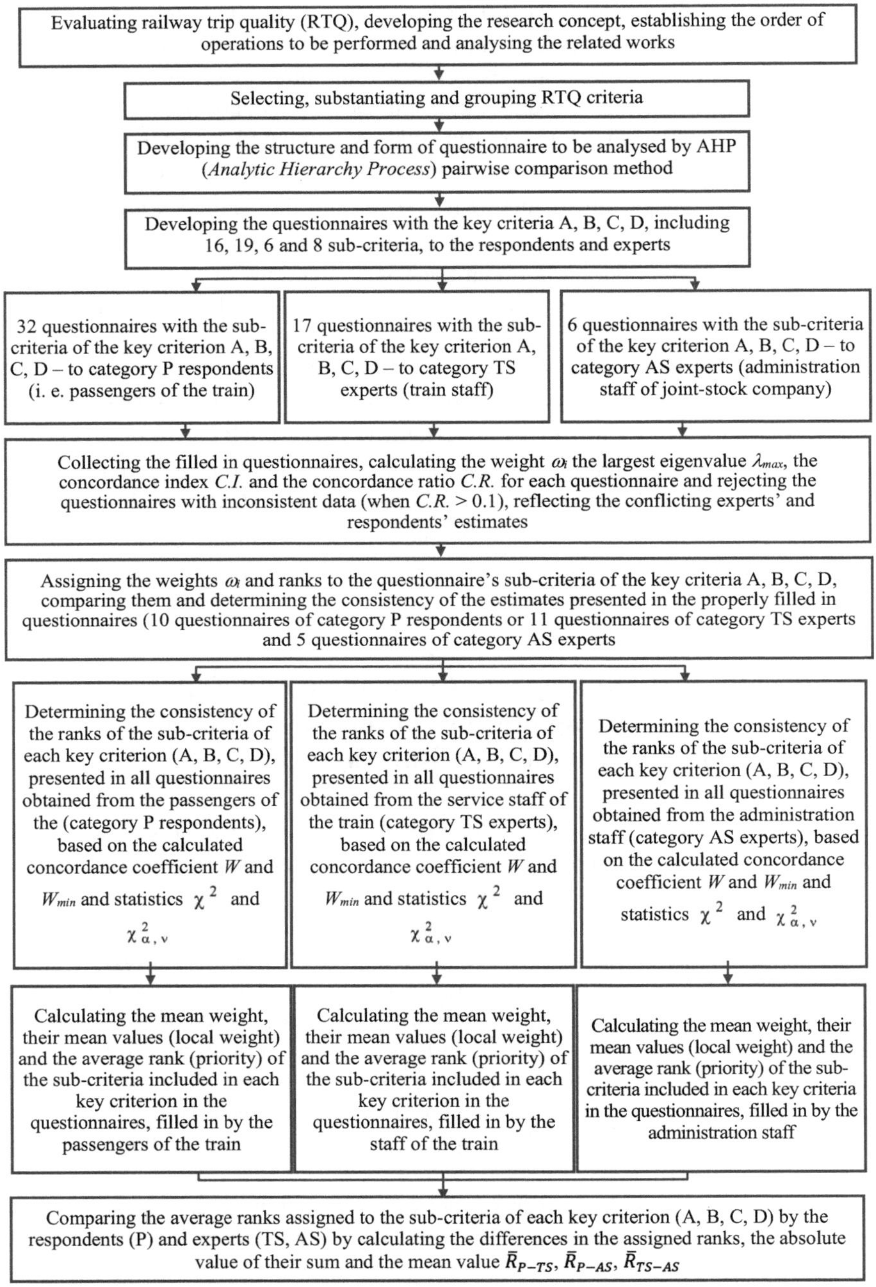


**Fig. 3.8** The algorithm for investigating and the railway trip quality

**Table 3.12** Differences in ranks assigned to the sub-criteria by the respondents and experts

| The sub-criterion No in the questionnaire $i = 1, 2, \ldots, m$ | Key criterion | | | | | | | | | | | |
|---|---|---|---|---|---|---|---|---|---|---|---|---|
| | A | | | B | | | C | | | D | | |
| | Pairs of particular category respondents and experts compared in the analysis | | | | | | | | | | | |
| | $\overline{R}_{P-TS}$ | $\overline{R}_{P-AS}$ | $\overline{R}_{TS-AS}$ | $\overline{R}_{P-TS}$ | $\overline{R}_{P-AS}$ | $\overline{R}_{TS-AS}$ | $\overline{R}_{P-TS}$ | $\overline{R}_{P-AS}$ | $\overline{R}_{TS-AS}$ | $\overline{R}_{P-TS}$ | $\overline{R}_{P-AS}$ | $\overline{R}_{TS-AS}$ |
| 1 | 1 | 10 | 9 | − 1 | 0 | 1 | − 1 | 0 | 1 | − 1 | 0 | 1 |
| 2 | − 6 | 0 | 6 | 2 | 2 | 0 | − 1 | 0 | 1 | − 2 | − 5 | − 3 |
| 3 | − 4 | 1 | 5 | − 7 | − 4 | 3 | 0 | 0 | 0 | 1 | 2 | 1 |
| 4 | 1 | 2 | 1 | − 7 | − 3 | 4 | 0 | 0 | 0 | 4 | 4 | 0 |
| 5 | − 3 | 0 | 3 | − 3 | − 6 | − 3 | 1 | 0 | − 1 | 2 | 2 | 0 |
| 6 | 2 | 1 | − 1 | − 3 | − 5 | − 2 | 1 | 0 | − 1 | − 4 | − 3 | 1 |
| 7 | − 1 | − 1 | 0 | 0 | 1 | 1 | | | | − 1 | − 2 | − 1 |
| 8 | 2 | − 5 | − 7 | − 3 | − 1 | 2 | | | | 1 | 2 | 1 |
| 9 | 1 | 6 | 5 | 3 | 4 | 1 | | | | | | |
| 10 | 0 | − 4 | − 4 | 4 | 2 | − 2 | | | | | | |
| 11 | 0 | − 3 | − 3 | − 2 | − 2 | 0 | | | | | | |
| 12 | − 8 | − 7 | 1 | 5 | 1 | − 4 | | | | | | |
| 13 | 10 | 1 | − 9 | − 3 | 0 | 3 | | | | | | |
| 14 | 0 | − 1 | − 1 | 3 | − 1 | − 4 | | | | | | |
| 15 | − 1 | − 4 | − 3 | 2 | 0 | − 2 | | | | | | |
| 16 | 6 | 4 | − 2 | 4 | 2 | − 2 | | | | | | |
| 17 | | | | 1 | 1 | 0 | | | | | | |
| 18 | | | | 2 | 8 | 6 | | | | | | |
| 19 | | | | 3 | 1 | − 2 | | | | | | |

(continued)

**Table 3.12** (continued)

| The sub-criterion No in the questionnaire $i = 1, 2, \ldots, m$ | Key criterion | | | | | | | | | | | |
|---|---|---|---|---|---|---|---|---|---|---|---|---|
| | A | | | B | | | C | | | D | | |
| | Pairs of particular category respondents and experts compared in the analysis | | | | | | | | | | | |
| | $\overline{R}_{P-TS}$ | $\overline{R}_{P-AS}$ | $\overline{R}_{TS-AS}$ | $\overline{R}_{P-TS}$ | $\overline{R}_{P-AS}$ | $\overline{R}_{TS-AS}$ | $\overline{R}_{P-TS}$ | $\overline{R}_{P-AS}$ | $\overline{R}_{TS-AS}$ | $\overline{R}_{P-TS}$ | $\overline{R}_{P-AS}$ | $\overline{R}_{TS-AS}$ |
| $\sum_{i=1}^{m} \left\|\overline{R}_{1-2}\right\|_i$ | 46 | 50 | 60 | 58 | 44 | 42 | 4 | 0 | 4 | 16 | 20 | 8 |
| $\overline{\overline{R}}_{1-2}$ | 2.88 | 3.125 | 3.75 | 3.053 | 2.32 | 2.21 | 0.67 | 0 | 0.67 | 2.00 | 2.50 | 1.00 |

**Table 3.13** A questionnaire for ranking key criterion A, B, C and D describing the quality of a railway trip

| The key criterion | A group and brief description of key criteria |
|---|---|
| A | Train elements and technical state of rails (unevenness of a track, train speed, structure and equipment for a passenger car ensuring the comfort of passengers) |
| B | Railway trip planning and technology (timely departure and arrival of the train, quality of services provided on train, the appearance of TS, their personal contact with passengers, knowledge of foreign languages, etc.) |
| C | Expenses of the trip (ticket price, medical care insurance abroad, visa, etc.) |
| D | Safe railway trip (availability of fire extinguishing and first medical care facilities, efficiency of wheelset control, fire alarm and manual braking systems, possibility of calling doctors and policemen to the train) |

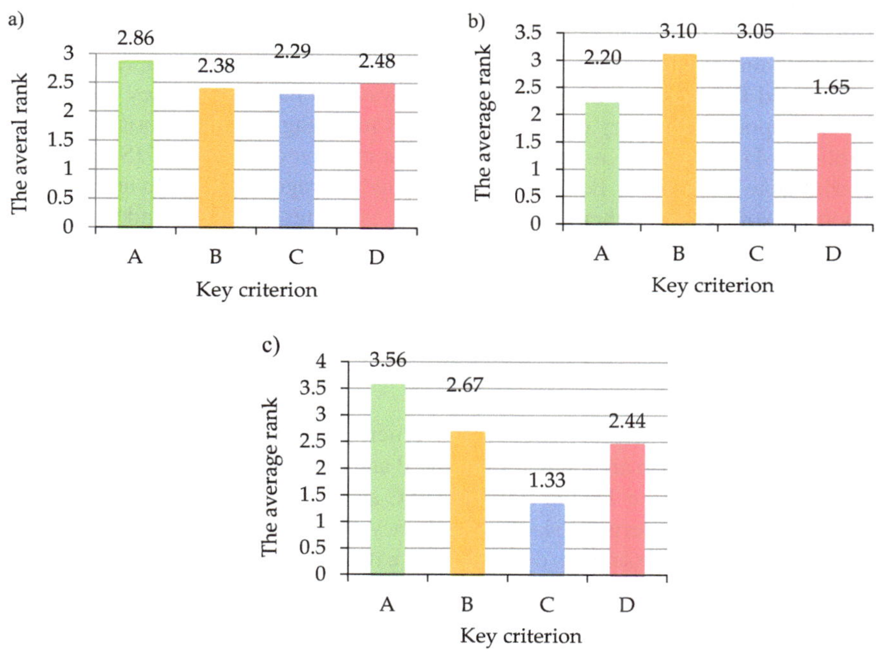


**Fig. 3.9** The diagram of average ranks $\overline{R}_f$ given to key criterion A, B, C and D describing the quality of a railway trip considering the position of the **a** the passengers'; **b** the train staff; **c** administrative staff of the JSC

(20 members of the train team and 9 staff members of the JSC). Based on the methods described above, the consistency of opinions expressed by the respondents and experts representing two different categories as well as the consistency of the estimates elicited from all respondents and experts (50 evaluators) (Table 3.14) was determined.

**Table 3.14** Concordance coefficient $W$, critical value $\chi^2_{\alpha,v}$ and the lowest concordance coefficient $W_{min}$ obtained for the estimates provided by the P, TS, AS

| The category of respondents and experts* | $m_e$ | Calculated value | | | |
|---|---|---|---|---|---|
| | | $W$ | $W_{min}$ | $\chi^2$ | $\chi^2_{3;0.05}$ |
| P | 21 | 0.038 | 0.124 | 2.37 | 7.81 |
| TS | 20 | 0.295 | 0.130 | 17.70 | 7.81 |
| AS | 9 | 0.501 | 0.289 | 13.53 | 7.81 |
| B (P, TS, AS) | 50 | 0.047 | 0.052 | 6.984 | 7.81 |

* The categories of respondents (P) and experts (TS and AS), B—both respondents and experts

The calculated average ranks $\overline{R}_f$ given to key criterion A, B, C and D describing the quality of a railway trip by P show that key criterion C is more significant than B, D and A, i.e. the following priorities order is obtained: C $\succ$ B $\succ$ D $\succ$ A. However, from the point of view of the TS, average ranks $\overline{R}_f$ given by them to key criterion A, B, C and D describing the quality of a railway trip indicate that key criterion D is more important than key criteria A, C and B implying that the priorities order is as follows: D $\succ$ A $\succ$ C $\succ$ B. From the point of view of the AS of JSC Lietuvos geležinkeliai—C $\succ$ D $\succ$ B $\succ$ A. The calculated average ranks $\overline{R}_f$ do not show the importance of one or another key criterion. The estimates of the respondents (P) and experts (TS, AS) are closer to each other (Table 3.15) and match with determining the priority of key criterion A and C with respect to the quality of the railway trip.

Concordance coefficient $W$, critical value $\chi^2_{v,\alpha}$ (taken from the distribution table with a respective significance level $\alpha = 0.05$) and the lowest value of concordance coefficient $W_{min}$ obtained for the estimates provided by the respondents and experts are given in Table 3.14.

The estimates of the preference of key criterion A, B, C and D describing the quality of a railway trip are elicited from the P and remain inconsistent, whereas the judgements made by the TS and AS of the JSC are consistent. A general opinion of all respondents and experts also lacks consistency. It can be assumed that the interests and requirements of passengers differ to great extent: for example, some of them are primarily interested in low ticket prices actually ignoring the level of comfort, while

**Table 3.15** The number of tied ranks obtained from the respondents and experts

| Key criterion | P = TS | P = AS | TS = AS |
|---|---|---|---|
| A | 0 | 1 | 0 |
| B | 0 | 0 | 0 |
| C | 0 | 1 | 0 |
| D | 0 | 0 | 0 |
| Total | 0 | 2 | 0 |

* The notation of respondent and expert categories: *P* passengers, *TS* train staff, *AS* administrative staff

others, on the contrary, give preference to comfort not paying much attention to the ticket price.

The significance indicator $Z_f$ of any key criterion obtained from the respondents and experts is calculated using formula (3.28). When the quality of an object is determined by an additive mathematical model used for calculating a complex (integrated) quality key criterion (allowing quality to be defined by a single number and compared to the quality of other objects), significance indicators $Z_f$ rather than average ranks $\overline{R}_f$ should be used, which does not indicate the significance level of one or another rank.

The significance of key criteria describing the quality of the object evaluated by experts can be determined normalizing them (i.e., making their sum equal to one) and calculating the significance indicator $Z_f$ of each criterion from the formula given in:

$$Z_f = \frac{(F+1) - \overline{R}_f}{\sum_{f=1}^{F} \overline{R}_f}, \tag{3.28}$$

where $F$ is the number of key criteria describing the quality (characteristics) of the considered object; $\overline{R}_f$ is the average rank of the $f$-th key criterion ($f = 1, 2, \ldots, F$) (Fig. 3.9).

The estimates of the significance of key criterion A determined by the respondents and experts of all three categories (P, TS, AS) take part in the survey, are expressed by the mean value of normalized subjective weight $\overline{Z}_A$ (when the number of the respondents and experts in each category is the same) and are calculated applying the formula

$$\overline{Z}_A = \frac{\sum_{e=1}^{E} Z_{Ae}}{E} = \frac{Z_{AP} + Z_{ATS} + Z_{AAS}}{3}, \tag{3.29}$$

where $Z_{Ae}$ is weight given by the respondents or experts of the $e$-th category to key criterion A; $E$ – is number of estimators categories; $Z_{AP}$ is weight assigned to key criterion A by passengers; $Z_{ATS}$ is weight assigned to key criterion A by train staff; $Z_{AAS}$ is weight assigned to key criterion A by administrative staff; ($e = 1, 2, \ldots, E$).

The mean values of normalized subjective weight $\overline{Z}_B, \overline{Z}_C, \overline{Z}_D$ of other key criterion (B, C, D) were calculated employing similar formulas (Table 3.16):

$$\overline{Z}_B = \frac{\sum_{e=1}^{E} Z_{Be}}{E} = \frac{Z_{BP} + Z_{BTS} + Z_{BAS}}{3}, \tag{3.30}$$

$$\overline{Z}_C = \frac{\sum_{e=1}^{E} Z_{Ce}}{E} = \frac{Z_{CP} + Z_{CTS} + Z_{CAS}}{3}, \tag{3.31}$$

$$\overline{Z}_D = \frac{\sum_{e=1}^{E} Z_{De}}{E} = \frac{Z_{DP} + Z_{DTS} + Z_{DAS}}{3}, \tag{3.32}$$

**Table 3.16** Significance (weight) and preference order (priority) of key criterion A, B, C and D describing the quality of the railroad trip determined by P, TS and AS

| Weights $Z_f, \overline{Z}_f, \overline{Z}_f^*$ | Key criterion describing the quality of the railway trip ($f = 1, 2, \ldots, F$) | | | | |
|---|---|---|---|---|---|
| | A | B | C | D | Total |
| **Passengers ($n_P = 21$)** | | | | | |
| $Z_{fP}$ | 0.2143 | 0.2619 | 0.2714 | 0.2524 | 1.000 |
| Priorities | 4 | 2 | 1 | 3 | 10 |
| **Train service staff ($n_{TS} = 20$)** | | | | | |
| $Z_{fTS}$ | 0.2800 | 0.1900 | 0.1950 | 0.3350 | 1.000 |
| Priorities | 2 | 4 | 3 | 1 | 10 |
| **Administrative staff of JSC ($n_{AS} = 9$)** | | | | | |
| $Z_{fAS}$ | 0.1444 | 0.2333 | 0.3667 | 0.2556 | 1.000 |
| Priorities | 4 | 3 | 1 | 2 | 10 |
| The average estimate of all experts and respondents ($n = 50$) $\overline{Z}_f$ | 0.2129 | 0.2284 | 0.2777 | 0.2810 | 1.000 |
| The average estimate of all evaluators (experts and respondents) taking into account their number in a category $\overline{Z}_f^*$ | 0.2280 | 0.2280 | 0.2580 | 0.2860 | 1.000 |

The significance of key criterion A determined by the respondents and experts of all three categories (P, TS, AS) is expressed by the mean value of weight $\overline{Z}_A^*$ (when the number of the respondents and experts in each category is not the same). This normalized subjective weight is calculated by the formula

$$\overline{Z}_A^* = \frac{\sum_{e=1}^{E} Z_f \cdot n_e}{\sum_{e=1}^{E} n_e} = \frac{Z_{AP} \cdot n_P + Z_{ATS} \cdot n_{TS} + Z_{AAS} \cdot n_{AS}}{n_P + n_{TS} + n_{AS}}, \quad (3.33)$$

where $Z_{AP}, Z_{ATS}, Z_{AAS}$ denote weight given by the respondents (experts) of categories P, TS and AS to key criterion A; $n_P$, $n_{TS}$, $n_{AS}$ are the numbers of the respondents (P) and experts (TS, AS). $E$ – number of estimators category ($e = 1, 2, \ldots, E$).

In the considered case, the number of the respondents and experts is not the same, and therefore formula (3.33) is used, which is more suitable because the respondents make a larger part—21 passengers. The performed research accepts the opinion of passengers to be of primary importance.

Mean weight $\overline{Z}_f^*$ showing the significance of key criterion B, C and D was determined by the respondents and experts of categories P, TS and AS, and therefore is calculated applying formulas:

$$\overline{Z}_B^* = \frac{Z_{BP} \cdot n_P + Z_{BTS} \cdot n_{TS} + Z_{BAS} \cdot n_{AS}}{n_P + n_{TS} + n_{AS}}, \quad (3.34)$$

**Table 3.17** Local and global normalised weights for the criteria of international passenger train travel

| Key criterion and weight $\overline{Z}_f^*$ | Sub-criterion | Local weight $\overline{Q}_{fi}$ | Local priority | The product $\overline{Z}_f^* \cdot \overline{Q}_{fi}$ | The product $\overline{Z}_f^* \cdot \overline{Q}_{fi} \cdot m_f$ | Global weight $\overline{\omega}_i$ | Global priority |
|---|---|---|---|---|---|---|---|
| The technical state of the railway track and the elements of the train (key criterion A) $\overline{Z}_A^* = 0.228$; $m_A = 16$ | A1 | 0.0536 | 10 | 0.0122208 | 0.1955328 | 0.016548 | 30 |
| | A2 | 0.1121 | 1 | 0.0255588 | 0.4089408 | 0.034609 | 7 |
| | A3 | 0.0310 | 16 | 0.007068 | 0.113088 | 0.009571 | 44 |
| | A4 | 0.0585 | 8 | 0.013338 | 0.213408 | 0.018061 | 27 |
| | A5 | 0.0578 | 9 | 0.0131784 | 0.2108544 | 0.017845 | 28 |
| | A6 | 0.1050 | 3 | 0.02394 | 0.38304 | 0.032417 | 9 |
| | A7 | 0.1061 | 2 | 0.0241908 | 0.3870528 | 0.032757 | 8 |
| | A8 | 0.0787 | 4 | 0.0179436 | 0.2870976 | 0.024297 | 14 |
| | A9 | 0.0635 | 7 | 0.014478 | 0.231648 | 0.019605 | 21 |
| | A10 | 0.0642 | 6 | 0.0146376 | 0.2342016 | 0.019821 | 19 |
| | A11 | 0.0339 | 15 | 0.0077292 | 0.1236672 | 0.010466 | 41 |
| | A12 | 0.0366 | 13 | 0.0083448 | 0.1335168 | 0.011300 | 39 |
| | A13 | 0.0450 | 11 | 0.01026 | 0.16416 | 0.013893 | 34 |
| | A14 | 0.0766 | 5 | 0.0174648 | 0.2794368 | 0.023649 | 15 |
| | A15 | 0.0359 | 14 | 0.0081852 | 0.1309632 | 0.011083 | 40 |
| | A16 | 0.0415 | 12 | 0.009462 | 0.151392 | 0.012812 | 36 |
| The trip planning and technology (key criterion B) $\overline{Z}_B^* = 0.228$; $m_B = 19$ | B1 | 0.1128 | 1 | 0.0257184 | 0.4886496 | 0.041355 | 2 |
| | B2 | 0.0279 | 14 | 0.0063612 | 0.1208628 | 0.010229 | 42 |
| | B3 | 0.0532 | 8 | 0.0121296 | 0.2304624 | 0.019504 | 23 |
| | B4 | 0.0323 | 13 | 0.0073644 | 0.1399236 | 0.011842 | 37 |
| | B5 | 0.0273 | 15 | 0.0062244 | 0.1182636 | 0.010009 | 43 |
| | B6 | 0.0444 | 11 | 0.01001232 | 0.1923408 | 0.016278 | 32 |
| | B7 | 0.0604 | 6 | 0.01137712 | 0.2616528 | 0.022144 | 17 |
| | B8 | 0.0227 | 17 | 0.0051756 | 0.0983364 | 0.008322 | 46 |
| | B9 | 0.0212 | 18 | 0.0048336 | 0.0918384 | 0.007772 | 47 |
| | B10 | 0.0429 | 12 | 0.0097812 | 0.1858428 | 0.015729 | 33 |
| | B11 | 0.0180 | 19 | 0.004104 | 0.077976 | 0.006599 | 49 |
| | B12 | 0.0250 | 16 | 0.0057 | 0.1083 | 0.009166 | 45 |
| | B13 | 0.0965 | 2 | 0.022002 | 0.418038 | 0.035379 | 5 |
| | B14 | 0.0540 | 7 | 0.012312 | 0.233928 | 0.019798 | 18 |
| | B15 | 0.0959 | 3 | 0.0218652 | 0.4154388 | 0.035159 | 6 |
| | B16 | 0.0530 | 9 | 0.012084 | 0.229596 | 0.019431 | 25 |
| | B17 | 0.0832 | 4 | 0.0189696 | 0.3604224 | 0.030503 | 11 |
| | B18 | 0.0498 | 10 | 0.0113544 | 0.2157336 | 0.018258 | 26 |

(continued)

**Table 3.17** (continued)

| Key criterion and weight $\overline{Z}_f^*$ | Sub-criterion | Local weight $\overline{Q}_{fi}$ | Local priority | The product $\overline{Z}_f^* \cdot \overline{Q}_{fi}$ | The product $\overline{Z}_f^* \cdot \overline{Q}_{fi} \cdot m_f$ | Global weight $\overline{\omega}_i$ | Global priority |
|---|---|---|---|---|---|---|---|
| | B19 | 0.0795 | 5 | 0.018126 | 0.344394 | 0.029146 | 13 |
| The expenses a trip by train and provided service (key criterion C) $\overline{Z}_C^* = 0.258$; $m_C = 6$ | C1 | 0.3324 | 1 | 0.0857592 | 0.5145552 | 0.043547 | 1 |
| | C2 | 0.1030 | 4 | 0.026574 | 0.159444 | 0.013494 | 35 |
| | C3 | 0.0547 | 6 | 0.0141126 | 0.0846756 | 0.007166 | 48 |
| | C4 | 0.1509 | 3 | 0.0389322 | 0.2335932 | 0.019769 | 20 |
| | C5 | 0.2722 | 2 | 0.0702276 | 0.4213656 | 0.035661 | 4 |
| | C6 | 0.0868 | 5 | 0.0223944 | 0.1343664 | 0.011372 | 38 |
| Internal safety trip by train (key criterion D) $\overline{Z}_D^* = 0.286$; $m_D = 8$ | D1 | 0.1524 | 3 | 0.0435864 | 0.3486912 | 0.029510 | 12 |
| | D2 | 0.1008 | 5 | 0.0288288 | 0.2306304 | 0.019518 | 22 |
| | D3 | 0.1007 | 6 | 0.0288002 | 0.2304016 | 0.019499 | 24 |
| | D4 | 0.1887 | 1 | 0.0539682 | 0.4317456 | 0.036539 | 3 |
| | D5 | 0.1208 | 4 | 0.0345488 | 0.2763904 | 0.023391 | 16 |
| | D6 | 0.0908 | 7 | 0.0259688 | 0.2077504 | 0.017582 | 29 |
| | D7 | 0.1606 | 2 | 0.0459316 | 0.3674528 | 0.031098 | 10 |
| | D8 | 0.0852 | 8 | 0.0243672 | 0.1949376 | 0.016498 | 31 |
| Total | – | 4.0000 | – | – | 11.816 | 1.000000 | – |

$$\overline{Z}_C^* = \frac{Z_{CP} \cdot n_P + Z_{CTS} \cdot n_{TS} + Z_{CAS} \cdot n_{AS}}{n_P + n_{TS} + n_{AS}}, \tag{3.35}$$

$$\overline{Z}_D^* = \frac{Z_{DP} \cdot n_P + Z_{DTS} \cdot n_{TS} + Z_{DAS} \cdot n_{AS}}{n_P + n_{TS} + n_{AS}}. \tag{3.36}$$

A general model for calculating the quality level indicator (QLI) $K$ of a passenger train, when the weight of key criterion was obtained by expert evaluation, is determined by the formula:

$$\begin{aligned} K &= \sum_{f=1}^{F} \left[ \overline{Z}_f^* \cdot \left( \sum_{i=1}^{m} \overline{Q}_{fi} \cdot x_{fi} \right) \right] \\ &= \overline{Z}_A^* \cdot \left( \sum_{i=1}^{m} \overline{Q}_{Ai} \cdot x_{Ai} \right) + \overline{Z}_B^* \cdot \left( \sum_{i=1}^{m} \overline{Q}_{Bi} \cdot x_{Bi} \right) \\ &\quad + \overline{Z}_C^* \cdot \left( \sum_{i=1}^{m} \overline{Q}_{Ci} \cdot x_{Ci} \right) + \overline{Z}_D^* \cdot \left( \sum_{i=1}^{m} \overline{Q}_{Di} \cdot x_{Di} \right), \end{aligned} \tag{3.37}$$

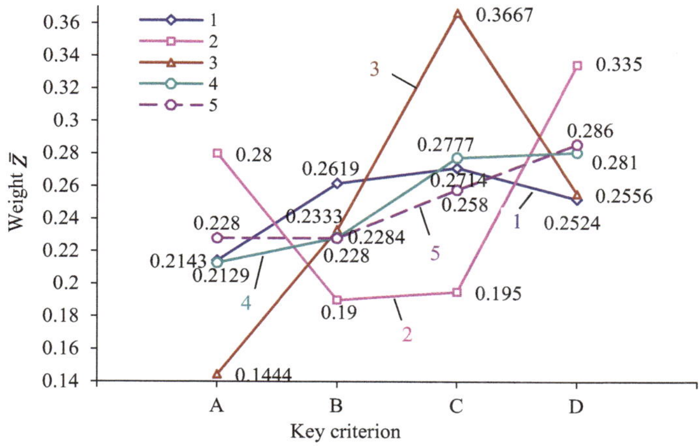


**Fig. 3.10** Lines showing the weight of a quality railway trip taking into account key criterion A, B, C and D given by: 1—P; 2—TS; 3—AS; 4—all experts and respondents based on their mean estimate $\overline{Z}_f$; 5—all experts and respondents based on mean estimate $\overline{Z}_f^*$ of all experts and respondents taking into account the number in a category

where $K$ is QLI of an international train ($K$ may be in the range from 0 to 1); $\overline{Q}_{Ai}$, $\overline{Q}_{Bi}$, $\overline{Q}_{Ci}$, $\overline{Q}_{Di}$ denote the local weights (Fig. 3.7) of the $i$-th sub-criteria of the $f$-th key criterion determined by the expert evaluation method ($f = 1, 2, \ldots, F$); $x_{Ai}$, $x_{Bi}$, $x_{Ci}$, $x_{Di}$ are the variables of the $i$-th sub-criteria of the $f$-th key criterion used for calculating the real key criterion value ranging from 0 to 1.

The calculation results of significance (weight) $Z_f$ and the priorities of key criterion A, B, C and D describing the quality of the railroad trip determined by P, TS and AS of the JSC are given in Table 3.16. The lines of significance (weight) $Z_f$ and mean weight $\overline{Z}_f$, $\overline{Z}_f^*$ are shown in Fig. 3.10.

Passengers think that key criterion C describing the quality of the trip by international train is the most important because $Z_{CP} = 0.2714$. Key criterion B seems to be less important because $Z_{BP} = 0.2619$ while key criterion A with $Z_{AP} = 0.2143$ is assessed as the least important (Fig. 3.10). For the train service staff, key criterion A ($Z_{ATS} = 0.2800$) followed by key criterion D ($Z_{DTS} = 0.3350$) is the most significant. The administrative staff of JSC believe that key criterion C with respective $Z_{CAS} = 0.3667$ is the most important while key criterion A is the least important due to the fact that, in this case, $Z_{AAS} = 0.1444$. The values of weight often differ or even contradict each other. The mean values of the weight of various key criteria are calculated using different formulas. Therefore, they show that key criterion D is more significant than C, while C is more important than B and B, in turn, is more important than A. The considered data allow for a conclusion that railway and train parameters (key criterion A) are the least important for all categories of the respondents and experts, whereas the parameters of traffic safety (key criterion D), on the contrary, are the most important.

The processed data obtained from the surveyed respondents (P) and experts (TS and AS) show that the problem of the quality of the railway trip allow the authors determining the weight of key criterion A, B, C and D and individual criteria describing the quality of the railway trip in various trains and expressing it in a single number.

The use of methods for expert evaluation and AHP technique was a difficult task of the conducted research. Differently from the respondents, the passengers of the train did not show enough initiative. The motivation and experience of the TS helped them with filling in the questionnaires. The P estimates of the significance of key criterions A, B, C and D describing the quality of the railway trip are not consistent. The needs of passengers vary to great extent: some are interested in the trip cost ignoring the range and quality of the provided services while the other part requires various high-quality services and comfort at any cost.

The opinions TS on key criterions A, B, C and D describing the quality of the railway trip are in a good agreement. These people are qualified evaluators. The estimates of the AS provided for key criterions A, B, C and D describing the quality of the railway trip are consistent. The opinions of P and AS about the significance (preference) of key criterions A, B, C and D describing the quality of the railway trip are similar.

The present investigation has determined the weight $\overline{Z}_f^*$ of the key criteria of the additive model (3.33)–(3.36) and the mean weight $Q_{fi}$ of the particular sub-criteria of these key criteria.

## 3.8 Calculation of Global Weights for Train Journey Quality Criteria

Where the object of study is characterised by hierarchically structured criteria divided into key criteria and sub-criteria (Figs. 2.2b and 3.3), local and global non-normalised weights [12] or normalised weights [16–18] are calculated for the sub-criteria. Once the normalised subjective weights $\overline{Z}_f^*$ and the normalised local weights $\overline{Q}_{fi}$ of the sub-criteria for the key criteria of passenger train journey quality have been established, it is possible to calculate the normalised global weights $\overline{\omega}_i$ for all the subcriteria.

The subjective local weights $\overline{Q}_{fi}$ calculated by through AHP method and obtained on the basis of the survey on the significance of the international passenger train journey quality criteria for P, TS and AS are presented in Tables 3.7, 3.8, 3.9, 3.10 and Fig. 3.7. They indicate that the sum of the local mean weights $\overline{Q}_{fi}$ for each of the four key criteria (A, B, C, D) is equal to 1. In other words, the sub-criteria for each key criterion have been compared against each other.

To calculate the normalised weights of all 49 sub-criteria without hierarchical structuring, i.e. without dividing them into 4 key criteria, the following formula is applied:

$$\overline{\omega}_i = \frac{\overline{Z}_f^* \cdot Q_{fi} m_f}{\sum_{f=1}^{E} \sum_{i=1}^{m} \overline{Z}_f^* \cdot \overline{Q}_{fi} \cdot m_f}, \tag{3.38}$$

where $\overline{Z}_f^*$ is the subjective normalised weight for the $f$-th key criterion, obtained by interviewing P, TS and AS ($f = 1, 2, \ldots, F$; $F = 4$; Table 3.16); $\overline{Q}_{fi}$ is the subjective normalised local weight of the $i$-th sub-criterion for the $f$-th main criterion ($i = 1, 2, \ldots, m_f$); $m_f$ is the number of sub-criteria that make up the $f$-th main criterion ($m_A = 16$; $m_B = 19$; $m_C = 6$; $m_D = 8$).

The results of the calculation are shown in Table 3.17. The table shows the global weights and global priorities for all 49 sub-criteria. For example, the subjective normalised weigh for the key criterion A is $\overline{Z}_A^* = 0.228$, the local weight for its sub-criterion A1 is $\overline{Q}_{Ai} = 0.0536$, and this key criterion is made up of $m_A = 16$ sub-criteria. The subjective normalised global weight calculated by applying formula (3.38) is:

$$\overline{\omega}_{Ai} = \frac{0.228 \times 0.0536 \times 16}{3.648 + 4.332 + 1.548 + 2.288} = 0.016548.$$

The global weight $\overline{\omega}_{Ai} = 0.016548$ of sub-criterion A1 for the key criterion A ranks 30th among the other 48 sub-criteria, i.e. it has a lower than average global priority.

The results of the calculations (Table 3.17) suggest that of the 49 sub-criteria that have the greatest impact on the quality of international train travel (in terms of global weight and priority) are sub-criteria C1 (ticket price), B1 (schedule and timing of train departures and arrivals), D4 (operational state of axle-box overheated and fire alarm system), and C5 (cost of visa), B13 (storage of passenger luggage and personal items), B15 (availability of a visa at the cross-border station), A2 (duration and speed of train travel-trip), A7 (required temperature inside a passenger car), A6 (operation of the ventilation, air-conditioning, cooling and lighting systems, with regard to their timely activation and deactivation), D7 (operational state of hand brake). These top 10 sub-criteria account for more than one-third (35.85%) of the significance of all 49 sub-criteria, expressed as a sum of global weights.

## References


1. Akgüngör AP, Demirel A (2008) Investigating urban traffic based noise pollution in the city of Kirikkale, Turkey. Transport 23(3):273–278. https://doi.org/10.3846/1648-4142.2008.23.273-278
2. Belov IV, Tereshina NP, Galaburda VG et al (2001) Ekonomika Zheleznodorozhnogo Transporta. Marshrut, Moskva. 600 s (in Russian)
3. Stoilova S (2018) Study of railway passenger transport in the European Union. Tehnički vjesnik 25(2):587–595
4. Saaty TL (1980) The Analytic Hierarchy Process. McGraw-Hill, New York, NY

5. Maskeliūnaitė L, Sivilevičius H, Podvezko V (2009) Research on the quality of passenger transportation by railway. Transport 24(2):100–112. https://doi.org/10.3846/1648-4142.2009.24.100-112
6. Maskeliūnaitė L, Sivilevičius H (2012) Expert evaluation of criteria describing the quality of travelling by international passenger train: technological, economic and safety perspectives. Technol Econ Dev Econ 18(3):544–566. https://doi.org/10.3846/20294913.2012.710178
7. Morkvėnas R, Bivainis J, Jaržemskis A (2008) Assessment of employee's knowledge potential in transport sector. Transport 23(3):258–265. https://doi.org/10.3846/1648-4142.2008.23.258-265
8. Sivilevičius H, Zavadskas EK, Turskis Z (2008) Quality attributes and complex assessment methodology of asphalt mixing plant. Baltic J Road Bridge Eng 3(3):161–166
9. Sivilevičius H (2011) Application of expert evaluation method to determine the importance of operating asphalt mixing plant quality criteria and rank correlation. Baltic J Road Bridge Eng 6(1):48–58. https://doi.org/10.3846/bjrbe.2011.07
10. Tavana M, Soltanifar M, Santos-Arteaga FJ (2021) Analytical hierarchy process: revolution and evolution Ann Oper Res 1–29 https://doi.org/10.1007/s10479-021-04432-2
11. Farhan J, Fwa TF (2009) Pavement maintenance prioritization using analytic hierarchy process. Transp Res Rec 2093(1):12–24. https://doi.org/10.3141/2093-02
12. Gudienė N, Banaitis A, Podvezko V, Banaitienė N (2014) Identification and evaluation of the critical success factors for construction projects in Lithuania: AHP approach. J Civ Eng Manag 20(3):350–359. https://doi.org/10.3846/13923730.2014.914082
13. Šakalys R, Sivilevičius H, Miliauskaitė L, Šakalys A (2019) Investigation And evaluation of main indicators impacting synchromodality using ARTIW and AHP methods. Transport 34(3):300–311. https://doi.org/10.3846/transport.2019.9718
14. Sivilevičius H, Maskeliūnaitė L (2010) The criteria for identifying the quality of passengers' transportation by railway and their ranking using AHP method. Transport 25(4):368–381. https://doi.org/10.3846/transport.2010.46
15. Saaty TL, Vargas LG, Dellmann K (2003) The allocation of intangible resources: the analytic hierarchy process and linear programming. Socio Econ Plan Sci 37(3):169–184. https://doi.org/10.1016/S0038-0121(02)0039-3
16. Byun D-H (2001) The AHP approach for selecting an automobile purchase model. Inf Manage 38(5):289–297
17. Chang S-C, Tsai P-H (2016) A hybrid financial performance evaluation model for wealth management banks following the global financial crisis. Technol Econ Dev Econ 22(1):21–46. https://doi.org/10.3846/20294913.2014.986771
18. Rabbani A, Zamani M, Yardani-Chamzini A, Zavadskas EK (2014) Proposing a new integrated model based on balanced scorecard (SBSC) and MCDM approaches by using linguistic variables for the performance evaluation of oil producing companies. Expert Syst Appl 41(16):7316–7327. https://doi.org/10.1016/j.eswa.2014.05.023

# Chapter 4
# Technical Parameters of the Railway Track and Train Elements

*The journey has its own lyrics*
*A duet of balanced motion*
*The rails and wheels in tune.*
Richard L. Ratliff

## 4.1 Technical Parameters of the Railway Track and Train Elements and Their Impact on the Quality of the Train Journey

Passenger comfort is at the heart of the manufacture and subsequent operation of rolling stock. The main objective is to ensure the quality of the passenger journey. However, comfort is a very subjective concept and hard to be directly and quantitatively measured. Travel comfort can be influenced by a variety of factors, which is why there are various measures to improve the quality of travel. The study provides an overview of the latest methods to assess and improve rail passenger comfort. The key factors that contribute to ride comfort are summarised primarily according to the sources of discomfort including static comfort, vibration comfort, noise comfort, aural pressure comfort, thermal comfort, and visual comfort. Technologies to improve each type of ride comfort are provided [1].

Urban transit agencies are working to meet the commuting needs of commuters and they aim to persuade commuters to switch from a private car for their daily journeys. This is a key factor contributing to a sustainable urban environment. Considerable attention is paid to the evaluation of travel time and cost, as well as customer satisfaction factors. Human aspects such as comfort are rarely valued. A study [2] recommends a quantitative method to assess the comfort of rail passengers by considering the average temperature in the carriage, vibration, $CO_2$ concentration, noise, and lighting.



H. Sivilevičius and L. Maskeliūnaitė, *Assessment of the Quality of Passenger Transportation by Train Using Multiple Criteria Decision Making Methods*, Lecture Notes in Intelligent Transportation and Infrastructure,
https://doi.org/10.1007/978-3-031-85877-2_4

The model for calculating the importance of the key criterion A factors (they refer to train elements and technical condition of the track) on the quality of international train travel is provided in formula (4.1):

$$K_A = \overline{Z}_A^* \cdot \sum_{i=1}^{m} \left(\overline{Q}_{Ai} \cdot x_{Ai}\right) = \overline{Z}_A^* \cdot \left(\overline{Q}_{A1} \cdot x_{A1} + \overline{Q}_{A2} \cdot x_{A2} + \cdots + \overline{Q}_{A16} \cdot x_{A16}\right), \tag{4.1}$$

where $K_A$ is a comprehensive quality index (CQI) related to the train elements and the technical state of railway track (ranging from 0 to $\overline{Z}_A^*$); $\overline{Z}_A^*$ is the mean weight, reflecting the respondents' and experts' estimates of the significance of the key criterion A (Table 3.16, $\overline{Z}_A^* = 0.228$); $\overline{Q}_{A1}, \ldots, \overline{Q}_{A16}$ denote the local weight values of $i$-th sub-criterion of A key criterion (1, …, $m$) determined by the expert evaluation method (Table 3.7, Fig. 3.7); $x_{A1}, \ldots, x_{A16}$ are the variables of $i$-th sub-criteria of the key criterion A (1, …, $m$), their estimates are used for determining the real sub-criterion value, ranging from 0 to 1.

The mean weight $\overline{Z}_A^*$ shows the significance of the key criterion A (when the number of the respondents and experts in each category differs). This weight is calculated by the Eq. (3.33).

To calculate the CQI, showing the significance of the train elements and technical state of rails (railway track), the variables of any sub-criterion $x_{A1}, \ldots, x_{A16}$, serving as a basis for calculating the real sub-criterion value (which ranges from 0 to 1), should be determined.

### *4.1.1 Quality of the Railway Track*

Infrastructure shapes mobility. No major change in transport will be possible without the support of an adequate network and more intelligence in using it. Overall, transport infrastructure investments have a positive impact on economic growth, create wealth and jobs, and enhance trade, geographical accessibility, and the mobility of people [3].

In the process of train operation, the interaction between train wheels and railway track underneath leads to changes in railway track geometry. The changes in track geometry result in track irregularity [4]. Ensuring the safety of railway traffic and the comfort of passengers is a significant management task for railway transportation enterprises, and they are greatly affected by track irregularities (or track geometry defects) [5]. A cost model is proposed to specify the cost-effective maintenance limits for track geometry maintenance [6]. Andrade and Texeira [7] explores a quantitative model with different maintenance strategies to control railway track geometry degradation in Portugal. The maintenance cost for railway infrastructure, in the case of traditional ballasted track, represent considerable amounts, around €25,000–30,000 per kilometre per year [8].

The quality of track geometry is assessed by the variation of seven track geometric parameters: track gauge, longitudinal level (or surface) (left/right), cross level (or cant), alignment (or horizontal alignment) (left/right) and twist [9].

Traditionally, railway track quality is quantified by a specialized car that measures the basic geometric parameters of the track, short wavelength irregularities, and absence of complete track settlement. Usually, the inspection measures and records the variation of the track gauge, vertical and lateral alignments, and cross-level (angular variation on a track section—cant or super elevation). The values measured are compared to standardized limits [10]. For managing the mean values of track irregularity, track quality index (TQI) is used to quantify track irregularity of 200 m long unit track sections.

When a train runs on uneven, rough tracks, vertical and horizontal oscillations occur in the passenger carriages, and this undermines travel comfort and safety. The higher the train speed, the greater the negative impact on comfort and safety caused by variations in track geometry. Therefore, countries tend to restrict the maximum allowable TQI values with reference to the maximum speed of trains: the higher the speed, the lower the allowable TQI values (Table 4.1).

Based on the allowable speed, the railway track of China has been classified into categories, each of which has its owner TQI management threshold, as listed in Table 4.1.

The TQI is a quantitative index of track irregularities, and it is a continuous variable in railway system. TQI is used for planning and scheduling railway track maintenance and guiding track maintenance work. The TQI management value (standard deviations $\sigma_g$) is based on the Chinese Standard 'Rules of railway track maintenance, as shown in Table 4.2. It shows that not only the TQI values are restricted, but also the standard deviations of each of the seven geometric parameters $\sigma_{g,\,st}$.

It is necessary to standardize the way that track geometry quality is assessed to permit safe and cost-effective railway traffic by focusing on the functional requirements of both track and vehicle.

**Table 4.1** TQI management threshold (or limits) [4]

| Allowable speed range $v_{max}$ (km/h) | $\leq 160$ | 160 200 | 200 250 | 300 350 |
|---|---|---|---|---|
| TQI management threshold (mm) | 15 | 10 | 8 | 5 |

**Table 4.2** TQI management values (unit: mm) [5]

| Speed $v_{max}$ | Indexes | | | | | | | |
|---|---|---|---|---|---|---|---|---|
| | Track gauge | Longitudinal level | | Alignment | | Cross-level | Twist | Total (TQI) |
| | | Left rail | Right rail | Left rail | Right rail | | | |
| $\leq$ 160 km h | 1.6 | 2.5 | 2.5 | 2.2 | 2.2 | 1.9 | 2.1 | 15.0 |
| > 160 km h | 1.1 | 1.5 | 1.5 | 1.6 | 1.6 | 1.3 | 1.4 | 10.0 |

Table 4.3 defines the limit values of longitudinal level and alignment for each track quality class (TQC) and for each speed range. The table is based on the track characteristics of the European Railway Networks participating in the European Track Quality Survey.

TQI allows to observe the changes in geometric track parameters and is used to describe track quality. The total TQI value is calculated for each kilometre of the railway track by summing up the TQI values of the geometric railway track parameter. The total TQI, describing the quality of the considered railway track in terms varying from "very good" to "poor", may range from 0 to 45 and more points (Table 4.4) [12]. When the estimate value of the railway track quality decreases, the train speed is also decreased. The account of TQI is provided by the track recording vehicle (TRV) program EMGraph. It consists of the considered railway track, as well as the considered parameters and their values.

Track recording vehicle (TRV) is self-propelled or hauled vehicle with fixed, dedicated, measuring equipment and systems, used for the measurement, assessment and recording of track geometry parameters under loaded conditions. It measures and produces consistent results with reference to the requirements of EN 13848-1-2019 [9].

The condition quality of the railway track may be described by seven geometric parameters, including track gauge (1), cross level (2), left and right longitudinal level (3, 4), left and right alignment (5, 6) and twist (7). All these parameters by TRV are measured in *millimetres*. Their values for 1 km and the total track length vary, i.e. are characterized by variation (spread). The higher the variation, the larger the deviation of each geometric parameter values from the nominal (specified) values and the closer they approach critical values. This increases the possibility of an accident, e.g. the

**Table 4.3** Track longitudinal level and alignment threshold (or limits) values of standard deviations [11]

| Track geometric parameter | Speed (in km/h) | Limit value of standard deviation (in mm) | | | | |
|---|---|---|---|---|---|---|
| | | Track quality class | | | | |
| | | A | B | C | D | E |
| Longitudinal level left and right rail | $80 < v \leq 120$ | < 0.75 | 1.10 | 1.80 | 2.50 | > 2.50 |
| | $120 < v \leq 160$ | < 0.65 | 0.85 | 1.40 | 1.85 | > 1.85 |
| Alignment left and right rail | $80 < v \leq 120$ | < 0.50 | 0.70 | 1.05 | 1.45 | > 1.45 |
| | $120 < v \leq 160$ | < 0.45 | 0.55 | 0.75 | 1.00 | > 1.00 |

**Table 4.4** Qualitative evaluation of track based on $TQI_{st}$ values

| Total $TQI_{st}$ ($mm^2$) | Track quality |
|---|---|
| 0–20 | Very good |
| 21–34 | Good |
| 35–44 | Satisfactory |
| ≥ 45 | Poor |

derailing of the train (or its cars). High variation of track geometric parameters gives rise to faster wearing of rails and wheelsets and increases railway car vibration, noise, and acceleration, thereby decreasing the level of comfort for passengers and the quality of travel by train [13, 14].

Ride comfort depends on human tolerance to vibration exposure over time. Presently, ISO 2631-5:2018 [15], EN 12299:2009 [16] and Sperling's method are the most common evaluation methods for assessing passenger comfort on trains [17].

The riding quality is a key index that affects the riding comfort for passengers. Based on the standards ISO 2631-5:2018, EN 12299, GB/T 5599-1985 [18], the Sperling Index $W_s$ is used to evaluate the riding quality, it can be written as

$$W_s = 0.896 \sqrt[10]{\frac{a^3}{f_v} \times F(f_v)}, \tag{4.2}$$

where $a$ is the vertical vibration (the amplitude) acceleration of the car body, cm/s$^2$; $f_v$ is the vertical vibration (excitation) frequency of the car body, Hz; $F(f_v)$ is the modifying coefficient (frequency weighting function) of the vibration frequency that expresses human vibration sensitivity.

Studies of Sperling's comfort index values show [19] that there are no huge differences in the significance of Sperling's comfort index for the dynamics in the vertical direction when comparing a passenger car with integral wheelsets and wheelsets with independently rotating wheels both when the vehicle runs in curves and in a straight section.

The variation of each geometric parameter of the railway track TRV each 1 km-long track section is measured by a mobile laboratory, i.e. on TRV. The sum $TQI_{f,200}$ of five 200 m sections of a 1 km track generates the value $TQI_{fs}$. The total of the variations of all 7 parameters shows the technical condition of the track expressed by TQI. TQI is the value that characterises track geometry quality of track section based on parameters and measuring methods compliant with EN [9] series. It is necessary to standardize the way that track geometry quality is assessed to permit safe and cost-effective railway traffic with the focus on both track and vehicle.

The smaller the real $TQI_f$ value, the higher the track guality. In the Instruction K/259 self-propelled TRV (or track geometry car) EM-140 [12], issued by order of the director of JSC "LTG Link", four levels of track quality (ranging from very good to poor), which correspond to various $TQI_{st}$ values, have been specified (Table 4.4).

The intervals of track quality levels based on non-dimensional TQI$_{st}$ values, as specified by the Instruction K/259, [12] differ to various extent. Thus, for "very good" (quality), the difference is 20 (20–0), for "good", the difference is 13 (34–21), for "satisfactory", the difference is 9 (44–35) and for "poor", the difference is not clear because $TQI_{st} \geq 45$. There is no indication in Table 4.4 as to how high the TQI value can be in a poor track condition. The experimental investigation performed by the authors of the present work has shown that the highest real (actual) $TQI_{fmax}$ value for 1 km track is 92.4, while the lowest $TQI_{fmin}$ value is equal to 7.49 [20].

The Instruction K/259 of Track recording vehicle EM-140 [12] erroneously states that TQI is a non-dimensional value, indicating track gauge deviation. Since all geometric parameters of the track are measured in millimetres (mm), their total, making the TQI value, also has a dimension. [4], presenting the modelling of the track quality index, states that the method used in China for defining TQI is based on the total of seven standard deviations (TQI = $\sum_{g=1}^{G} \sigma_g$, where $G = 7$ ($g = 1, 2, \ldots, G$) denotes the number of the geometric parameter), given in mm. In this work, the variation of any geometric parameter (including track gauge, cross level, left and right longitudinal level, left and right alignment, and twist) is expressed by the standard deviation $\sigma_g$. The authors think, however, that to sum up standard deviations according to mathematical rules is not correct. In this case, it is more appropriate to use variances $\sigma_g^2$ [20].

Also, without a graphical standard deviation model with a probability density curve for a normal or lognormal distribution, it is almost impossible to understand whether $\sigma_g$ is the spread measure of a geometric track parameter used in the work [4]. We think that the TQI is the sum of variances $\sigma_g^2$ of individual track geometric parameters, expressed in mm$^2$.

In the absence of the highest fixed $TQI_{max}$ value, or $TQI_{limit}$ value (which means that trains are not allowed to run even at the lowest speed), it is hardly possible to calculate the variable $x'_{A1}$ of the first sub-criterion of the key criterion A based on the relationship between the actual values of $TQI_f$ and maximal $TQI_{max}$ or $TQI_{limit}$ subtracted from one.

Given that the track with $x'_{A1} = 1$ for each 1 km track length is the best, while the track with $x'_{A1} = 0$ is the worst, and the specified quantitative variation of its quality is uniform (making 0.25), the third-degree regression equation (4.3), defining non-linear dependence of $x'_{A1}$ on $TQI_{st}$, has been solved and the curve has been plotted (see Fig. 4.1).

$$x'_{A1} = -3 \cdot 10^{-6} \cdot TQI_{st}^3 - 2 \cdot 10^{-5} \cdot TQI_{st}^2 - 0.0106 \cdot TQI_{st} + 0.9999. \quad (4.3)$$

A relationship close to functional dependence has been found between the standard index of track quality $TQI_{st}$ determined by the TRV and the variable variable $x'_{A1}$ of the first sub-criterion of the key criterion A, whose coefficient of determination $R^2 = 0.9984$.

The value of $TQI_{st}$, corresponding to $x'_{A1} = 0$, which was obtained by using the Eq. (4.3), reached about 51–52 mm$^2$. For the real track of very poor quality, this value may be several times larger. Given that $x'_{A1}$ values cannot be negative, and the real $TQI_f$, determined by a TRV may exceed 52 mm$^2$, four additional points, extending the area of poor track quality TQI from 45 to 90, have been introduced (Fig. 4.1).

The increase of TQI values from 60 mm$^2$ to 90 mm$^2$ by the 10 mm$^2$ step caused the decrease of $x'_{A1}$ values (solid points) so that the curve of the function $x'_{A1} = f$ ($TQI$) could run smoothly from $TQI_{st} = 0$, when $x'_{A1} = 1$, through all standard $TQI_{st}$ values, and asymptotically approach $x'_{A1} = 0$. The standardized $TQI_{st}$ values and the

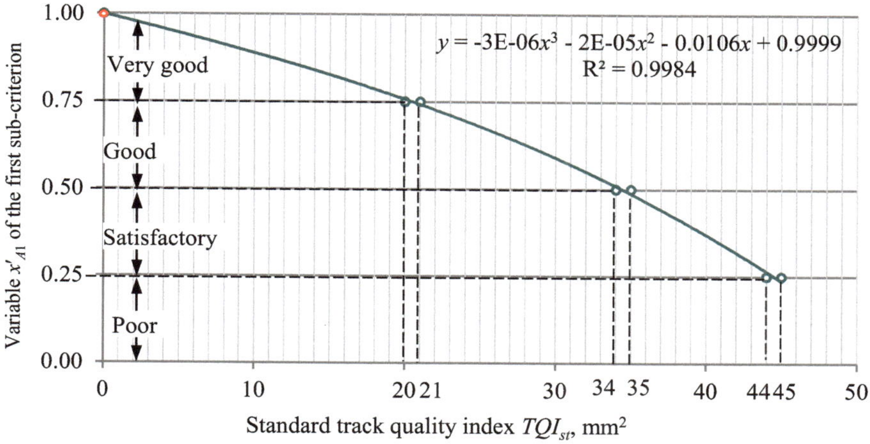


**Fig. 4.1** The relationship between the variable $x'_{A1}$ of the key criterion A describing the dependence of the railway trip quality on the condition of the railway track and $TQI_{st}$

modelled $TQI_{mod}$ values (solid points) were described by various regression equations, such as linear ($y_1$), quadratic ($y_2$), cubic ($y_3$) and fourth degree ($y_4$) regression equations (Fig. 4.2).

The coefficients of determination of the regression equations $R^2$ (Fig. 4.2) increase as the equation power increases. The coefficient of the fourth-degree regression

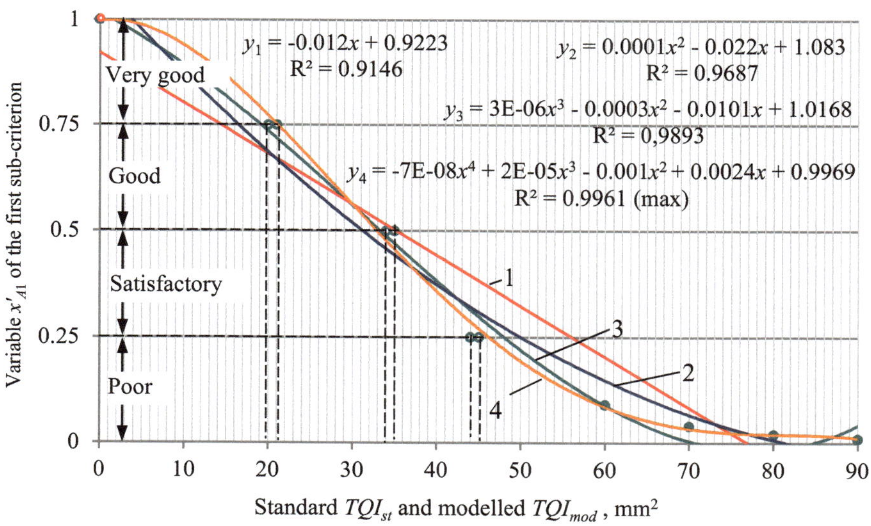


**Fig. 4.2** Various regression equations describing the values of standard track quality index $TQI_{st}$ and modelled $TQI_{mod}$

equation is $R^2 = 0.9961$, while its regression coefficients of higher accuracy for $TQI_{fs}$ allow us to obtain the final expression, showing how the real quality index $TQI_{fs}$ of the $s$-th track section ($s = 1, 2, \ldots, N_{km}$) is related to the variable $x'_{A1}$, showing the quality of this track:

$$x'_{A1} = 0.997 + 0.00245 \cdot TQI_{fs} - 0.000962 \cdot TQI_{fs}^2 + 1.532 \cdot 10^{-5} \cdot TQI_{fs}^3 - 6.99 \cdot 10^{-8} \cdot TQI_{fs}^4. \tag{4.4}$$

The curve (Fig. 4.3) obtained based on the regression equation (4.4) does not completely comply with some standard $TQI_{st}$ values. The smallest $x'_{A1}$ value, which is $x'_{A1} = 0.5$, corresponds to $TQI_{fs} \approx 33$ mm$^2$ (see $TQI_{st} = 34$ mm$^2$ in Table 4.4 and Fig. 4.3), while the largest poor quality track value $x'_{A1} = 0.25$ corresponds to $TQI_{fs} \approx 46$ mm$^2$ ($TQI_{st} = 45$ mm$^2$ in Table 4.4 and Fig. 4.3). Some limiting standard values of various track quality levels, which are used now, could be slightly corrected because they are conditional and do not reflect the essential parameters determining comfort of travel by train for passengers as well as limiting acceleration values, accident risk, etc.

The railway track is the total of all $s$-th kilometre lengths with various $TQI_{fs}$ values determined by a TRV. Therefore, the quality of the whole track may be determined in terms of the compliance of the average $\overline{TQI_f}$ for its 1 km lengths with the variable $x_{A1}$ of the first sub-criterion of key criterion A obtained from the Eq. (4.4) or the curve (Fig. 4.3). In the regression equation, the average value of the length of one kilometre of track is used instead of the $TQI_{fs}$:

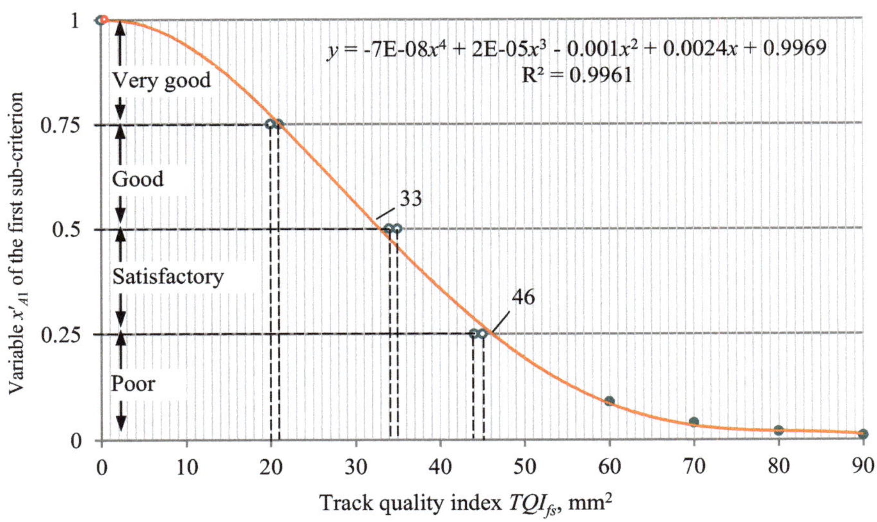


**Fig. 4.3** The regression equation and the curve used for determining $x'_{A1}$ value for the first sub-criterion of key criterion A, when the real $TQI_{fs}$ is given

$$\overline{TQI}_f = \frac{\sum_{g=1}^{G} \sigma_{gfs}^2}{N_{km}} = \frac{\sum_{s=1}^{N_{km}} TQI_{fs}}{N_{km}}, \tag{4.5}$$

where $\overline{TQI}_f$ is the average track quality index of evaluated railway track the length of which is $N_{km}$; the numerical value is equal to the total variance $\sigma_{tot}^2 = \sum_{g=1}^{G} \sigma_{gfs}^2$, mm$^2$; $TQI_{fs}$ is quality index of $s$-th one-kilometre track length, the numerical value of which is equal to the total variance $\sigma_{gfs}^2$ of the particular $g$-th geometric parameters, mm$^2$ ($g = 1, 2, \ldots, G$) and $N_{km}$ is the track length, km ($s = 1, 2, \ldots, N_{km}$).

Thus, track quality is shown by the compliance of the variable $x_{A1}$ of the first sub-criterion of the key criterion A, which was calculated from formula (4.6) having the average $\overline{TQI}_f$, with one of the standard quality levels (very good, good, satisfactory, poor) given in Table 4.4.

$$x_{A1} = 0.997 + 0.00245 \cdot \frac{\sum_{s=1}^{N_{km}} TQI_{fs}}{N_{km}} - 0.000962 \cdot \left(\frac{\sum_{s=1}^{N_{km}} TQI_{fs}}{N_{km}}\right)^2 + 1.532 \cdot 10^{-5} \cdot \left(\frac{\sum_{s=1}^{N_{km}} TQI_{fs}}{N_{km}}\right)^3 - 6.99 \cdot 10^{-8} \cdot \left(\frac{\sum_{s=1}^{N_{km}} TQI_{\text{fs}}}{N_{km}}\right)^4. \tag{4.6}$$

The total track variance $\sigma_{total}^2$ for all seven geometric parameters (their variances denoted as $\sigma_g^2$, $g = 1, 2, \ldots, G$), measured by a TRV, is used as TQI, from which $x_{A1}$ is calculated. The value of $x_{A1}$ obtained from the regression equation (4.6) is used in the model (4.1) to evaluate the quality of travel by train, according to railway track quality.

### 4.1.2 Trip Duration of the Train

For night train passengers, it is the time they reach their destination in the morning that counts, not how long the journey takes. Infrastructure access charges account for a large share of total costs. Lower charges for night trains would significantly improve the competitive position of night trains and would contribute to climate protection objectives [21].

To ensure safe railway traffic, and considering such parametres as the surface railway structures, embankments and their condition, as well as various types of trains and their operation according to the timetable, and also in compliance with the technical rules of railway operation, the maximum permissible speed of trains and locomotives on the railway tracks near the stations and on the railway tracks between the stations has been determined [22, 23]. Sub-criterion A2 is the most important of all the 16 sub-criteria of the key criterion A: Its priority (rank) $\overline{R}_{A2} = 1$, and local weight $\overline{Q}_{A2} = 0.1121$ (Table 3.7). The variable $x_{A2}$ of the CQI $K_A$ which determines the influence of the train speed (time of travel) on the quality travel by train, is calculated by the formula:

$$x_{A2} = \frac{\sum_{l=1}^{n_{stage}} \left(1 - \frac{|v_{f,l} - v_{max,a}|}{v_{max,a}}\right) \cdot l_l}{\sum_{l=1}^{n_{stage}} l_l}, \tag{4.7}$$

where $x_{A2}$ is the variable of the 2nd sub-criterion of the key criterion A; $n_{stage}$ is the number of railway track sections between stations ($l = 1, 2, \ldots, n_{stage}$); $v_{f,l}$ is the real (measured) and the maximum permissible speed difference of a train at the $l$-th section of the track, km/h; $v_{max,a}$ is the maximum permissible speed of a train at the $l$-th section of the track, km/h; $l_l$ is the denotes the lengths of the sections of the track, km.

### *4.1.3 The Condition of the Passenger Carriage Exterior*

Trains must be parked on platforms for passenger boarding in a washed condition. The bodies of carriages get dirty during operation (with dust, smoke from coal-burning carriages in winter, freight trains passing on the road). This results in degraded aesthetics, accelerated carriage body surface wear and corrosion. Dirty carriage windows do not make passengers happy. Vandalism of rolling stock is not uncommon, with graffiti painted on the body of carriages and windows of carriages smashed on the outside. There are also cases of breaking carriage windows from the inside. This is a costly global rail transport problem.

Vandalism and graffiti on the railway have a major impact on railway authorities, the development of railway services, costs, and the timeliness of services. There are also important social costs which stem from passengers feeling unsafe, not to mention the environmental costs of removing graffiti and repairing vandalism [24]. The trend, known as "graffiti tourism", is known abroad where gangs from different countries link up and travel to each other's home city. Irish Rail has spent more than €2 million on the cleaning of rail carriages in the last two years [25].

Racist, homophobic, anti-Semitic, rude, or offensive graffiti is a big problem. They must be removed immediately if they are noticed or reported. There have been graffiti on political issues on the body of a train carriage. The collapse of the Soviet Union led to the specific situation of Kaliningrad region. After the restoration of the independent Baltic States, Kaliningrad was separated from the rest of Russia and turned into an exclave, and after Poland and Lithuania joined the European Union, Kaliningrad became an enclave of the EU [26]. In Russia, attempts have been made to escalate this issue with notes written on the body of carriages of the Lithuanian international train.

Maintenance and repair of rail carriages is carried out at maintenance depots and wagon depots. The technical maintenance of rail carriages involves not only checks on the condition of units and components, their depreciation and their dimensions, the technical condition of chassis (bogies), the functioning of brakes and automatic couplings, the condition of braking platforms, special steps, and railings, but also on the technical condition of the rail carriage bodies. Repairs include repairing bodywork damage and painting. The exterior of the trains is washed in rolling stock washing facilities. To protect against vandalism, railway companies in some countries cover their rolling stock with anti-graffiti film, which not only protects against graffiti (a layer of film is peeled off and a new layer reapplied), but also in the event of window breakage. In the presence of the film, the glass cracks on impact, but does not shatter or inflict injuries on passengers [27]. Anti-graffiti solutions can also be used to clean graffiti.

The condition of the passenger carriage exterior (cleanness, deformations, damages, etc.) is evaluated in points as follows: 0 means that the carriage is not clean and has some deformations or damages; 1 shows that the carriage is not clean, but no deformations or damages can be observed, or it is clean, but there are some damages; 2 indicates that the carriage is clean and has no damages. The variable $x_{A3}$ of the CQI $K_A$, which, determines the influence of the carriage cleanness and damages on the quality travel by train, is calculated by the formula:

$$x_{A3} = \frac{\sum_{c=1}^{n_{car}} \left( \frac{CE_f}{CE_{max}} \right)_c}{n_{car}}, \tag{4.8}$$

where $x_{A3}$ is the variable of the 3rd sub-criterion of the key criterion A; $CE_f$ is the real condition (cleanness, damages) of the $c$-th carriage ($c = 1, 2, \ldots, n_{car}$) expressed in points (0, 1 or 2 points); $CE_{max}$ is the highest estimate value of the carriage in points (2 points); $n_{car}$ is the number of carriages in the train.

### 4.1.4 Noise Reduction Insulation

Rolling stock intended for the carriage of passengers must be designed, constructed, and operated in such a way that people travelling by train do not suffer discomfort from excessive noise levels. For passenger train rolling stock, the equivalent sound pressure level in the carriage compartment shall not exceed $NL_{max,a} = 70$ dBA. For passenger rolling stock equipped with sleeping berths, the equivalent sound pressure level in the compartment shall not exceed $NL_{max,a} = 65$ dBA [28]. These threshold values indicate the worst acceptable level of travel quality (variable $x_{A2} = 0$). The ideal $x_{A2} = 1$ can be taken as the factual zero equivalent sound pressure level $NL_f = 0$ dBA, which is not achievable in practice.

Reality is better represented by the principle that the ideal sound pressure level is that of a quiet rural area, i.e. $NL_{min} = 30$ dBA. To improve the quality of the train journey, an even lower sound pressure level of 20 dBA (whispering, rustling leaves) or 10 dBA (breathing) can be used as a possible minimum sound pressure level $NL_{min}$. The average ($x_{A4} = 0.5$) can therefore be taken as either $NL_f = \frac{70+30}{2} = 50$ dBA

**Fig. 4.4** The dependence (correlation) of variable $x_{A4}$ of sound pressure level in the rail carriage compartment and quality of passenger transport

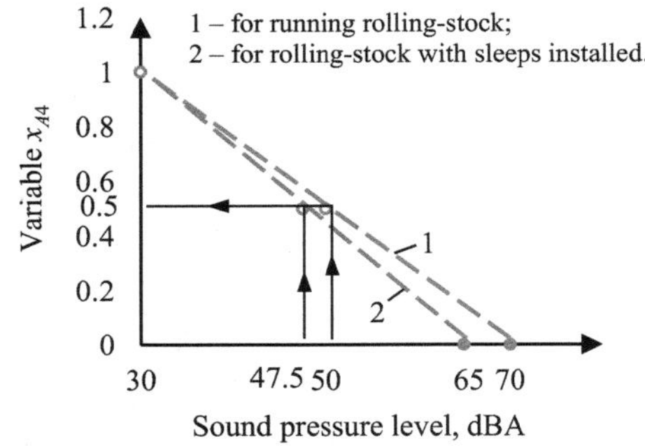


or $NL_f = \frac{65+30}{2} = 47.5$ dBA factual equivalent sound pressure level in the carriage compartment when passenger train rolling stock is in motion (Fig. 4.4).

Snoring, children crying or running around the carriage is a common problem, especially in the sleeping carriages. Around 45% of people snore occasionally and 25% snore regularly. In the US, around 37 million people snore regularly. Moderate snoring is between 50 and 65 dB(A). However, it can reach around 90 dB(A). In the UK, 111.6 dB(A) snoring has been registered [29]. Trains are considered the best mode of transport for travelling with children, as there is more space to play and move around, and there are working toilets. Having a restaurant car on board makes it easy to feed your child. Children aged between 9 months to 6 years old can have a crying level between 99 and 120 dB(A) [30]. It must be admitted that snoring train passengers and those travelling with small children negatively affect the quality of the journey for other passengers. However, just because a person snores or travels with a child, he or she cannot be banned from travelling. Rail companies are the ones who are supposed to find solutions to these problems. On the French International (Intercities) train (Paris-Limoges-Tolouse line) children can play in the "Espace Kid" area (Travelling by train … 2023). It has everything a child needs: pillows, a play rug, special seats for children aged 1–4 to keep them safe. There are also baby changing tables, sockets for feeding bottle warmers and areas to store prams, pushchairs, or strollers. Train staff could, in an exceptional case, provide a snoring passenger with a free compartment (if available).

The factual sound pressure level $NL_f$, measured in the carriage compartment has been compared to the maximum permissible $NL_{max}$. The variable $x_{A4}$ calculated from formula (4.9) indicates the comparative effectiveness of noise insulation measures on board the train which can be used to improve the quality of train travel.

$$x_{A4} = \frac{\sum_{c=1}^{n_{car}} \left( \frac{NL_{max,a} - NL_{fc}}{NL_{max,a} - NL_{min}} \right)_c}{n_{car}}, \tag{4.9}$$

where $x_{A4}$ is the variable of the 4th sub-criterion of the key criterion A; $NL_{f,c}$ is the real noise level in the $c$-th carriage, dBA; $NL_{max,a}$ is the maximum permissible noise level in the $c$-th carriage, dBA (65 or 70 dBA), $n_{car}$ is the number of carriages in the train ($c = 1, 2, \ldots, n_{car}$); $NL_{min}$ is the minimal sound pressure level. The target point of the noise level in the $c$-th carriage should be 30 dBA.

### 4.1.5 The Interior of a Passenger Carriage

Today passengers expect from train travels more than just being transported from point A to point B. They want entertainment and aesthetics, comprehensive information, and safety. Interior design plays a significant role in the design of new rail carriages or the refurbishment of the old ones. Around the world, railways are increasingly working on how to further improve the aesthetics and functionality of rail interiors [31]. The interior design of passenger trains is an important aspect of improving the quality of service in rail transport. Nowadays, interior design is still focused on improving function and structure. There is room for improvement considering the psychological needs and visual experience of passengers. The interior design of passenger trains is a significant issue of improving the quality of service in rail transport [32].

This is important because it contributes to creating a favourable impression about the company, which provides passenger transportation services. Moreover, a pleasant interior design ensures that passengers and train staff feel comfortable during the journey. Passenger carriage interior is evaluated in points as follows: 0 is given to an old (not modernized) carriage, 1 is assigned to an old (modernized) carriage and 2 is given to a new carriage. The variable $x_{A5}$ of the CQI $K_A$ which assesses the carriage interior and influences the quality travel by train, is calculated by the formula:

$$x_{A5} = \frac{\sum_{c=1}^{n_{car}} \left( \frac{CI_{f,c}}{CI_{max,p}} \right)_c}{n_{car}}, \tag{4.10}$$

where $x_{A5}$ is the variable of the 5th sub-criterion of the key criterion A; $CI_{f,c}$ is the real estimate of the $c$-th carriage interior in points (0, 1 or 2 points); $CI_{max,p}$ is the maximum possible carriage interior estimate value in points (2 points); $n_{car}$ is the number of carriages in the train ($c = 1, 2, \ldots, n_{car}$).

### 4.1.6 Operation of Ventilation, Air Conditioning, Cooling and Lighting Systems in Terms of Their Timely On/Off Switching

To ensure passenger comfort, it is essential that the heating, ventilation, air-conditioning, and lighting systems in the train carriage work perfectly. The efficiency of heating, ventilation, and air conditioning (HVAC) systems has become particularly important with the onset of the COVID-19 pandemic. Recommendations have been made for train operators to ensure proper maintenance and operation of HVAC systems on trains to reduce the risk of virus transmission [33].

A series of experiments were carried out in an intercity train carriage with the aim of obtaining a "Proof of Concept" by applying three methods to improve the understanding of airflow and the dispersion of exhaled droplets. Each experiment was successful in providing practical insights into the risk of airborne contamination. The air in the carriage did not appear to be well mixed along the length of the carriage but it was likely to be well mixed along the height and width of the carriage. There is also a discussion of the appropriateness of fresh air supply rates for UK train carriages based on the $CO_2$ concentrations measured in these experiments [34].

The internal and external environments of a high-speed train are connected by a ventilation system, the structure of which is complex and large. Therefore, a one-dimensional assumption can be applied. In this work, a high-speed train carriage is chosen for the study. The calculated results were compared with the design. One-dimensional modelling was found to be suitable for the investigation of the reliability of the ventilation system in HST carriages [35].

Modern railway air-conditioning systems use fluorinated refrigerants. Since 2016, they have had to be phased out due to their negative impact on the climate. Air-assisted air-conditioning systems, known as air cycle systems, use compressed outdoor air for cooling and do not require any fluorinated refrigerants. These air-conditioning systems, which have been in operation on trains for some time, have been proven to be environmentally friendly and cost-effective [36].

Safety ensurance of employees and passengers is a top priority for every railway undertaking. Not only the operation of the heating, ventilation, and air-conditioning systems, but also the lighting of the carriage is important for passenger comfort. On a Eurostar train from France to London, the effect of bright lights on passengers was monitored. The evaluation results were unfavourable compared to the one with softer light on previous generation trains on this route.

Many airlines also make clever use of lighting to improve the perception of space, calm passengers and create the right mood and, in the COVID-19 timeframe, cleanliness [37]. There are numerous measures to help ensure good lighting, including modern LEDs. The new interior accent lighting on the train provides greater comfort and enjoyment for passengers. The parameter-controlled indoor lighting adapts to natural lighting conditions and can thus have a positive effect on the human body. The lights are automatically regulated, and lighting automatically adjusts according to season and time of day [38].

The operation of ventilation, air conditioning, cooling, heating, and lighting systems is evaluated in points as follows: 0 means that none of these systems operate in the carriage, 1 shows that one of these systems operates in the carriage, 2 indicates that two systems operate in the carriage, 3 implies that three systems operate in the carriage and 4 means that all the above-mentioned systems operate in the carriage. This sub-criterion A6 is the third among the most relevant sub-criteria of key criterion A. Its priority $\overline{R}_{A6} = 3$, and local weight $\overline{Q}_{A6} = 0.1050$ (Table 3.7). The variable $x_{A6}$ of the CQI $K_A$ which, determines the influence of carriage ventilation, air conditioning, heating, and lighting systems on the quality travel by train, is calculated by the formula:

$$x_{A6} = \frac{\sum_{c=1}^{n_{car}} \left( \frac{VCL_{f,c}}{VCL_{max,p}} \right)_c}{n_{car}}, \tag{4.11}$$

where $x_{A6}$ is the variable of the 6th sub-criterion of the key criterion A; $VCL_f$ is the estimate (in points) of the operation of the $c$-th carriage systems (0, 1, 2, 3 or 4 points); $VCL_{max}$ is the highest possible estimate of the systems operating in the carriage (4 points); $n_{car}$ is the number of carriages in the train ($c = 1, 2, \ldots, n_{car}$).

### *4.1.7 Temperature Control System for Passenger Carriage Interior*

The factual temperature of the carriage compartment is influenced by the efficiency of the carriage heating system (the amount of heat generated), the quality of the carriage thermal insulation and the temperature of the ambient air on the outside of the train. The carriage compartment temperature is regulated automatically or manually by the train staff.

The optimum temperature $T_0$ of the carriage compartment (interior), or the tolerances for deviations thereof, may be regulated by nationally adopted standards, either by dividing the period of the year into a winter (cold) season and a summer (warm) season, or by having it uniform throughout the year. In the first case, the temperature in $T_{OW}$ is lower in the cold season than in the warm season $T_{OS}$ (Fig. 4.5a). In the second case, where the year is not divided into cold and warm periods, the optimum temperature of the carriage compartment $T_0$ is equal to the average of the minimum $T_L$ and maximum $T_U$ allowable temperatures (Fig. 4.5b). By taking the average temperature measured in the compartments of all carriages of the train $\overline{T}_{f,c}$ and comparing it with the standard temperature $T_L$ and $T_U$, the variable $x_{A7}$ is calculated to show the effect of the internal temperature on the quality of train travel.

The system for maintaining the internal temperature of the carriage compartment is the second most important sub-criterion of the core criterion A. Its priority (rank) $\overline{R}_{A7} = 2$, and local weight $\overline{Q}_{A7} = 0.1061$ (Table 3.7).

#### 4.1.7.1 Temperature Assessment with Distinction Between Cold and Warm Periods

The water heating system in winter should provide comfortable conditions for passengers and service staff on the train (and on a particular carriage). The air temperature in a passenger carriage should be at least + 18 °C or at most 22 °C, even if the outdoor temperature reaches – 40 °C and the train speed $v_{max}$ < 160 km/h (Fig. 4.5a). The carriages are heated by burning solid fuel. When the water in the boiler is heated

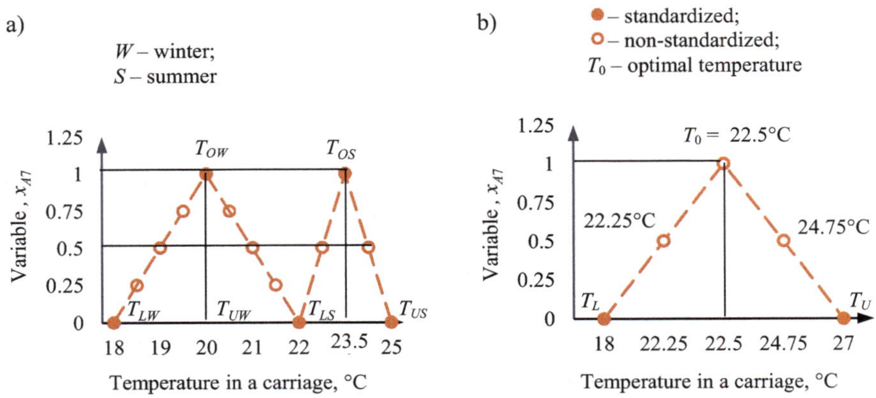


**Fig. 4.5** The relationship between the standardized temperature (lower $T_L$ and upper $T_U$) in a carriage and variable of quality level $x_{A7}$: **a** in winter (when the outdoor temperature is not lower than—40 °C, train speed $v_{max}$ < 160 km/h) and in summer (when air conditioning is used); **b** when the year is not divided into cold and warm periods [when air cooling (conditioning) is used]

by electric heaters (from the high voltage overhead contact system of 3000 V), the temperature in the passenger carriage is automatically maintained in the range of (20 ± 2) °C.

In summer, the temperature is maintained in the range of (22 to 25) °C due to the use of air-conditioning systems. If in winter the average real temperature $\overline{T}_f$ in the carriage is $T_{OW}$ + 20 °C (Fig. 4.5 a), this is considered to correspond to the highest quality level ($x_{A7} = 1$). When the real measured average temperature $\overline{T}_f$ in winter is equal + 19 °C or + 21 °C, then, $x_{A7} = 0.5$. When the real temperature $\overline{T}_f$ is equal to the lowest permissible value of the winter temperature $T_{LW}$ = + 18 °C or to the highest permissible value $T_{UW}$ = + 22 °C, then, $x_{A7} = 0$. If, in summer, the average temperature $\overline{T}_f$ matches the optimal temperature $T_{OS}$ in this season which equals to + 23.5 °C, then, $x_{A7} = 1$. When the real temperature $\overline{T}_f$ corresponds to the lowest permissible value in summer, $T_{LS}$ = 22 °C, or to the highest permissible temperature value $T_{US}$ = + 25 °C, then, there is the lowest quality level ($x_{A7} = 0$) in the carriage. When the air temperature in a passenger carriage $\overline{T}_f$ is equal to + 22.75 °C or + 24.25 °C, then, $x_{A7} = 0.5$. When $\overline{T}_f > T_{UW}$ and $\overline{T}_f < T_{LW}$, then $x_{A7}$ is considered equal to 0. The standards do not regulate the temperature in the carriage higher than + 26 °C and lower than + 22 °C in summer and higher than + 22 °C in winter [39].

The variable $x_{A7}$ of the CQI $K_A$ which assesses the temperature inside the carriage and influences the quality travel by train in winter, is calculated by the formula:

$$x_{A7} = 1 - \frac{\sum_{c=1}^{n_{car}} \left| \frac{2(T_{OW} - \overline{T}_f)}{T_{UW} - T_{LW}} \right|_c}{n_{car}}, \tag{4.12}$$

where $T_L$ is the lowest permissible carriage temperature in °C. In winter, when solid fuel is burned and the outside temperature is not lower than – 40 °C and not higher than + 10 °C, $T_{LW}$ = 18 °C. When water is heated by electricity, $T_{LW}$ = 20 °C (the temperature of (22 ± 2) °C is automatically maintained). In summer, when air-conditioning is used, the temperature of (24 ± 2) °C is automatically maintained, therefore, $T_{LS}$ = 22 °C [39], $T_{US}$ is the highest permissible temperature inside a passenger carriage (+ 26 °C). In winter, when the outside temperature is not lower than − 40 °C and solid fuel is used for heating, $T_{UW}$ = + 22 °C. When water is heated by electricity, $T_{UW}$ = 24 °C (the temperature of (22 ± 2) °C is automatically maintained); $n_{car}$ is the number of passenger carriages in the train ($c = 1, 2, \ldots, n_{car}$); $\overline{T}_f$ is the average temperature in the $c$-th carriage measured $n_T$ times at a height of 1.10 m above the floor of the carriage during the trip:

$$\overline{T}_f = 1 - \frac{\sum_{t=1}^{n_T} T_{ft}}{n_T}, \tag{4.13}$$

where $T_{ft}$ is the temperature obtained in the $t$-th measurement, °C; $n_T$ is the number of temperature measurements ($t = 1, 2, \ldots, n_T$).

#### 4.1.7.2 Temperature Assessment When Not Divided into Cold and Warm Periods

The current *Lithuanian Hygiene Standard* HN 128:2013 [28] requires that the temperature in the carriage compartment be at least 18 °C ($T_L$) and not higher than 27 °C ($T_U$), regardless of the season (cold or warm), i.e., the standard provides only limit values. (Fig. 4.5b). The optimum temperature $T_0$ in a carriage compartment c can be taken as the average temperature $\frac{T_L + T_U}{2}$ = 22.5 °C, which corresponds to the highest level of comfort with variable $x_{A7} = 1$. When the average temperature of the carriage compartment is at $\overline{T}_f = T_L$ (or below) or at $\overline{T}_f = T_U$ (or above), the quality level variable $x_{A7c} = 0$. The average quality level ($x_{A7c} = 0.5$) corresponds to the factual temperature $\overline{T}_f$ of the carriage compartment, which is equal to the average of the minimum and medium authorised temperatures $\frac{T_L + T_0}{2}$ or $\overline{T}_f = \frac{T_0 + T_U}{2}$. The impact of the factual carriage $c$-th compartment ($c = 1, 2, \ldots, n_{car}$) temperature $\overline{T}_f$ on the quality level variable $x_{A7c}$ can be calculated according to the formulae:

$$x_{A7c} = \frac{\overline{T}_f - T_L}{\frac{T_L + T_U}{2} - T_L}, \quad \text{when} \quad \overline{T}_f < T_0, \tag{4.14}$$

$$x_{A7c} = \frac{T_U - \overline{T}_f}{T_U - \frac{T_L + T_U}{2}}, \quad \text{when} \quad \overline{T}_f > T_0 \tag{4.15}$$

The variable $x_{A7}$, which indicates the quality level of the train in terms of the compliance of the factual temperature of the carriage compartment with the requirements of the *Hygiene Standards* HN 128:2013 [28], can be calculated from formulas (4.16) or (4.17):

$$x_{A7} = \frac{1}{n_{car}} \sum_{c=1}^{n_{car}} \left( \frac{\overline{T}_f - T_L}{T_0 - T_L} \right)_c, \quad \text{when} \quad \overline{T}_f < T_0; \tag{4.16}$$

and

$$x_{A7} = \frac{1}{n_{car}} \sum_{c=1}^{n_{car}} \left( \frac{T_U - \overline{T}_f}{T_U - T_0} \right)_c, \quad \text{when} \quad \overline{T}_f > T_0. \tag{4.17}$$

The worst case ($x_{A7} = 0$) can be observed, when $\overline{T}_f$ is by 2 °C smaller than the lowest permissible value $T_L$ or when $\overline{T}_f$ is by 2 °C higher than the highest permissible value $T_U$. If $\overline{T}_f$ is lower than $T_L$ by more than 2 °C or is by 2 °C higher than $T_U$, a negative value of $x_{A7}$ is obtained, which is considered equal to zero.

In this model the impact of the speed of air movement in the carriage compartment, the relative air humidity, and the level of artificial lighting on the quality of the train journey is not assessed. The *Lithuanian Hygienic Standard* HN 128:2013 [28] requires that the velocity of air movement in the cold season be not less than 0.05 m/s and not more than 0.2 m/s, and in the warm season not more than 0.9 m/s. Relative air humidity shall not exceed 70%. The overall level of artificial lighting on the surface of the table and, where there is no table, on the surface of the passenger seating or sleeping area, shall be not less than 100 lx.

### *4.1.8 Train Toilets Type and Condition*

The type, design and condition of train toilets (WC) have an impact on sanitation, hygiene, and ecology. These parameters are therefore relevant to the quality of the journey. They are not always up to high standards.

The train WC can form a barrier for those wishing to travel by train as it is perceived as being dirty, and therefore its use as being unpleasant. In addition, Dutch train WC users have the additional issue of storing their hand luggage in the toilet's confined space. It investigates what kind of hand luggage train travelers have and finally what travelers do with their hand luggage when using the toilet [40]. Decisions on the use of WC on trains and the legal regulation of the number of WC on the train are analysed. The issue of the number of WC on trains and their level of maintenance is discussed [41].

Lack of cleanliness on Indian trains, especially in toilets, has always been a major problem. Until now, trains have not had a system in place to handle human excrement. Passengers would use the non-vacuum WC, where excrement would fall onto the rails. For 24 years, Indian Railways has been trying to find a solution to the 3980 tonnes—497 lorry loads (8 tonnes per lorry)—of excrement that is dumped every day from trains onto the rail tracks. A recent effort has been the installation of the so-called "bio-digesters" under the train WC, where bacteria in the composting chamber digest faeces. However, this type of WC has got many disadvantages. Smouldering and non-functioning toilets on India Railways—one of the world's largest rail networks—may soon be a memory. Since 2018, the National Carrier has been installing biovacuum WC in its train carriages. These biovacuum WC do not smell bad. Water consumption has been reduced by more than 20 times. Water saving is a high priority for the railways. A biological vacuum WC requires about half a litre of water per flush and all waste is disposed off by vacuuming. The initiative by the railways to upgrade the existing bio-toilets has come after regular complaints from passengers that the toilets are clogged with waste such as plastic bottles, paper, hygiene packs, nappies, etc., which are thrown down the toilet [42, 43]. However, biovacuum WC also have disadvantages. Sometimes vacuum WC cause additional noise in train carriages. Gunaki and Devaraj [44] explains about the issues produced by bio-toilets and steps to convert human waste from Indian Railway sanitary system to energy, nutritious product for the agricultural sector using value chain concept.

In the carriages of long-distance trains and local passenger trains, two vacuum WC should be equipped, while one WC should be provided in the luggage carriage and one in the dining car.

A WC is evaluated in points as follows: 0 means that WC in a passenger carriage are out of order, 1 shows that there are ordinary WC in the carriage, which are clean and in good condition, 2 shows that, in a passenger carriage, there is one ordinary WC and one vacuum WC, which are clean and in good order, 3 implies that there are vacuum WC in the carriage, but they are not clean and not in good condition, 4 means that there are vacuum WC in the carriage, which are clean and in good condition. The variable $x_{A8}$ of the CQI $K_A$ which assesses the quality of travel by train with reference to the availability of WC on the passenger carriage, is calculated by the equation:

$$x_{A8} = 1 - \frac{\sum_{c=1}^{n_{car}} \left( \frac{WC_f}{WC_{max}} \right)_c}{n_{car}}, \tag{4.18}$$

where $x_{A8}$ is the variable of the 8-th sub-criterion of the key criterion A; $WC_f$ is the estimate of the real condition of lavatories in the $c$-th carriage, expressed in points (0, 1, 2, 3 or 4 points); $WC_{max}$ is the highest possible estimate (in points) for the $c$-th carriage (4 points); $n_{car}$ is the number of carriages in the train ($c = 1, 2, \ldots, n_{car}$).

### 4.1.9 The Structure of Berths and Special Facilities for People with Disabilities

Austrian Federal Railways plans to launch new, modern, and comfortable passenger trains in 2022. The berths on the new night trains will be slightly longer, even though the European gauge of 1435 mm does not allow berths of 2 m [21].

The structure of berths (the safety belts on upper level berths) and special facilities for the disabled passengers are evaluated in points as follows: 0 means that the upper level berths are not provided with safety belts and a passenger carriage is not adapted for passengers with disabilities, 1 implies that the safety belts are available on the upper level berths, but the carriage is not adapted for the disabled passengers, 2 shows that the upper level berths have safety belts and the carriage is adapted for the disabled passengers, 3 shows that a passenger carriage has only lower level berths (therefore, safety belts are not necessary), but it is not adapted for the disabled passengers, 4 means that a passenger carriage has only lower level berths and is adapted for the disabled passengers. The variable $x_{A9}$ of the CQI $K_A$ which assesses the quality of travel by train with rference to the structure of the berths in the carriage (the availability of safety belts on the upper-level berths) and special facilities for people with disabilities is calculated by the equation:

$$x_{A9} = \frac{\sum_{c=1}^{n_{car}} \left( \frac{BD_f}{BD_{max}} \right)_c}{n_{car}}, \quad (4.19)$$

where $x_{A9}$ is the variable of the 9-th sub-criterion of the key criterion A; $BD_f$ is the estimate in points (0, 1, 2, 3 or 4 points) of the berth structure in the $c$-th carriage (the availability of safety belts on the upper level berths) and the adaptation of a passenger carriage for the disabled passengers; $BD_{max}$ is the highest possible estimate (4 points) of the berth structure (the availability of safety belts on the upper level berths) and the adaptation of a passenger carriage for the disabled passengers; $n_{car}$ is the number of carriages in the train ($c = 1, 2, \ldots, n_{car}$).

### 4.1.10 The Availability of Shower Facilities that Operate on a Regular Basis

Anyone travelling for more than a few hours has questions about how to maintain good hygiene. Many passengers are concerned about whether trains have on-board showers. The new rolling stock is expected to be comfortable, as passengers travelling in the higher classes want separate toilets and showers. Even those travelling in economy class do not like "sharing" bathrooms with strangers. Therefore, innovations such as "mini apartments" are expected to appear on trains soon. There are also pricing innovations that make trains affordable for single travellers or couples [21].

The availability of shower facilities in a passenger carriage is evaluated in points as follows: 0 means that shower equipment is not installed in the carriage, 1 implies that shower facilities are available in the carriage at fixed hours and are in good working condition, 2 shows that shower facilities are available in the carriage at any time. The variable $x_{A10}$ of the CQI $K_A$ which assesses the quality of travel by train with reference to the availability of shower equipment availability in a passenger carriage, is calculated by the equation:

$$x_{A10} = \frac{\sum_{c=1}^{n_{car}} \left( \frac{SE_f}{SE_{max}} \right)_c}{n_{car}}, \quad (4.20)$$

where $x_{A10}$ is the variable of the 10-th sub-criterion of the key criterion A; $SE_f$ is the estimate of shower equipment availability and its work hours in the $c$-th carriage in points (0, 1 or 2 points); $SE_{max}$ is the highest possible estimate of shower equipment availability and its work hours (2 points); $n_{car}$ is the number of carriages in the train ($c = 1, 2, \ldots, n_{car}$).

### 4.1.11 Bicycle and Scooter Storage

Many cyclists prefer travelling by train, but sometimes it can cause inconvenience. This is due to train stations that are not cycling-friendly, of confusing information in the seat reservation system, or the absence of the facility to carry a bike on board. For these reasons, it can be problematic, if not impossible, to combine cycling with train travel. The train is a relatively fast, comfortable means of transport which, when combined with a bicycle, is a great blend for sustainable mobility. Regional and local trains mostly provide transportation for bicycles. However, transporting a bicycle on high-speed and long-distance trains is problematic. Transport service providers apply different rules and regulations for the carriage of bicycles. This makes it complicated for cyclists to plan their train and bike journeys.

Recently, a significant step has been taken towards a more bicycle-friendly future on trains in Europe. In 2020, the European Cyclists' Federation (ECF) launched the "Cyclists Love Trains" campaign. As of 2025, new and upgraded rolling stock is expected to have at least four-cycle spaces per train. But there is still much progress to be made. In a recent report published by the European Cyclists' Federation (ECF), 69 railway companies and services have been ranked and analysed according to their bicycle-friendliness. The focus has been on long journeys. Of the 69 companies, 15 (22%) do not accept unfolded and disassembled bicycles. In contrast, the Intercity Berlin trains of the NS-DB joint venture between Berlin and Amsterdam rank highest for their friendliness and flexibility towards users. Other highly rated train companies are Swiss Federal Railways (SBB), the Belgian National Railway Company (SNCB/NMBS), Deutsche Bahn (DB) and Hungarian State Railways (MÁV-START). France and Austria are also making positive developments regarding cycling, which could serve as a benchmark for other rail companies [45].

It would not be rational to have a special compartment in every passenger carriage for the transport of bicycles, so it was considered that there should be at least one carriage or a special compartment in the train for the transport of bicycles and scooters. The possibility of transporting bicycles on the train is evaluated in points as follows: 0 means that there is no special carriage or compartment on the train for carrying bicycles, therefore, their transportation is prohibited, 1 denotes that there is no special carriage or compartment on the train for transporting bicycles, but they may be carried for a fixed payment, when disassembled and packed, 2 implies that there is no special carriage or compartment on the train for transporting bicycles, but they may be carried free of charge, when disassembled and packed, 3 shows that there is a special carriage (or compartment) for transporting bicycles for a fixed payment, 4 means that there is a special carriage (or compartment) for transporting bicycles free of charge. The variable $x_{A11}$ of the CQI $K_A$ which assesses the quality of travel by train with reference to the availability of conditions to carry bicycles and scooters on the train is calculated by the equation:

$$x_{A11} = \frac{BS_f}{BS_{max}}, \tag{4.21}$$

where $x_{A11}$ is the variable of the 11-th sub-criterion of the key criterion A; $BS_f$ is the means the presence or absence of a special carriage or compartment for carrying bicycles and scooters in the train and the evaluation of transportation conditions in points (0, 1, 2 or 4 points); $BS_{max}$ is the highest estimate of the conditions for carrying bicycles ans scooters in the train (4 points).

### *4.1.12 Availability of the Smoking Places*

Smoking on public transport is a pressing issue which is addressed in different ways.

Researchers led by the Chinese Tobacco Control Association conducted a study on four ordinary trains. Three trains had smoking areas between the carriages and on one train smoking was banned. The results of the study have shown that the concentration of carcinogenic particles in train carriages where passengers could smoke exceeded 500 μg per cubic metre. According to the Beijing Environmental Monitoring Centre, when the average concentration of PM 2.5—airborne particles with a diameter of 2.5 microns or less—is above 150 mcg/cubic metre, the air is hazardous to human health. The air is even more toxic in the areas between carriages where smokers congregate. The study revealed that the highest concentration of PM 2.5 reached was more than 1000 mcg/cu.m. Cigarette smoke can easily enter the train carriages and spread through the train, posing a health risk to many passengers. The researchers also interviewed 94 passengers on four trains and learned that 77 of them would be in favour of a total ban on smoking on ordinary (slow) trains. Twelve male smokers opposed the proposal. The remaining five respondents did not give a specific answer but said they would stick to the no-smoking rule. All high-speed railway networks in China ban smoking, yet the ban does not apply to slow trains, despite the years-long urging by tobacco control experts. Passengers on slow trains are allowed to smoke in designated areas between carriages. In 2017, a passenger on a Chinese ordinary train sued the railway company for installing smoking areas and ashtrays on board. The company was ordered to abolish all smoking facilities on the train within 30 days of the ruling [46]. Many European countries have passed laws that ban smoking on trains and severely restrict smoking in train stations. Railway stations generally have designated smoking areas, which may not always follow the uniform rules in different countries.

Effective implementation of no-smoking regulations requires provision of training for railway staff, posting of no-smoking signs and display of information to train and railway station passengers [47].

Despite the health risks of smoking, there are still passengers who refuse to give up this harmful habit. On long train journeys, they feel like smoking. It is therefore recommended to offer such passengers the facility to smoke without causing harm to other non-smokers.

This criterion is evaluated in points as follows: 0 means that smoking in the train is prohibited, 1 smoking is allowed in special vestibules, 2 implies that there is a special carriage in the train for smokers. The variable $x_{A12}$ of the CQI $K_A$ which assesses the quality of travel by train with reference to the availability of conditions for smoking while on the train, is calculated by the equation:

$$x_{A12} = \frac{SP_f}{SP_{max}}, \tag{4.22}$$

where $x_{A12}$ is the variable of the 12th sub-criterion of the key criterion A; $SP_f$ is the estimate of smoking places in the train in points (0, 1 or 2 points); $SP_{max}$ is the highest possible estimate of smoking places in the train (2 points).

### 4.1.13 *Radio Broadcasting Unit and Its Centralized Operation*

During the journey, the radio broadcasts information about the route, the names of the stations (the stations where border and customs controls are carried out), the arrival of the train at the final station, information about the services provided, as well as the rules of conduct for passengers on the train and the fire safety regulations that they must observe on board. Passengers are also informed about the delay of the train, the causes of the delay and reduction of the stopping time [48]. A radio broadcasting unit and its centralized operation (switching on/off) are evaluated in points as follows: 0 means that a radio broadcasting unit is not available in the carriage (it is either not installed or is out of order), 1 implies that a radio broadcasting unit in the carriage can be switched on/off by a switch located in the conductor's compartment, 2 shows that a radio broadcasting unit in the carriage can be switched on/off by switches found in each compartment, 3 denotes that a radio broadcasting unit in the carriage can be switched on/off by switches found in each compartment and in the corridor, 4 means that any passenger, who would like to listen to the radio, can use special headphones. The variable $x_{A13}$ of the CQI $K_A$ which assesses the quality of travel by train with reference to the availability of radio broadcasting unit and its centralized on/off switching, is calculated by the equation:

$$x_{A13} = \frac{\sum_{c=1}^{n_{car}} \left( \frac{RB_f}{RB_{max}} \right)_c}{n_{car}}, \tag{4.23}$$

where $x_{A13}$ is the variable of the 13th sub-criterion of the key criterion A; $RB_f$ is the estimate in points (0, 1, 2, 3 or 4 points) of the operation of the radio broadcasting unit in the $c$-th carriage and its centralized on/off switching; $RB_{max}$ is the highest possible estimate of the radio broadcasting unit operation and centralized on/off switching (4 points); $n_{car}$ is the number of carriages in the train.

### *4.1.14 Availability of the Dining-Car*

On the trains of European railway companies, there is usually a dining-car, which is open to all passengers, regardless of a ticket price. For example, one of the countries with high quality and reasonably priced meals on trains is the Czech Republic. Most Swiss trains have dining-cars serving high quality cuisine, including national dishes, however, these are quite expensive. The German railway operator (DB) focuses on the use of local ingredients in its dining-cars. The food menu on the train aims to offer healthy and nutritious meals. One disadvantage of DB is that most meals are heated (cooked) using microwaves. The comfortable dining-cars are a distinctive advantage. UK train routes are known for their high-quality food, prepared by top chefs. There is a trend towards stricter food standards in the rail industry. Most rail operators, such as Thalys, Alleo, Trenitalia, SNCF, and Eurostar, are now sourcing local products and they also offer vegan food [49].

A dining-car is a carriage on a passenger train with a small kitchen for food preparation and a dining area for passengers. Due to space constraints, both the dining and kitchen areas of these carriages are designed to make the most efficient use of the available space. In the 19th and early twentieth centuries, the dining-car on the train offered passengers high-quality food in elegant settings. Since the beginning of the twenty-first century, many restaurant carriages have been replaced by snack bar (buffet) carriages. In the 19th and early twentieth centuries, when long-distance train journeys were commonplace, a dining-car was a practical necessity. The main function of the dining-car is obvious from its name. This is a place on the train where travellers can enjoy cooked meals. Passengers do not need to take their own food on board the train if there is a dining-car. Dining-cars offer train passengers a space where they can interact with each other. They can feature expensive china, ornate chandeliers, and impeccably dressed and mannered waiters. Nowadays, in the US, Europe, and other countries where rail passenger transport is popular, more and more sophisticated dining-cars are being built. This is aimed at attracting customers. Since the beginning of the twenty-first century, cars and planes have made long-distance rail travel a fact of the past. In many countries, passengers travel short distances on commuter trains. They commute mostly for work purposes. Most commuter trains do not have a dining car. Railway companies that offer food often do so in a snack bar/buffet car [50]. When there is no dining-car or snack bar/buffet car on board, snacks, pastries, and drinks are usually offered to passengers by train attendants.

The availability of the dining-car and the service and quality provided are scored as follows: 0 means that there is no dining car on the train and the conductors do not offer any snacks or drinks, 1 means that there is no dining car on the train but the conductors offer tea, coffee, etc. to the passengers, 2 means that the dining-car is only available at weekends, strong drinks are not offered, there are no waiters in this car and the conductors do not offer any drinks or snacks to the passengers, 3 means that the dining- car is only available at weekends, strong drinks are not served and there are no waiters in this car, but conductors do serve tea, coffee, etc., 4 means that the dining-car is only available at weekends, strong drinks are served, there is a waiter in this car and conductors do serve tea, coffee, etc., 5 means that there is a permanent dining-car on the train, strong drinks are not served, there are no waiters in this car and conductors do not serve drinks or snacks to passengers, 5 means that there is a dining-car on the train at all times, strong drinks are not served, there is no waiter and conductors do not offer passengers drinks or snacks, 6 means that there is a dining-car on the train at all times, strong drinks are not served, there is no waiter but conductors do offer passengers tea, coffee, etc., 7 means that there is a permanent dining-car on the train, where strong drinks are served, but there are no waiters in this car, and conductors serve tea, coffee, etc. to passengers. 8 means that there is a permanent dining-car on the train, where strong drinks are served, there is a waiter in this car, and conductors serve tea, coffee, etc. to passengers. The variable $x_{A14}$ of the CQI $K_A$ which assesses the quality of travel by train with respect to the availability of the dining-car on the train, is calculated by the equation:

$$x_{A14} = \frac{DC_f}{DC_{max}}, \tag{4.24}$$

where $x_{A14}$ is the variable of the 14th sub-criterion of the key criterion A; $DC_f$ is the estimate of the dining-car availability and the provided services in points (0, 1, 2, 3, 4, 5, 6, 7 or 8 points); $DC_{max}$ is the highest possible estimate of the dining -car availability and the provided services (8 points).

### *4.1.15 Calling Conductor to a Passenger Compartment in an Emergency*

First-class soft carriages are often equipped with emergency call buttons. There are conflicting views among staff and passengers on the purpose of the emergency call button both on trains and planes. Often, staff say that the emergency call button is only for emergencies and should not be used for requests such as getting a drink or a blanket. Flight attendants are of the opinion that the emergency call button should never be used during the landing or take-off of an aircraft because of the risk of injury to crew [51]. In Europe and North America, it is expected that the emergency call button will only be used in cases of emergency or serious necessity, such as illness. Also, in cases where you want to inform the service staff of something that the staff member should know immediately: a noticed malfunction, an extraneous sound, etc. Business and first-class carriages/airplanes offer more services, so passengers can feel free to use the call button but are advised not to disturb a staff member if he/she is busy at the time. Some carriers, especially in East and South-East Asia, encourage passengers to use call buttons to reduce walking through the carriage [52]. However, the call button has long been a source of concern for both passengers and crew members. It has even been named an "ancient torture device". Sometimes children start playing with the call button on their journey [53]. The compartment of a train carriage may not be equipped with an emergency call button. It may not be working. If the attendant is not at the service area, he or she may not hear the call (if he or she is in the shower, toilet or has left for some reason). There is no such thing as a perfect system. All measures need to be taken to ensure the safety of passengers and the quality of service, whether this is just a request for extra attention on the part of a passenger or a necessity.

It is comfortable for passengers to have an opportunity to call the conductor of a carriage to the compartment when they need some information or help in the case of emergency. They can do it by pressing a button. The possibility of calling the conductor to a passenger compartment in the case of emergency is evaluated in points as follows: 0 means that there is no button (switch) for calling the conductor to a passenger's compartment in the carriage, 1 implies that a button (switch) for calling the conductor to a passenger's compartment is installed in every compartment, though it is out of order (e.g. a special magnetic card is required for it to start operating, which is not available); 2 means that a properly operating button (switch) for calling the conductor to the compartment is available in the carriage. The variable $x_{A15}$ of the CQI $K_A$ which assesses the quality of travel by train with respect to calling the conductor to a passenger compartment in emergency is calculated by the equation:

$$x_{A15} = \frac{\sum_{c=1}^{n_{car}} \left( \frac{CC_f}{CC_{max}} \right)_c}{n_{car}}, \tag{4.25}$$

where $x_{A15}$ is the variable of the 15th sub-criterion of the key criterion A; $CC_f$ is the estimate in points (0, 1 or 2 points) of the possibility to call the conductor to a passenger compartment of the $c$-th carriage in emergency; $CC_{max}$ is the highest possible estimate of the possibility to call the conductor to a passenger compartment in emergency (2 points); $n_{car}$ is the number of carriages in the train ($c = 1, 2, \ldots, n_{car}$).

### 4.1.16 Use of Electrical Equipment on Board the Train

This possibility is rated in points as follows: 0 means that passengers cannot use hairdryers, irons or other electrical appliances on the train, 1 means that passengers can use their own hairdryers on the train, although there is no iron for ironing clothes, 2 means that there is a special compartment on the train where passengers can use a hairdryer and an iron for a fixed payment, 3 means that there is a special compartment on the train where passengers can use a hairdryer and iron free of charge, 4 means that there is a beauty salon where a passenger can have his/her hair done and a special compartment on the train where passengers' clothes are washed and ironed free of charge. The variable $x_{A16}$ of the CQI $K_A$ which assesses the quality of travel by train with reference to the possibility to use a hairdryer and an iron on the train, is calculated by the equation:

$$x_{A16} = \frac{EA_f}{EA_{max}}, \tag{4.26}$$

where $x_{A16}$ is the variable of the 16-th sub-criterion of the key criterion A; $EA_f$ is the estimate of the possibility for passengers to use electrical appliances (hairdryer and an iron) on the train expressed in points (0, 1, 2, 3 or 4 points); $EA_{max}$ is the highest estimate of the possibility for passengers to use a electrical appliances (hairdryer and an iron) on the train (4 points).

## 4.2 Additive Model for the Sub-criteria of the Key Criterion a

The comprehensive quality index $K_A$, describing the sub-criteria associated with the train elements and railway track condition is calculated by the additive equation:

$$K_A = \overline{Z}_A^* \cdot \Big[\overline{Q}_{A1} \cdot \Big(0.997 + 0.00245 \cdot \overline{TQI}_f - 0.000962 \cdot \overline{TQI}_f^2 + 1.532 \cdot 10^{-5}$$

$$\cdot \overline{TQI}_f^3 - 6.99 \cdot 10^{-8} \cdot \overline{TQI}_f^4\Big) + \overline{Q}_{A2} \cdot \frac{\sum_{l=1}^{n_{stage}} \left(1 - \frac{|v_{f,l} - v_{max,a}|}{v_{max,a}}\right) \cdot l_l}{\sum_{l=1}^{n_{stage}} l_l}$$

$$+ \overline{Q}_{A3} \cdot \frac{\sum_{c=1}^{n_{car}} \left(\frac{CE_f}{CE_{max}}\right)_c}{n_{car}} + \overline{Q}_{A4} \cdot \frac{\sum_{c=1}^{n_{car}} \left(\frac{NL_{max,a} - NL_f}{NL_{max,a} - NL_{min}}\right)_c}{n_{car}}$$

$$+ \overline{Q}_{A5} \cdot \frac{\sum_{c=1}^{n_{car}} \left(\frac{CI_f}{CI_{max}}\right)_c}{n_{car}} + \overline{Q}_{A6} \cdot \frac{\sum_{c=1}^{n_{car}} \left(\frac{VCL_f}{VCL_{max}}\right)_c}{n_{car}}$$

$$+ \overline{Q}_{A7} \cdot \frac{1}{n_{car}} \sum_{c=1}^{n_{car}} \left(\frac{T_f - T_L}{T_0 - T_L}\right)_c + \overline{Q}_{A8} \cdot \frac{\sum_{c=1}^{n_{car}} \left(\frac{WC_f}{WC_{max}}\right)_c}{n_{car}}$$

$$+\overline{Q}_{A9}\cdot\frac{\sum_{c=1}^{n_{car}}\left(\frac{BD_f}{BD_{max}}\right)_c}{n_{car}}+\overline{Q}_{A10}\cdot\frac{\sum_{c=1}^{n_{car}}\left(\frac{SE_f}{SE_{max}}\right)_c}{n_{car}}$$

$$+\overline{Q}_{A11}\cdot\frac{BS_f}{BS_{max}}+\overline{Q}_{A12}\cdot\frac{SP_f}{SP_{max}}+\overline{Q}_{A13}\cdot\frac{\sum_{c=1}^{n_{car}}\left(\frac{RB_f}{RB_{max}}\right)_c}{n_{car}}$$

$$\left.+\overline{Q}_{A14}\cdot\frac{DC_f}{DC_{max}}+\overline{Q}_{A15}\cdot\frac{\sum_{c=1}^{n_{car}}\left(\frac{CC_f}{CC_{max}}\right)_c}{n_{car}}+\overline{Q}_{A16}\cdot\frac{EA_f}{EA_{max}}\right]. \quad (4.27)$$

All symbols of the module (4.27) are given in Sect. 4.1 for the analysis of each sub-criterion from A1 to A16.

The variable $x_{A7}$ in the mathematical model (4.27) which is multiplied by $\overline{Q}_{A7}$, is calculated according to formula (4.16) when $\overline{T}_f < T_0$. In the case where $\overline{T}_f > T_0$, the variable $x_{A7}$ is calculated according to formula (4.17).

The additive equation (4.27) was verified by solved example with real data [54]. Consequently, the results are in line with the principles of the proposed method for the determination of the quality of rail travel and are real. ($K_A = 0.1341$, i.e., it is $0 < K_A < \overline{Z}_A^*$). According to the key criterion A, the best train quality can be achieved when $K_A = 0.228$.

In this monograph, the mathematical model (4.27) has been upgraded due to changes in the requirements of the norms, revisions to the formulae for the calculation of the variables $x_{A4}$ ir $x_{A7}$ and additional experimental studies. The model adequacy, i.e., its full correspondence to reality, is demonstrated by the example it resolves.

The developed original additive mathematical model (4.27) is used to calculate the values of each of the 16 sub-criteria based on the normalised weight of the main criterion A multiplied by its variable. The correspondence between the variable of the sub-criterion and the specified values, the critical values, the highest values, the acceptable values, or the best values is determined. For this purpose, the value indicating the closeness of a given subcriterion describing the train journey to the best level (if its value is equal to 1), the average level (if its value is about 0.114) and the worst (unacceptable) level ($\overline{Z}_A^* = 0.228$, its value is about 0) is calculated by comparing the real (actual) values of the subcriterion variable with the specified, critical, highest, admissible and best values.

## 4.3 Practical Application of the Key Criterion a Model and Adequacy Assessment

Based on the considered model (4.27) and real research data, as well as the acceptable or best values of each sub-criterion, the importance of key criterion A for the evaluation of the quality of passenger rail transport can be obtained by a single value.

Using the formulas suggested in the paper, all CQI components (16 sub-criteria) and their significance can be calculated based on the real data taken from standards, technical specifications, and other reliable sources. The testing was performed on the IPT No. 6.

***Quality of the railway track.*** The railway transportation enterprise uses a track recording vehicle (TRV) EM-140 made in 2009 by the Plasser & Theurer Company (Austria) for measuring geometrical parameters of the railway track. The geometrical railway track parameters measured by TRV and stored on a storage device are as follows: track gauge, left longitudinal level, right longitudinal level, cross-level, left alignment, right alignment, and twist. All geometrical railway track parameters are obtained by precise measuring of the spatial curve of the track.

In Lithuania, the variation in geometric parameters of all the railways in the country, expressed in $TQI_{fs}$, is measured once a month on the TRV EM-140. The $TQI_{fs}$ values for the 1-km sections of each road are shown in the diagram (example Fig. 4.6) and compared with the allowable $TQI_{fs}$ (Table 4.4) to determine the track quality level. Figure 4.6 shows that the $TQI_{fs}$ values for the nine railway tracks are spread over a range from 12 to 72 mm$^2$. Only a small percentage of the railway sections are in poor condition ($TQI_{fs}$ > 45 mm$^2$) and in need of repair.

The main objective of track geometry interventions is to maintain or restore the initial geometric condition in such a way that the geometry degrades at a rate considered acceptable from a life cycle perspective. Typical maintenance actions to restore the quality of the track geometry include tamping, ballasting, measured ballasting, track lining, track lifting, rail and sleeper bracing and replacement of rail fasteners, sleepers, pads, and spring elements [55].

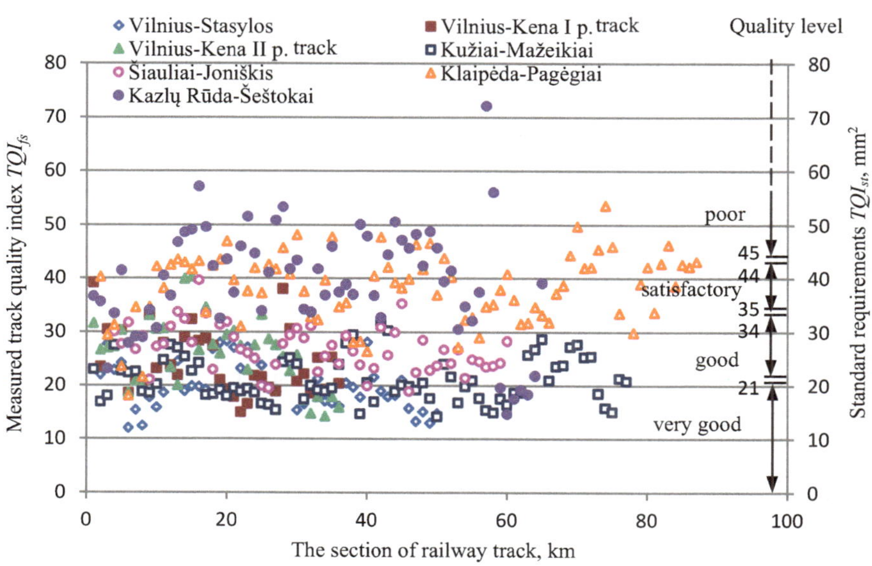


**Fig. 4.6** Variation of the TQI measured in March using the EM-140 on the nine railway track gauges of the Republic of Lithuania

On the route of the IPT No. 6, the total length of which is 944 km, the railway double track, running on the territory of Lithuanian Republic from Vilnius to the frontier with Belarus at Kena, makes 36 km. The average $\overline{TQI}_f$, measured by a TRV for the railway track running through Vilnius–Kena state frontier, was calculated by the formula (4.5) based on $TQI_{fs}$ of all one-kilometre track lengths (Table 4.5).

The data on the variation of the geometric parameters measured by the TRV on the territory of foreign countries for the calculations $\overline{TQI}_f$ could not be obtained.

The total TQI of all track lengths making 36 km on the route Vilnus–Kena state frontier is $\sum_{s=1}^{36} TQI_{fs} = 903.2\,\text{mm}^2, N_{km} = 36\,\text{km}, \overline{TQI}_f = 25.09\,\text{mm}^2$. The smallest $TQI_{fmin}$ value obtained for 1 km track length was $TQI_{fmin} = 14.95$ mm$^2$, while the largest value was $TQI_{fmax} = 64.04$ mm$^2$.

Based on the standard $TQI_{st}$ (from 0 to 45 mm$^2$) and modelled $TQI_{mod}$ (from 45 to 90 mm$^2$) values, the correlation (with determination coefficient $R^2 = 0.9961$) between the real $TQI_f$ and the variable $x_{A1}$ of the first sub-criterion of key criterion A, showing the effect of the variation of geometric track parameters on the track and trip (by train) quality, has been obtained:

$$
\begin{aligned}
x_{A1} &= 0.997 + 0.00245 \times \overline{TQI}_f - 0.000962 \times \overline{TQI}_f^2 \\
&\quad + 1.532 \times 10^{-5} \times \overline{TQI}_f^3 - 6.99 \times 10^{-8} \times \overline{TQI}_f^4.
\end{aligned} \tag{4.28}
$$

Given the average $\overline{TQI}_f = 25.09$ mm$^2$ of the track measured by a geometry car, the value of the variable $x_{A1} = 0.667$, obtained from the regression equation (4.28), corresponds to a good quality of train travel, depending on the track parameters that determine its quality.

The study shows that the effect of variations in individual geometric parameters on the TQI value is not uniform (Table 4.5). The largest part of $\overline{TQI}_f = 25.09$ mm$^2$ is due to the variation of the track gauge (20.7%), and the smallest variation of the left rail lonitudinal plane (10.9%) is almost twice as small. It would be of interest to practitioners and academics to have a study that would identify and assess the importance (weighting) of each individual geometric parameter on the TQI value.

***Trip duration of the train.*** The railway track on the route Vilnius–Kena–State border consists of 4 sections of a journey (representing the lengths between the stations) (Fig. 4.7). The total length of the railway track is $L = 36$ km. The lengths between intermediate stations are $l_1 = 9$ km, $l_2 = 8$ km, $l_3 = 12$ km, and $l_4 = 7$ km ($L = l_1 + l_2 + \ldots + l_l$). The highest permissible train speed on the route (railway track) Vilnius–State border is $v_{max} = 120$ km/h [22]. The engine driver had a written instruction that, on May 17, the train speed on the route Kena–State border had to be limited to $v_{max} = 40$ km/h. The real (measured) train speed on the sections $v_{f,l}$ of the considered railway track was $v_{f1} = 43$ km/h, $v_{f2} = 77$ km/h, $v_{f3} = 57$ km/h and $v_{f4} = 51$ km/h (the average speed was recorded based on the data provided by the control panel screen of the railway carriage). The CQI variable $x_{A2}$ was calculated by formula (4.7) and $x_{A2} = 0.531$ was obtained. In the ideal case, the train could move at the highest permissible speed $v_{max,a}$ ($x_{A2} = 1$) on all railway track sections. In the

**Table 4.5** Geometric parameters of the railway track on the route Vilnius–Kena (the main track 1)

| Number of 1 km length railway track section | Railway track geometrical parameter ($mm^2$) | | | | | | | |
|---|---|---|---|---|---|---|---|---|
| | Track gauge | Longitudinal level left rail | Longitudinal level right rail | Alignment left rail | Alignment right rail | Cross level | Twist | $TQI_{fc} = \sum_{g=1}^{G} \sigma_{gc}^2$ |
| 1 | 4.60 | 6.35 | 6.52 | 5.21 | 4.32 | 4.75 | 4.94 | 36.69 |
| 2 | 4.28 | 2.84 | 3.07 | 3.22 | 3.04 | 3.22 | 4.24 | 23.91 |
| 3 | 3.28 | 3.99 | 3.75 | 3.47 | 3.79 | 3.26 | 4.44 | 25.98 |
| 4 | 5.20 | 6.35 | 6.52 | 4.31 | 4.28 | 4.75 | 4.94 | 36.35 |
| 5 | 4.28 | 2.84 | 3.07 | 3.22 | 3.04 | 3.22 | 4.24 | 23.91 |
| 6 | 3.28 | 3.99 | 3.75 | 3.47 | 3.79 | 3.26 | 4.44 | 25.98 |
| 7 | 3.68 | 2.63 | 2.92 | 2.28 | 2.31 | 2.69 | 3.15 | 19.66 |
| 8 | 3.87 | 2.51 | 2.95 | 2.41 | 3.03 | 2.07 | 2.75 | 19.59 |
| 9 | 3.51 | 3.89 | 4.26 | 4.48 | 3.93 | 3.03 | 3.79 | 26.89 |
| 10 | 8.35 | 2.93 | 3.04 | 3.84 | 2.75 | 2.67 | 3.11 | 26.69 |
| 11 | 6.54 | 2.52 | 2.71 | 3.79 | 2.71 | 2.54 | 3.66 | 24.47 |
| 12 | 4.53 | 3.02 | 3.45 | 3.02 | 3.26 | 2.90 | 3.74 | 23.92 |
| 13 | 10.84 | 2.00 | 2.05 | 3.84 | 2.15 | 2.94 | 4.61 | 28.43 |
| 14 | 6.96 | 2.63 | 2.88 | 3.45 | 3.61 | 3.17 | 3.91 | 26.61 |
| 15 | 12.07 | 2.95 | 2.66 | 3.71 | 4.23 | 3.70 | 5.07 | 34.39 |
| 16 | 9.23 | 2.71 | 2.62 | 3.41 | 3.52 | 3.40 | 4.59 | 29.48 |
| 17 | 6.66 | 2.78 | 3.14 | 3.53 | 3.71 | 3.33 | 4.35 | 27.50 |
| 18 | 3.59 | 2.95 | 3.00 | 4.06 | 3.86 | 3.28 | 4.13 | 24.87 |
| 19 | 5.89 | 1.62 | 1.70 | 3.19 | 3.69 | 2.23 | 3.05 | 21.37 |
| 20 | 5.89 | 1.62 | 1.70 | 3.19 | 3.69 | 2.23 | 3.05 | 21.37 |
| 21 | 5.71 | 1.82 | 2.03 | 2.45 | 2.55 | 2.42 | 3.14 | 20.12 |
| 22 | 3.37 | 1.89 | 2.07 | 4.65 | 2.74 | 2.05 | 2.58 | 19.35 |
| 23 | 2.80 | 1.65 | 1.62 | 2.55 | 2.33 | 1.85 | 2.15 | 14.95 |
| 24 | 2.83 | 2.21 | 2.29 | 2.53 | 2.39 | 2.02 | 2.20 | 16.47 |
| 25 | 5.77 | 2.12 | 2.26 | 3.71 | 3.33 | 2.66 | 3.95 | 23.80 |
| 26 | 5.98 | 2.12 | 2.19 | 2.26 | 2.60 | 2.44 | 3.93 | 21.52 |
| 27 | 5.97 | 2.03 | 2.11 | 2.36 | 2.13 | 2.45 | 3.69 | 20.74 |
| 28 | 4.16 | 2.10 | 1.97 | 3.03 | 2.90 | 2.32 | 2.53 | 19.01 |
| 29 | 5.44 | 3.58 | 3.54 | 5.76 | 5.71 | 2.98 | 3.55 | 30.56 |
| 30 | 4.83 | 2.16 | 2.47 | 2.91 | 2.57 | 2.34 | 3.48 | 20.76 |
| 31 | 2.98 | 1.77 | 2.12 | 2.21 | 2.41 | 2.06 | 2.87 | 16.42 |
| 32 | 4.73 | 2.90 | 3.06 | 2.53 | 2.37 | 3.30 | 4.37 | 23.26 |
| 33 | 2.70 | 2.19 | 2.48 | 2.49 | 2.56 | 2.72 | 3.21 | 18.35 |
| 34 | 3.19 | 2.88 | 3.18 | 2.77 | 2.59 | 3.82 | 4.70 | 23.13 |

(continued)

**Table 4.5** (continued)

| Number of 1 km length railway track section | Railway track geometrical parameter ($mm^2$) | | | | | | | |
|---|---|---|---|---|---|---|---|---|
| | Track gauge | Longitudinal level left rail | Longitudinal level right rail | Alignment left rail | Alignment right rail | Cross level | Twist | $TQI_{fc} = \sum_{g=1}^{G} \sigma_{gc}^2$ |
| 35 | 4.58 | 2.03 | 2.28 | 2.48 | 1.83 | 2.37 | 48.47 | 64.04 |
| 36 | 4.98 | 2.00 | 2.12 | 3.71 | 3.30 | 2.67 | 3.88 | 22.66 |
| Total | 186.55 | 98.57 | 103.55 | 119.5 | 113.02 | 103.11 | 178.9 | 903.2 |
| Average | 5.18 | 2.74 | 2.88 | 3.32 | 3.14 | 2.86 | 4.97 | 25.09 |
| Percent | 20.7 | 10.9 | 11.5 | 13.2 | 12.5 | 11.4 | 19.8 | 100 |

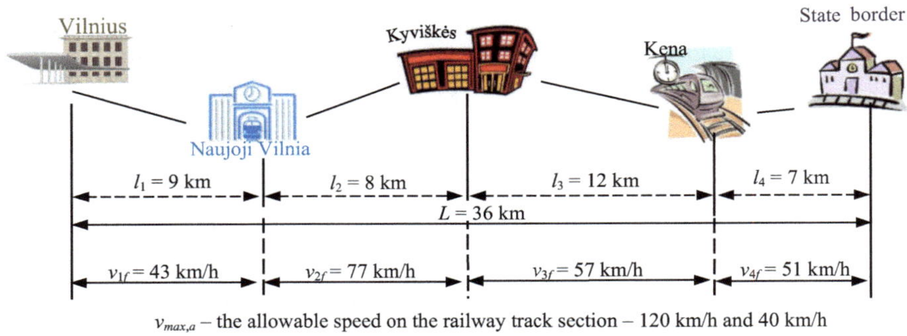


**Fig. 4.7** Lengths between intermediate stations on the Vilnius-State border route and speed of IPT No. 6 train on this route

worst case, it could either exceed the highest permissible speed twice or more times which is risky (taking into account traffic safety), or move very slowly. Therefore, it is assumed that $x_{A2} = 0$.

***The condition of the passenger carriage exterior (cleanness, deformations, damages, etc.).*** There are seven railway carriages in the train (Fig. 4.8). All carriages are washed before the train leaves the station where it is formed. They are therefore clean. There are no damages on carriage bodies either. The CQI variable $x_{A2}$ was calculated by formula (4.8) and the maximum $x_{A3} = 1$ was obtained.

***Noise reduction insulation.*** Train formation is carried out in accordance with the *Technical Regulations for the Use of Railways* (1996), the schedule of train runnings and the train formation plan. Train carriages are coupled in such a way as to ensure a high quality of train travel, depending on the noise level in the carriage compartment. Train formation is the responsibility of station staff. Train schedules and formation plans shall contain train weight and length standards for each section. They shall be compatible with the capacity of the locomotives used on these sections, the usable length of the approach and departure tracks at the stations, the profile of the track and, on electrified sections, the power supply conditions. The standards for the weight

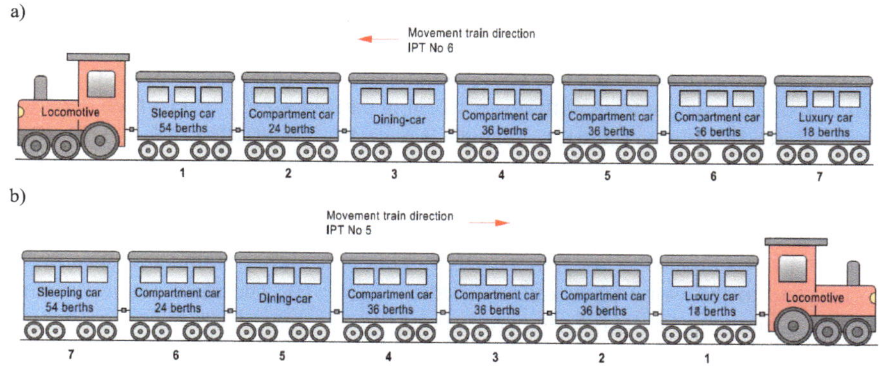


**Fig. 4.8** Procedure for the formation of the following trains: **a** IPT No. 6; **b** IPT No. 5 and carriage categories

and length of passenger trains and the arrangement of carriages on trains are laid down in the passenger train timetables [23].

The number of carriages in a train formation can change. Depending on the season of the year and the number of passengers, there may be more or fewer train carriages. The number and type of carriages a train is made up of depends on the category of train. In addition to the passenger carriages, a train may also have mail, baggage, "staff" and dining-carriages. The "staff" carriage is equipped with a compartment for the manager of the train, which contains a desk and a radio unit. This carriage also includes compartments for other staff serving the train, such as the train electrician and the train dining-car staff.

The train in our study is IPT No. 6 (Fig. 4.8a, b). It is made up of 7 passenger carriages: a second-class sleeper (54 berths), 4 s-class quadruple compartment sleeper carriages (3 carriages – 36 berths, 1 "staff" carriage – 24 berths), a dining-car, and a first-class soft double compartment carriage (18 berths).

The most expensive train tickets are in the first-class soft double compartment, which is why there is usually only one of these carriages in the train formation. If the route is not popular and the train is not a high-speed train, this carriage may not exist. The cheapest tickets are for the second-class sleeping car, which is also the one with the highest number of passengers. Additional carriages in this category are often attached. The proper train pattern ensures a comfortable ride for passengers. When the train IPT No. 6 departs from Vilnius, the first-class soft double compartment carriage is located at the rear of the train (Fig. 4.8a), and on IPT No. 5 to Vilnius the first-class soft double compartment carriage is at the front of the train formation, i.e., near the locomotive (Fig. 4.8b). The position of this carriage in the train formation contributes to the safety of the passengers travelling in the carriage, as passengers from other carriages can not pass through the carriage. The dining-car and the "staff" carriages are in the middle of the train formation to be easily accessible for passengers of all carriages. Typically, additional sleeping carriages ("open space sleepers") are attached to the sleeping carriages and sleeping quadruple compartments ("coupé cars") are attached to the coupé carriage group.

The factual sound pressure level $NL_{f,c}$ was measured in the compartment of each c-carriage of a passenger train with seven sleeping carriages ($n_{car} = 7$) which departed from Vilnius IPT No. 6. For each carriage, it was: $NL_{f,1} = 49$ dBA, $NL_{f,2} = 51$ dBA, $NL_{f,3} = 59$ dBA, $NL_{f,4} = 56$ dBA, $NL_{f,5} = 57$ dBA, $NL_{f,6} = 53$ dBA, $NL_{f,7} = 48$ dBA. By applying formula (4.9), the noise level variable $x_{A4}$ was calculated (Table 4.6). For the train consisting of all sleeping carriages, the maximum allowable sound pressure level shall be taken as $NL_{max,a} = 65$ dBA. This and higher sound pressure levels correspond to the worst train journey quality ($x_{A4} = 0$). The ideal sound pressure level is assumed to be the one that corresponds to $NL_{min} = 30$ dBA.

The CQI component was calculated by Eq. (4.9) and $x_{A4} = 0.335$ was obtained. In the ideal case, the noise level is equal to 30 dB (A) ($x_{A4} = 1$), while, in the worst case, when the noise level exceeds the limiting allowable noise exposure value $NL_{max,a}$, a negative value is obtained. It is considered that $x_{A4} = 0$.

***The interior of a passenger carriage.*** The renewal of the rolling stock of the company "LTG" Link (in particular, passenger carriages) is carried out in two ways: (a) the service life of the used passenger carriages is extended (by their modernisation); (b) new passenger coaches are purchased. To validate the proposed passenger car interior assessment model and demonstrate its applicability, it was tested in IPT 6. The train consists of 7 coaches, of which 4 are new ($CI_{f,c} = 2$) and 3 are modernised ($CI_{f,c} = 1$). The maximum possible estimation value for the carriage interior is ($CI_{max,p} = 2$). The CQI variable $x_{A5}$ component was calculated using Eq. (4.10) and $x_{A5} = 0.786$ was obtained.

***Operation of ventilation, air conditioning, cooling, and lighting systems in terms of their timely on/off switching.*** There are seven railway carriages in the train. Air conditioning systems are installed in all railway carriages, running on international routes. In six carriages, all systems, including air conditioning, ventilation, heating, and lighting systems, are operating, while, in one sleeping carriage with 54 berts (No. 1), all systems, except for air conditioning system, which has not been installed yet, are operating. The CQI variable $x_{A6}$ was calculated by Eq. (4.11) and $x_{A6} = 0.964$ was obtained.

***Temperature control system for passenger carriage interior.*** In winter season in March, the temperature in the cabins of all carriages of IPT No. 6 was measured from 4:00 p. m. to 7:00 a. m.

Before the train departs from Vilnius and before it reaches the fifth intermediate station, the carriages are heated in furnaces burning solid fuel (coal). Electric heaters

**Table 4.6** Measured noise level in train IPT No. 6 passenger carriages

| Carriage number | 1 | 2 | 3 | 4 | 5 | 6 | 7 | Average variable $x_{A4}$ |
|---|---|---|---|---|---|---|---|---|
| Measured noise level $NL_{f,c}$ | 49 | 51 | 59 | 56 | 57 | 53 | 48 | |
| Variable $x_{A4,c}$ | 0.457 | 0.400 | 0.171 | 0.257 | 0.229 | 0.343 | 0.486 | 0.335 |

powered by a 3000 V high voltage contact network are switched on when the train runs from the fifth intermediate station to its destination on a foreign railway, and the temperature in the interior of each carriage is automatically maintained at (22 ± 2) °C.

Sixty-one measurement was conducted in each carriage of the train in randomly selected compartments from 1 to 9 with the interval of 15 min. The indoor temperature in the carriage was measured by a digital thermometer (TA20) used for measuring indoor and outdoor temperature. The ambient air temperature was recorded based on the data displayed on the screen of the control panel of the carriage. The change of the outdoor and indoor temperature of first carriage is shown in Fig. 4.9.

At the time of measurement (Fig. 4.10), the average outside temperature was 7.23 °C i.e., above –40 °C and the average inside temperature was $\overline{T}_{f,1} = 23.00$ °C. It is assumed that the date of the measurement belongs to the winter period.

$\overline{T}_{f,c}$ is the average temperature in the *c*-th carriage ($c = 1, 2, \ldots, 7$), measured $n_T$ times at various places of the carriage during the trip and is calculated by formula (4.13).

The average inside temperature in the first carriage measured on March 22 was 23 °C (formula 4.13) $\left(\overline{T}_f = \frac{1402.7}{61} = 23\right)$ (Fig. 4.10b).

The average temperatures $\overline{T}_{f,c}$ measured and calculated for the other six carriages are shown in Table 4.7. It varied from 20.82 °C to 23.61 °C in all carriages, with the optimum value $T_{OW} = 22$ °C. The standard deviations $\sigma_T$ of the temperature in each carriage are also presented here. They show that the temperature of the carriage fluctuated during the train journey. The standard deviation $\sigma_T$ of the 61 factual temperature measurements $T_{ft}$ of the carriage compartment, measured every 15 min, varied between 0.89 °C and 1.94 °C, i.e. it varied by a factor of about two. The wide variation in the individual temperature values $T_{ft}$ shows that the average temperature of the carriage compartment $\overline{T}_{f,c}$, while close to the optimum temperature $T_{OW} = 22$ °C, may deviate significantly during certain periods of the train journey and reduce the comfort level.

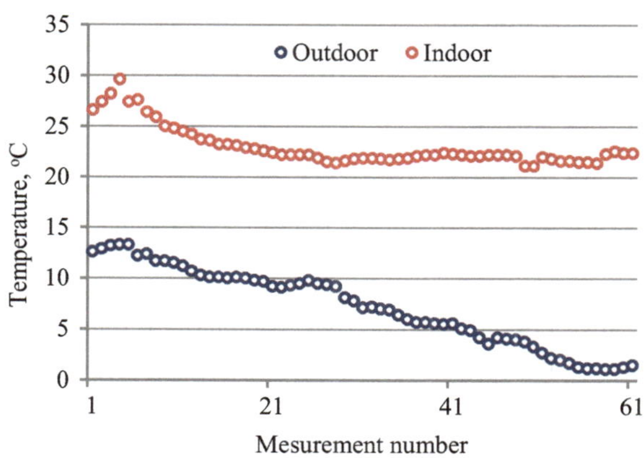


**Fig. 4.9** The change of the outside and inside temperature in the passenger carriage compartment 1

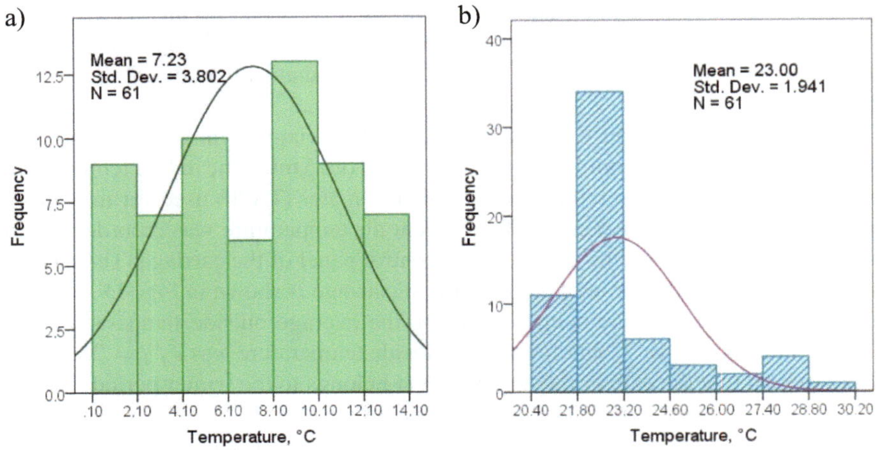


**Fig. 4.10** The histograms and statistical data of the outside (**a**) and inside (**b**) temperature (in the passenger carriage compartment 1)

**Table 4.7** Average temperature measured in the compartments of the carriages of an IPT No. 6 train and calculated value of the variable $x_{A7}$

| Indicator | Number of carriage ($c = 1, 2, \ldots, n_{car}$) (Fig. 4.8) | | | | | | | Total | Mean | Variable $x_{A7}$ |
|---|---|---|---|---|---|---|---|---|---|---|
| | 1 (1) | 2 (9) | 3 – | 4 (3) | 5 (7) | 6 (4) | 7 (6) | | | |
| Average $\overline{T}_{f,c}$ | 23.00 | 22.14 | 21.05 | 22.73 | 23.61 | 21.96 | 20.82 | – | – | – |
| Stand. dev. $\sigma_T$ | 1.94 | 0.89 | 1.65 | 1.48 | 1.12 | 1.36 | 0.98 | – | – | – |
| $x_{A7}$ (4.12) *num* | 1 | 1 | 0.475 | 1 | 1 | 0.02 | 0.59 | 5085 | 0.726 | 0.274 |
| $x_{A7c}$ (4.14, 4.15) | 0.889 | 0.920 | 0.678 | 0.949 | 0.753 | 0.880 | 0.627 | 5.696 | 0.814 | 0.814 |

The results of the study (Table 4.7) show that the factual temperature $\overline{T}_f$ of the seven carriages of the train, in terms of compliance with the different standards, shows a difference in the quality of travel. The calculated value of the variable $x_{A7}$ according to formula (4.12) equals 0.274, i.e., the quality of the train journey is about a quarter of the highest quality. The factual temperature $\overline{T}_f$ in the carriage compartments is more in line with the requirements of the new *Lithuanian Hygiene Standard* HN 128:2013 [28]. The value of the variable $x_{A7}$ calculated according to formula (4.16) and formula (4.17), equals 0.814, and indicates that the factual temperature of the compartments of the carriages is more than 80% of the required optimum temperature and that the quality of the journey is close to the best possible quality. For the integrated assessment of the quality of the trip IPT No. 6, the variable $x_{A7} = 0.814$ calculated in accordance with the new *Lithuanian Hygiene Standard* HN 128:2013 [28] is considered.

***Train toilets type and condition.*** At present, 70% of the carriages operated by the company "LTG Link" are equipped with vacuum WC. It is planned to install ecological toilets in all carriages in 2–4 years. The train has seven railway carriages. In six carriages, there are vacuum WC, which are properly maintained (the train manager controls their condition), while the toilets in the first carriage are ordinary, but clean ($WC_{f,1} = 1$, $WC_{f,2-7} = 4$, $WC_{max} = 4$). The CQI variable $x_{A8}$ was calculated by Eq. (4.18) and $x_{A8} = 0.893$ was obtained.

***The structure of berths and special facilities for people with disabilities.*** In some coaches of JSC "LTG Link", the upper berths are equipped with special brackets, while in ordinary coaches there are safety belts for passengers. Some passenger diesel and electric trains are adapted to the special needs of disabled passengers. The trains have special places for their wheelchairs, as well as toilets adapted for disabled people and enlarged doorways. The number of such passenger carriages is constantly increasing because, when new carriages are purchased or the old ones are modernized, special needs of disabled passengers are considered. Wheelchairs with disabled persons are raised to the level of the carriage entrance by mobile hoists, which are available at large railway stations.

The international passenger train No 6 has six carriages (not including the dining-car). In five passenger carriages the upper layer berths are equipped with safety belts, though these carriages are not adapted for the disabled ($BD_{f,1-6} = 1$, $BD_{max} = 4$). The first (No 7, Fig. 4.8a) carriage (luxury car with 18 berths) has only lower berths, therefore, safety belts are not needed there. This carriage is not adapted for disabled passengers either ($BD_{f,7} = 3$). The CQI variable $x_{A9}$ was calculated by Eq. (4.19) and $x_{A9} = 0.333$ was obtained.

***The availability of shower facilities that operate on a regular basis.*** Currently, no shower facilities are available in the passenger carriages of the JSC "LTG Link" ($SE_{f,c} = 0$, $SE_{max} = 2$), though they are urgently needed for both passengers and service staff of international long-distance trains. The CQI variable $x_{A10}$ was calculated by Eq. (4.20) and the critically low $x_{A10} = 0$ was obtained.

***Bicycle and scooter storage.*** The international trains of the JSC "LTG Link" do not have any carriages equipped to carry bicycles. However, the regulations [56] allow to carry one engineless knocked-down and packed bicycle if there is room for it in a special compartment for hand luggage. In this case, the service is provided to passengers free of charge ($BS_f = 2$, $BS_{max} = 4$). The CQI variable $x_{A11}$ was calculated by Eq. (4.21) and $x_{A11} = 0.50$ was obtained.

***Availability of the smoking places.*** The Lithuanian law of tobacco control prohibits smoking of cigarettes (or other tobacco products) in public transport (except for long-distance trains, which should have carriages for smokers and non-smokers). In passenger carriages of international trains of the JSC "LTG Link", smoking is allowed only in special vestibules ($SP_f = 1$, $SP_{max} = 2$). Some passengers object to it because smoke often penetrates through poorly sealed or open doors and gets into the corridor and compartments of the carriage. The experiment was made in the IPT No. 6 to assess the travel quality with reference to smoking, according to the above-mentioned regulations. The CQI variable $x_{A12}$ was calculated by Eq. (4.22) and $x_{A12} = 0.50$ was obtained.

***Radio broadcasting unit and its centralized operation.*** Volume controls for the train radio are available in the compartments and corridors of the carriages. At present, passengers can also use headphones in the compartments of new carriages manufactured at the Tver Railway Carriage Plant. If there are no headphones in a carriage (because the conductor does not give them to passengers), passengers can listen to the radio without the headphones. In these carriages, the radio is switched on and off using the button in the conductor's compartment, but it cannot be used to adjust the volume. When the train departs from the main station, passengers often ask the conductor to turn off the radio because it is often too loud, and they do not hear the information about the train's route given by the train manager. The train has six carriages (not including the dining-car). In the three carriages ($c = 1, 2, 4$) the radio is switched on/off and controlled by a switch in the train attendant's compartment ($RB_{f,\,1,\,2,\,4} = 1$). In the other three carriages ($c = 5, 6, 7$) the radio is controlled by switches located in each compartment and in the corridor ($RB_{f,\,5,\,6,\,7} = 3$). The CQI variable $x_{A13}$ was calculated by Eq. (4.23) and $x_{A13} = 0.50$ was obtained.

***Availability of the dining-car.*** All international trains of the JSC "LTG Link" have dining-cars. Strong drinks are served there. For economic reasons, it is rare to find a waiter in the dining-car. Usually, it is self-service, or the conductor of a dining-car provides the services of a waiter. Conductors of passenger cars offer tea, coffee, soft drinks, and pastries to passengers. If a passenger requests it, the conductor should bring the requested food or drink from the dining-car ($DC_f = 7$, $DC_{max} = 8$). The CQI variable $x_{A14}$ was calculated by Eq. (4.24) and $x_{A14} = 0.875$ was obtained.

***Calling conductor to a passenger compartment in an emergency.*** At present, not all carriages in the train have buttons for calling a conductor to the compartment in an emergency. All new carriages manufactured in the Tver plant have the possibility of calling a conductor to the compartment. However, the pushbuttons that connect to the conductor only work if a magnetic card is inserted in the magnetic card reader. These cards are still not available, so passengers cannot call a conductor to the compartment. The train has six carriages (not including the dining-car) (Fig. 4.8) ($CC_{f,\,1,\,2} = 0$, $CC_{f,\,4\text{–}6} = 1$, $CC_{f,\,7} = 2$, $CC_{max} = 2$). The CQI variable $x_{A15}$ was calculated by Eq. (4.25) and $x_{A15} = 0.417$ was obtained.

***Use of electrical equipment on board the train.*** The trains made up in Lithuania do not have any compartments, where passengers could use a hairdryer or an iron. On this train you can use your own personal hairdryer, although there is no access to a clothes iron ($EA_f = 1$, $EA_{max} = 4$). The CQI variable $x_{A16}$ was calculated by Eq. (4.26) and $x_{A16} = 0.25$ was obtained.

The quality level of the key criterion A, and the respective variable $x_{Ai}$ value based on the real data are presented in Table 4.8. The quality of train represented by the estimate of the variable $x_{Ai}$ is equal to 1, when it is based on the highest values of the real data, it is equal to 0.5, when based on the average values, and to 0, when based on the lowest values of the data.

The highest quality level determined by using the real data was obtained for the sub-criterion A3 (the state of the carriage exterior) ($x_{A3} = 1$). The lowest quality level determined by using the real data was obtained for the sub-criterion A10 (availability of regularly operating shower) ($x_{A10} = 0$). The average quality level was determined for A11 (special compartments for transporting bicycles) ($x_{A11} = 0.50$), A12

**Table 4.8** Quality level of the sub-criteria of the key criterion A and the respective value of the variable $x_{Ai}$ based on real data

| Variable $x_{Ai}$ of the key criterion A (formula) | Real (actual) vale of the component | The calculated value $x_{Ai}$ |
|---|---|---|
| $x_{A1}$ (4.6) | $N_{km} = 36$ km; $\sum_{s=1}^{N_{km}} TQI_{fs} = 903.2$ mm$^2$; $\overline{TQI}_f = 25.09$ mm$^2$; $TQI_{fmin} = 14.95$ mm$^2$; $TQI_{fmax} = 64.04$ mm$^2$ | 0.667 |
| $x_{A2}$ (4.7) | $n_{stage} = 4$; $l_1 = 9$ km; $l_2 = 8$ km; $l_3 = 12$ km; $l_4 = 7$ km; $L = 36$ km; $v_{f1} = 43$ km/h; $v_{f2} = 77$ km/h; $v_{f3} = 57$ km/h; $v_{max,a\ 1\text{–}3} = 120$ km/h; $v_{max,a\ 4} = 40$ km/h | 0.531 |
| $x_{A3}$ (4.8) | $n_{car} = 7$; $CE_{f,\ 1\text{–}7} = 2$ (each); $CE_{max} = 2$ | 1.000 |
| $x_{A4}$ (4.9) | $n_{car} = 7$; $NL_{f,1} = 49$ dBA; $NL_{f,2} = 51$ dBA; $NL_{f,3} = 59$ dBA; $NL_{f,4} = 56$ dBA; $NL_{f,5} = 57$ dBA; $NL_{f,6} = 53$ dBA; $NL_{f,7} = 48$ dBA; $NL_{max,a} = 65$ dBA; $NL_{min} = 30$ dBA; | 0.335 |
| $x_{A5}$ (4.10) | $n_{car} = 7$; $CI_{f,1\text{–}4} = 2$; $CI_{f,5\text{–}7} = 1$; $CI_{max,p} = 2$ | 0.786 |
| $x_{A6}$ (4.11) | $n_{car} = 7$; $VCL_{f,1\text{–}6} = 4$; $VCL_{f,7} = 3$; $VCL_{max,p} = 4$ | 0.964 |
| $x_{A7}$ (4.12, 4.16, 4.17) | $n_{car} = 7$; $\overline{T}_{f,1} = 23.00$ °C; $\overline{T}_{f,2} = 22.14$ °C; $\overline{T}_{f,3} = 21.05$ °C; $\overline{T}_{f,4} = 22.73$ °C; $\overline{T}_{f,5} = 23.61$ °C; $\overline{T}_{f,6} = 21.96$ °C; $\overline{T}_{f,7} = 20.82$ °C; $T_{OW} = 20$ °C; $T_{LW} = 18$ °C; $T_{UW} = 22$ °C; $T_o = 22.5$ °C; $T_L = 18$ °C; $T_U = 27$ °C | 0.274<br>0.814 |
| $x_{A8}$ (4.18) | $n_{car} = 7$; $WC_{max} = 4$; $WC_{f,1} = 1$; $WC_{f,2} = 4$; $WC_{f,3} = 4$; $WC_{f,4} = 4$; $WC_{f,5} = 4$; $WC_{f,6} = 4$; $WC_{f,7} = 4$ | 0.893 |
| $x_{A9}$ (4.19) | $n_{car} = 6$; $BD_{max} = 4$; $BD_{f,1} = 1$; $BD_{f,2} = 1$; $BD_{f,4} = 1$; $BD_{f,5} = 1$; $BD_{f,6} = 1$; $BD_{f,7} = 3$; | 0.333 |
| $x_{A10}$ (4.20) | $n_{car} = 7$; $SE_{max} = 2$; $SE_{f,1\text{–}7} = 0$ | 0 |
| $x_{A11}$ (4.21) | $BS_{max} = 4$; $BS_f = 2$ | 0.50 |
| $x_{A12}$ (4.22) | $SP_{max} = 2$; $SP_f = 1$ | 0.50 |
| $x_{A13}$ (4.23) | $n_{car} = 6$; $RB_{max} = 4$; $RB_{f,1,2,4} = 1$; $RB_{f,5\text{–}7} = 3$ | 0.50 |
| $x_{A14}$ (4.24) | $DC_{max} = 8$; $DC_f = 7$ | 0.875 |
| $x_{A15}$ (4.25) | $n_{car} = 6$; $CC_{max} = 2$; $CC_{f,1,2} = 0$; $CC_{f,4\text{–}6} = 1$; $CC_{f,7} = 2$ | 0.417 |
| $x_{A16}$ (4.26) | $EA_{max} = 4$; $EA_f = 1$ | 0.25 |

(smoking places) ($x_{A12} = 0.50$) and A13 (radio broadcasting unit and its centralized operation) ($x_{A13} = 0.50$) based on the real data. First, the real values of the sub-criteria, whose variables $x_{Ai} = 0$ or are about zero, should be improved. The developed model may be applied to determining the quality or effectiveness of other objects or processes, which can be described by sets of criteria.

Using the final Eq. (4.27), the comprehensive quality index $K_A$ of the key criterion A describing the train elements and the technical state of rails (railway track) is calculated as follows:

$$K_A = 0.228 \begin{pmatrix} 0.0536 \times 0.667 + 0.1121 \times 0.531 \\ +0.0310 \times 1.0000 + 0.0585 \times 0.335 \\ +0.0578 \times 0.786 + 0.1050 \times 0.964 \\ +0.1061 \times 0.814 + 0.0787 \times 0.893 \\ +0.0635 \times 0.333 + 0.0642 \times 0 \\ +0.0339 \times 0.50 + 0.0366 \times 0.50 \\ +0.0450 \times 0.50 + 0.0766 \times 0.875 \\ +0.0359 \times 0.417 + 0.0415 \times 0.25 \end{pmatrix}.$$

$$= 0.228 \times 0.6205 = 0.1415$$

The sum $\sum_{i=1}^{16} \overline{Q}_{Ai} \times x_{Ai} = 0.6205$ of multiples $\overline{Q}_{Ai} \times x_{Ai}$ of the weights of the 16 sub-criteria $\overline{Q}_{Ai}$ of the main criterion A (Table 3.7, Fig. 3.7a) and of the variables $x_{Ai}$ (Table 4.8) indicates that the quality of the journey on the IPT No. 6 is slightly above average in terms of the technical condition of the train elements and the track.

For this key criterion, the calculated products of the local weights $Qai_{Ai}$ of each of the 16 sub-criteria and of the variables $x_{Ai}$ of the train under study (Fig. 4.11) vary from 0 (A10) to 0.1012 (A6). A higher than the average influence of all sub-criteria (average $\overline{Q}_{Ai} \times x_{Ai} = 0.0388$) on the quality of the train trip depends on speed of the train (trip duration) (A2), the interior of a passenger carriage (A5), operation of ventilation, air conditioning, cooling and lighting systems in terms of their timely switching on/off (A6), the system for maintain temperature inside a passenger carriage (A7), type and condition of the train toilets (A8), and availability of the dining-car (A14). The following have the least impact on improving the quality of travel on this train: the availability of regulary operating shower equipment (A10), the possibility of using electrical appliances in the train (A16), the possibility of calling the conductor to a passengers' compartment in emergency case (A15), special compartments for transporting bicycles ans scoots (A11), availability of the smoking places (A12), noise reduce insulation (A4), structure of berths and special feasibilities' for the disabled (A9), and radio broadcasting unit and its centralized operation (A13). On this train, the condition of the exterior of passenger carriages (A3) and quality (roughness) of the railway track (A1) were close to average.

Taking into consideration the weight of the core criterion A ($\overline{Z}_A = 0.228$) and the sum (0. 6205) of the multiples of all 16 sub-criteria ($\overline{Q}_{Ai} \times x_{Ai}$) the effect of this key criterion on the integrated quality indicator $K$ (Eq. 3.1) has been calculated. The influence on the quality of the train travel elements and the technical condition of

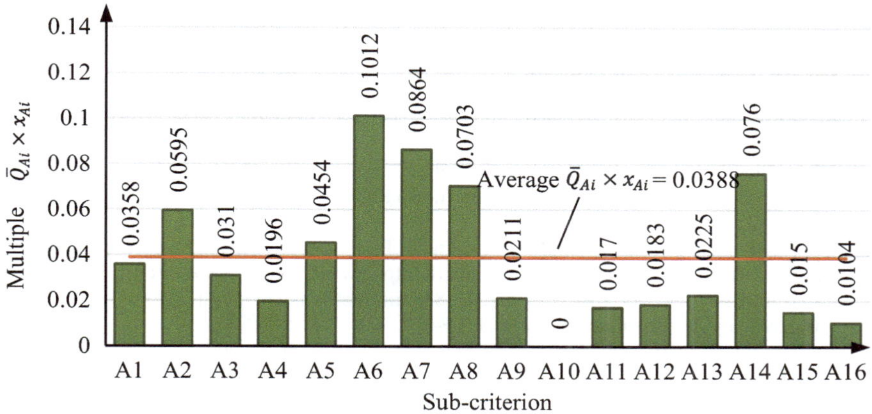


**Fig. 4.11** Results of the calculations of the impact of the sub-criteria of the key criterion A on the quality of the trip on the international passenger train under consideration

the railway track of the train studied is shown by the integrated quality index $K_A = 0.1415$. The other part of the Quality Level Indicator (QLI) $K$ consists of the B, C and D estimated CQI $K_B$, $K_C$, $K_D$, which have been investigated for the train IPT No. 6, and which are presented in other chapters of this monograph.

The resolved example shows that the additive mathematical model (4.27) is correct. The use of normalised subjective local weights $\overline{Q}_{Ai}$ given by the experts for the sub-criteria and the actual train parameters determined experimentally leads to a result between 0 and $\overline{Z}_A^*$. It can therefore be used in practice to assess the quality of a train journey.

## References

1. Peng Y, Zhou J, Fan C, Wu Z, Zhou W, Sun D, Lin Y, Xu D, Xu Q (2022) A review of passenger ride comfort in railway: assessment and improvement method. Transp Saf Environ 4(2):1–17. https://doi.org/10.1093/tse/tdac016
2. Mohammadi A, Amador-Jimenez L, Nasiri E (2020) A multi-criteria assessment of the passengers' level of comfort in urban railway rolling stock. Sustain Cities Soc 53:1–13. https://doi.org/10.1016/j.scs.2019.101892
3. White Paper on Transport (2011) Roadmap to a Single European Transport Area—Towards a Competitive and Resource-Efficient Transport System European Commission. Directorate-General for Mobility and Transport. Publications Office of the European Union, Luxembourg. https://doi.org/10.2832/30955
4. Xu P, Sun QX, Liu RK, Wang FT (2011) A short-range prediction model for track quality index. Proc Inst Mech Eng Part F J Rail Rapid Transit 225(3):277–285. https://doi.org/10.1177/204
5. Bai L, Liu R, Sun Q, Wang F, Wang F (2016) Classification learning-based framework for predicting railway track irregularities. Proc Inst Mech Eng Part F J Rail Rapid Transit 230(2):598–610. https://doi.org/10.1177/0954409714552818

6. ArastehKhouy I, Larsson-Kråik P-O, Nissen A, Kumar U (2016) Cost-effective track geometry maintenance limits. Proc Inst Mech Eng Part F J Rail Rapid Transit 230(2):611–622. https://doi.org/10.1177/0954409714542859
7. Andrade AR, Teixeira PF (2016) Exploring different alert limit strategies in the maintenance of railway track geometry. J Transp Eng 142(9):04016037. https://doi.org/10.1061/(ASCE)TE.1943-5436.0000867
8. López-Pita A, Teixeira RF, Casas C, Bachiller A, Ferreira PA (2008) Maintenance costs of high-speed lines in Europe state of the art. Transp Res Rec J Transp Res Board 2043(1):13–19. https://doi.org/10.3141/2043-02
9. EN 13848–1:2019. Railway applications. Track. Track geometry quality—Part 1: Characterization of track geometry. European Standard. https://www.google.com/search?q=EN13848&rlz=1C1GCEA_enLT892LT892&oq=EN13848&aqs=chrome..69i57j0i13i19i51217j0i13i19i3012.6952j0j4&sourceid=chrome&ie=UTF-8
10. Barbosa R (2016) Evaluation of railway track safety with a new method for track quality identification. J Transp Eng 142(11):04016053. https://doi.org/10.1061/(ASCE)TE.1943-5436.0000855
11. EN 13848-6:2014+A1:2020. Railway applications-Track-Track geometry quality-Part 6: characterisation of track geometry quality. European Standard. https://standards.iteh.ai/catalog/standards/cen/9c85a71e-0f1f-4825-b983-08b9aef34d6e/en-13848-6-2014a1-2020
12. Savaeigio kelio matavimo vagono EM-140 matuojamų geležinkelio kelio geometrinių parametrų įvertinimo instrukcija (K/259) (2012) Patvirtinta AB „Lietuvos geležinkeliai" generalinio direktoriaus 2012 m. sausio 4 d. įsakymu Nr. Į-10. Vilnius: UAB „Geležinkelių produkcijos atitikties vertinimo centras", 18 p. (in Lithuanian)
13. Javaid O, Choi D-H (2020) Effect of track irregularities on the response of two-way railway tracks. Appl Sci 10(1):11. https://doi.org/10.3390/app10010011
14. Yang Y, Zeng W, Qiu W-S, Wang T (2016) Optimization of the suspension parameters of a rail vehicle based on a virtual prototype Kriging surrogate model. Proc Inst Mech Eng F J Rail Rapid Transit 230(8):1890–1898. https://doi.org/10.1177/0954409715617213
15. ISO 2631-5:2018. Mechanical vibration and shock—Evaluation of human exposure to whole-body vibration—Part 5: Method for evaluation of vibration containing multiple shocks. Available from Internet: https://www.iso.org/standard/50905.html
16. EN 12299:2009. Railway applications—Ride comfort for passengers—Measurement and evaluation. European Standard. https://standards.iteh.ai/catalog/standards/cen/1fb5a39d-ef33-4d55-b91f-25037435497e/en-12299-2009
17. Jiang Y, Chen BK, Thompson C (2019) A comparison study of ride comfort indices between Sperling's method and EN 12299. Int J Rail Transp 7(4):1–18. https://doi.org/10.1080/23248378.2019.1616329
18. GB/T 5599-1985. Railway vehicles. Specification for evaluation testo the dynamic performance and accreditation test. The Standard measure of Railways. https://www.chinesestandard.net/PDF/BOOK.aspx/GBT5599-1985
19. Vaičiūnas G, Steišūnas S (2021) Sperling's comfort index study in a passenger car with independently rotating wheels. Transp Probl 16(2):121–130. https://doi.org/10.21307/tp-2021-028
20. Maskeliūnaitė L, Sivilevičius H (2014) The model for evaluating the criteria describing the quality of the trip by international train. Technol Econ Dev Econ 20(3):484–506. https://doi.org/10.3846/20294913.2014.949333
21. Cerny P (2021) 10 questions about night trains that you always wanted, but never dared, to ask. Heinrich Böll Stiftung. Brussels. https://eu.boell.org/en/10-questions-night-trains
22. Dėl didžiausio leistino traukinių ir lokomotyvų greičio geležinkelio stočių keliais ir tarpstočiais (2009) AB „Lietuvos geležinkeliai" generalinio direktoriaus 2009 m. lapkričio 3 d. įsakymas Nr. Į-843. Vilnius, 95 p (in Lithuanian)
23. Techninio geležinkelių naudojimo nuostatai (1996) Lietuvos Respublikos susisiekimo ministerija. Patvirtinta Lietuvos Respublikos susisiekimo ministro 1996 m. rugsėjo 20 d. įsakymu Nr.297. Informacijos ir leidybos centras, Vilnius, 126 p (in Lithuanian)

24. Thompson K, Offler N, Hirsch L, Every D, Thomas MJ, Dawson D (2012) From broken windows to a renovated research agenda: a review of the literature on vandalism and graffiti in the rail industry. Transp Res Part A Policy Pract 46(8):1280–1290. https://doi.org/10.1016/j.tra.2012.04.00
25. O'Keeffe C (2022) International graffiti gangs travelling to Ireland to deface trains. Irish Examiner. https://www.irishexaminer.com/news/arid-40869728.html
26. Ivanauskas V, Keršanskas V, Kasčiūnas L (2017) Kaliningrado veiksnys Lietuvos ir Rusijos santykiuose: reikšmė Lietuvos saugumui. Lietuvos metinė strateginė apžvalga 2016–2017, 15 tomas. 131–158. file:///C:/Users/40507/Downloads/lmsa-15-1-7_ivanauskas_kersanskas_kasciunas15%20(1).pdf (in Lithuania)
27. Sorrentino S (2018) Anti-graffiti film: a new layer of defence from vandalism. Comment by Railway-News. https://railway-news.com/anti-graffiti-film-a-new-layer-defence-vandalism/
28. HN 128:2013. „Keleiviniai geležinkelių riedmenys. Sveikatos saugos reikalavimai". Lietuvos higienos norma, patvirtinta Lietuvos Respublikos sveikatos apsaugos ministro 2013 m. birželio 14 d. įsakymu Nr. V-628. https://e-seimas.lrs.lt/portal/legalAct/lt/TAD/TAIS.451868?jfwid=-9dzqnu8op (in Lithuanian)
29. Rakicevic M (2020) 36 alarming snoring statistics and factsfor. https://disturbmenot.co/snoring-statistics/
30. Calderon LE, Carney LD, Kavanagh KT (2016) The cry of the child and its relationship to hearing loss in parental guardians and health care providers. J Evid Inf Social Work 13(2):198–205. https://doi.org/10.1080/23761407.2015.1018031
31. The inside track on rail interiors (2010) Railway technology. https://www.railway-technology.com/analysis/feature84864/
32. Zhang B, Guo W, Wang Y, Li S, Huang Y, Xu J (2016) Train's interior form design oriented visual image of passenger research. J Mech Eng 52(4):199–205. https://doi.org/10.3901/JME.2016.04.199
33. COVID-19 Rail Protocol. Recommendations for safe resumption of railway services in Europe (2020) European Union Agency for Railways, pp 1–19. file:///C:/Users/40507/Downloads/COVID-19%20Rail%20Protocol%20(1).pdf
34. Woodward H, Fan S, Bhagat RK, Dadonau M, Wykes MD, Martin E, Hama S, Tiwari A, Dalziel SB, Jones RL, Kumar P, Linden PF (2021) Air flow experiments on a train carriage—towards understanding the risk of airborne transmission. Atmosphere 12(10):1267, 1–19. https://doi.org/10.3390/atmos12101267
35. Zhu Y, Xu Y, Chen X (2015) Study on the one-dimensional carriage and ventilation system of high-speed train. In: Proceedings of the 2015 International Power, Electronics and Materials Engineering Conference (IPEMEC 2015), pp 947–951. https://doi.org/10.2991/ipemec-15.2015.173
36. Aigner R, Krawanja A, Luger C, Wollweber M, Claus P, Hörth L, Streng M (2019) Environmentally friendly air conditioning for trains—field data measurement and analysis on the ICE 3 Air-Cycle System. Final report. Umwelt Bundesamt, 167 p. https://www.umweltbundesamt.de/sites/default/files/medien/1410/publikationen/2019-10-14_texte_120-2019_air-condition-trains_0.pdf
37. Horan-Stear F (2022)Glow Up: interior lighting is getting smarter. Runway Girl Network. https://runwaygirlnetwork.com/2022/04/technology-means-interior-lighting-is-getting-smarter/
38. Deutsche Bahn honors Siemens as most innovative supplier (2016) Siemens. https://press.siemens.com/global/en/pressrelease/deutsche-bahn-honors-siemens-most-innovative-supplier
39. Egorov V (1987) Elektrooborudovanie Passazhirskikh Vagonov. Transport, Moskva. 312 s. UDK 629.45.064.5 (075.32) (in Russian)
40. Loth M, Molenbroek JFM, van Eijk DJ (2018) Hand luggage in the train toilet. WORK J Prev Assess Rehabil 59(3):387–399. https://doi.org/10.3233/WOR-182689
41. Lewandowski K (2020) Toilet in the train. Transp Overview Przeglad Komunikacyjny 2020(5):36–54. https://doi.org/10.35117/A_ENG_20_05_03

42. Make Way for Aircraft-Type Bio-Vacuum Toilets on Indiand Railways from January 2018 (2017) Banega Swasth India. https://swachhindia.ndtv.com/make-way-aircraft-type-bio-vacuum-toilets-indian-railways-january-2018-15619/
43. Sharma A, Unnikrishnan MK, Madaan A (2012) Toilet technology: improve sanitation on India's railways. Nature 489:33. https://doi.org/10.1038/489033e
44. Gunaki P, Devaraj S (2021) Value chain model for Indian Railway Sanitary System. Mater Today Proc 45(1):236–239
45. Tornow A (2021) Transporting bicycles on trains—inter-modality mini-series 2/3. Euro Velo the European Cycle Route Network. https://en.eurovelo.com/news/2021-07-16_transporting-bicycles-on-trains-inter-modality-mini-series-2-3
46. Xiaoyu W (2019) Health experts, researchers call for cigarette ban on slow trans. China Daily Society. http://global.chinadaily.com.cn/a/201901/11/WS5c37ef88a3106c65c34e3d74.html
47. Saleh SB, Faruque GM, Billah MM, Rahman MA, Shajib SH, Akhter A, Rayhan MA, Sujon AI, Reza MF, Munna K (2021) Passengers views on smoke-free provisions of Tobacco Control Law in trains and rail stations. In: 13th Asia Pacific Conference on Tobacco or Health. Apact 2021, september 3–4, vol 19(1), p A93. Tobacco Induced Diseases, Thailand. https://doi.org/10.18332/tid/141027
48. Tarptautinio susisiekimo keleivinio traukinio viršininko instrukcija (2003) Gelspa, Vilnius, 16 p (in Lithuanian)
49. Brandhuber S (2021) Discover the best train food services. Rail Europe. https://www.raileurope.com/en-us/blog/best-train-food-services
50. Huguelet MC (2023) What is a Dining Car? Wikimotors.com. https://www.wikimotors.org/what-is-a-dining-car.htm
51. Godfrey K, Sun T (2022) Passenger shamed by flight attendant for using the call button. Escape. Travel Advice. https://www.escape.com.au/travel-advice/passenger-shamed-by-flight-attendant-for-using-the-call-button/news-story/5e85829dbc1aabd99be6cce2c4000cd0
52. Walton J (2018) Plane insider: When should you press that cabin crew call button? Lonely Planet. Travel News. https://www.lonelyplanet.com/news/press-cabin-crew-call-button
53. Elliott C (2016) Pressing buttons: the most annoying things passengers do on planes. The Sydney Morning Herald. Traveler. https://www.traveller.com.au/pressing-buttons-the-most-annoying-things-passengers-do-on-planes-gnf4hj
54. Sivilevičius H, Maskeliūnaitė L (2014) The numerical example for evaluating the criteria describing the quality of the trip by international train. E+M Econ Manag 17(2):73–86. https://doi.org/10.15240/tul/001/2014-2-006/1648
55. Xin T, Famurewa SM, Gao L, Kumar U, Zhang Q (2016) Grey-system-theory-based model for the prediction of track geometry quality. Proc Inst Mech Eng F J Rail Rapid Transit 230(7):1735–1744. https://doi.org/10.1177/0954409715610603
56. Pagalbinė Nepriklausomų Valstybių Sandraugos šalių, Estijos Respublikos, Latvijos Respublikos, Lietuvos Respublikos geležinkelių administracijų tarpusavio susitarimo dėl atskirų normų taikymo medžiaga (2003) OP SMPS, Vilnius (in Lithuanian)

# Chapter 5
# Organisation and Technology of the Train Travel Process

*Trains are beautiful. They take people to places they've never been, faster than they could ever go themselves. Everyone who works on trains knows they have personalities, they're like people. They have their own mysteries.*
Sam Starbuck "The Dead Isle"

## 5.1 Elements of the Model to Assess the Impact of Train Travel Organisation and Technology on the Quality of Travelling

The aim of this chapter is to propose a multi-criteria mathematical model for assessing the importance of 19 sub-criteria that describe the organisation and technology of train travel. It also aims to determine the impact of the criteria weights on the CQI. In addition the mathematical model is validated using actual data from IPT 6.

Passenger trains on different routes are made up of different types of carriages, which may vary in their technical condition. They are operated by staff with different levels of experience and training. In the absence of quantitative estimates, the quality of train passenger services can only be determined roughly, subjectively and intuitively. In order to measure the quality of train travel more precisely, a method has been developed and an indicator $K$ has been presented to calculate the level of quality. This indicator $K$ is composed of 4 elements ($K_A$, $K_B$, $K_C$, $K_D$), the sum of which can vary between 0 and 1.

The model for calculating the significance (influence) of the key criterion B factors which characterise the organisation and technology of the train travel process on the quality of international train travel is given in the formula [1]:



H. Sivilevičius and L. Maskeliūnaitė, *Assessment of the Quality of Passenger Transportation by Train Using Multiple Criteria Decision Making Methods*, Lecture Notes in Intelligent Transportation and Infrastructure,
https://doi.org/10.1007/978-3-031-85877-2_5

**Table 5.1** The significance (weight) of the key criterion B describing travel by train

| Category of respondents and experts | Number of questionnaires | Weight |
|---|---|---|
| Passengers (P) | $n_P = 21$ | $Z_{B,P} = 0.2619$ |
| Train service staff (TS) | $n_{TS} = 20$ | $Z_{B,TS} = 0.1900$ |
| Administration staff (AS) | $n_{AS} = 9$ | $Z_{B,AS} = 0.2333$ |
| The average estimate value of all respondents and experts in their categories | 50 | $\overline{Z}_B^*= 0.228$ |

$$K_B = \overline{Z}_B^* \cdot \sum_{i=1}^{m} \left(\overline{Q}_{Bi} \cdot x_{Bi}\right) = \overline{Z}_B^* \cdot \left(\overline{Q}_{B1} \cdot x_{B1} + \overline{Q}_{B2} \cdot x_{B2} + \cdots + \overline{Q}_{B19} \cdot x_{B19}\right), \tag{5.1}$$

where $K_B$ is comprehensive quality index related to the railway trip planning and technology (ranging from 0 to $\overline{Z}_B^*$); $\overline{Z}_B^*$ is the mean weight, reflecting respondents' and experts' estimates of the significance of key criterion B (Table 3.16), $\overline{Z}_B^* = 0.2280$); $\overline{Q}_{B1}$, ...$\overline{Q}_{B19}$ denote local weight values of $i$-th sub-criterion ($i = 1, 2, \ldots, m$), determined by expert evaluation method (Table 3.8 and Fig. 3.7); $x_{B1}$, ..., $x_{B19}$ are the variables of $i$-th sub-criterion of key criterion B, whose estimates are used for determining the real sub-criterion value, ranging from 0 to 1.

The weights of 19 sub-criteria belonging to key criterion B, which describe the organization and technology of travel by train, are found from the survey of assessors, when three categories of respondents (P) and experts (TS and AS), provided their judgements. The mean weight $\overline{Z}_B^*$ shows the significance of key criterion B (when the number of the respondents and experts in each category differs). This weight is calculated by the formula (3.40).

The significance estimates (in normalized weights), provided to the key criterion B by P, TS and AS are presented in Table 5.1.

The sub-criteria variables, obtained (allowable) comparing their actual (real) and the specified permissible, as well as the best and the worst values, are calculated by using the above model. In the present monograph, the considered principle is applied to the evaluation of the quality of travel by trip.

To calculate the CQI, evaluating the significance of organization and technology of travel by train, the variables of any sub-criterion $x_{B1}, \ldots, x_{B19}$, serving as a basis for calculating the real sub-criterion value, ranging from 0 to 1, should be determined.

### *5.1.1 Departure and Arrival of Trains at the Scheduled Time*

Train departure and arrival on the scheduled time is the second most significant of the 49 sub-criteria for respondents (passengers) and experts (train staff and administrative staff) (Table 3.17). Train traffic is organised in accordance with timetables to be followed by all railway divisions.

Timetables play an important role in the organisation of rail traffic. To test the skills of train timetable managers, an assessment of the quality of timetabling is carried out. Timetables are assessed in the light of infrastructure features and traffic flows. The assessment also includes subjective data, such as expert scores. Results can sometimes be unreliable, as experts may have different opinions. It is proposed to use a relative efficiency index for train tracks, i.e. train routes are treated as production units. Each unit uses specific transport resources and generates feedback. The DEA (Data Envelopment Analysis) model is used to calculate the efficiency of the train track. Two statistical functions of train path efficiency are used to evaluate the timetabling quality [2].

Punctuality could be an advantage of rail travel, yet trains are often late. Tiong et al. [3] proposes a set of equations conditional on the train location for analyzing train arrival delay factors at stations. Accurate predictions of train arrival and departure delays improve the passenger service quality and are essential for real-time railway traffic management to minimise their further spreading [4]. To identify train delay patterns and predict train delays [5], records on the meteorological conditions, train delays and train timetables over three months period were collected and analysed. It has been noted that most train delays are caused by bad weather conditions. In normal weather conditions, delays are most often caused by historical train delays and the frequency of delays. By identifying the factors closely related to train delays, a model was developed to predict the delay time of an individual train at each station. The model not only benefits passengers who want to plan their journeys more reliably, it also assists railway operators in developing more efficient train schedules and pricing plans.

Passengers make their choice of the most preferable train depending on their travel preferences, pricing and timetables. The Railway Department, meanwhile, is constantly making adjustments to the train timetable according to the distribution of passenger flows throughout the day. The aim is to ensure optimum operating costs and energy consumption. This raises the question of how to respond to the different travel needs of passengers and achieve the sustainable objectives of service providers? This is why it is crucial to develop a demand-driven and environmentally friendly high-speed rail timetable. Huang et al. [6] formulates an optimisation of the train timetable for a given high-speed rail corridor that satisfies the interests of both passengers and the transport department.

The timing of train arrivals and departures has a direct impact on the performance of the passenger service. However, the existing station system is sometimes unable to accurately track the arrival time of a delayed train. Then the information can be sent to the passenger service system and fused with the information from the transporting dispatching management information system [7]. The system proposed in this paper presents a kind of low-cost solution for railway monitoring, which can effectively solve the problem of how to accurately get the arrival time through the linkage of the video monitoring system and of the passenger service system.

Adherence to the timetable is one of the most important sub-criteria describing the quality of railway operation. The priority (rank) $\overline{R}_{B1} = 1$, of this sub-criterion indicates that it is the most important of the 19 sub-criteria of key criterion B (Table 3.8). In

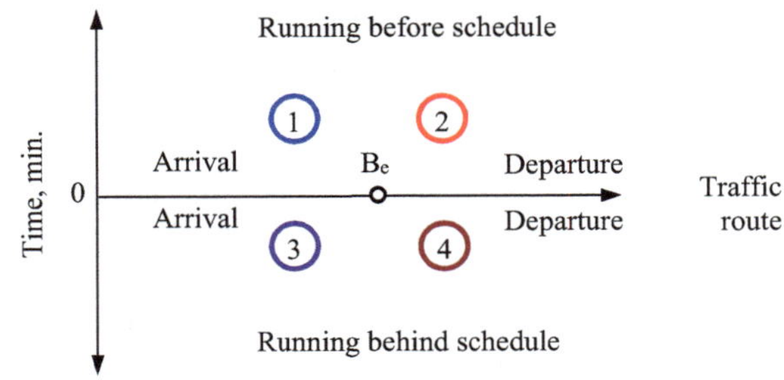


**Fig. 5.1** The model of passenger train's keeping to schedule at the intermediate station $B_e$ ($e = 1, 2, \ldots, b$)

the case of a failure to meet the schedule due to a technical malfunction or natural calamity, the staff members of all departments should take measures to ensure that the delaying passenger and freight locomotives would be running to schedule again because traffic safety depends on it [8]. Trains should arrive on time not only to the final destination point but to intermediate stations as well. If the train is up to 5 min behind schedule, the report is not written. Such cases are considered to be permissible. In terms of schedule, any station is considered from the perspective of four schedule maintenance alternatives (for every intermediate station) (Fig. 5.1):

1. The train arrives at the station before schedule.
2. The train departs from the station before it has had a full standing time (when the stoppage time of the train is cut down because of its arrival at the station behind schedule). It is possible, but not permitted.
3. The train arrives at the station behind schedule.
4. The train departs from the station behind schedule.

The train cannot depart from the station A before scheduled time, though it can depart behind scheduled time (Fig. 5.2).

The train can arrive at the destination station C a little behind or before scheduled time because of various problems on the way. With respect to the schedule there can be four combinations of train arrival at an intermediate station $B_e$ and departure from it (Fig. 5.3).

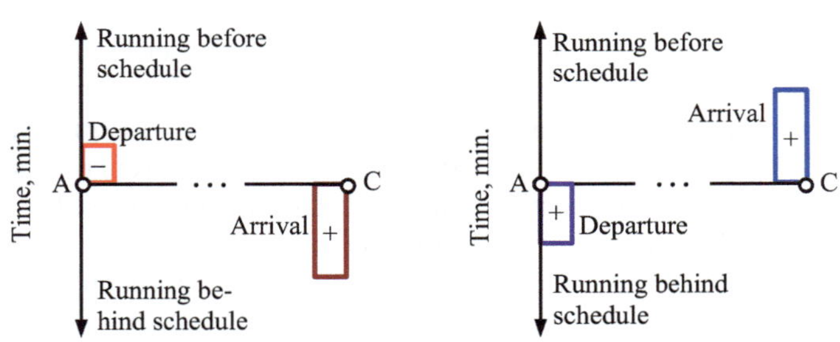


**Fig. 5.2** The model of passenger train adhering to the schedule, when departing from the station A and arriving at the station C

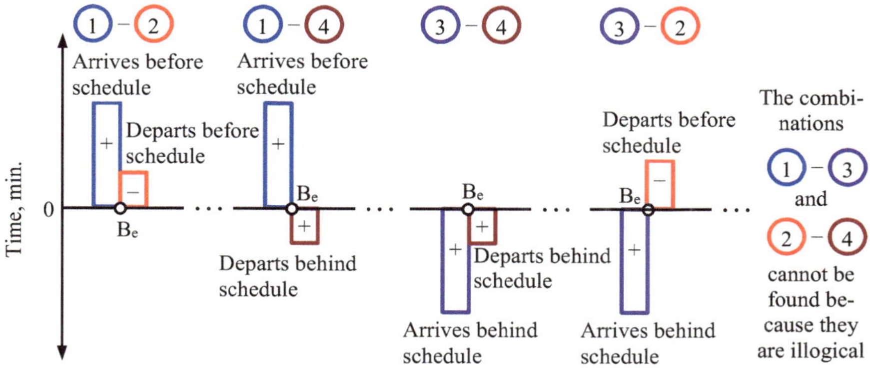


**Fig. 5.3** Model for arrival and departure of a passenger train ahead of schedule and arrival and departure behind schedule to/from intermediate station B

1–2 denote the arrival before schedule and departure before schedule (not permitted).

1–4 denote the arrival before schedule and departure behind schedule.

3–4 denote the arrival behind schedule and departure behind schedule.

3–2 denote the arrival behind schedule and departure before schedule (not permitted).

Co-ordination of bus, train, aircraft and ship timetables is essential for the integration of transport modes. The absence of this coordination can cause train delays and become a disruption to a passenger's plans for a multi-modal journey. This can seriously affect the quality of the journey.

The available combinations of the train's keeping to schedule over the whole route (distance from the departure to the destination station) are shown in Fig. 5.4.

When a train is from 60 to 120 min behind schedule or when the run is cancelled, passengers can require compensation from the company, which is provided for in

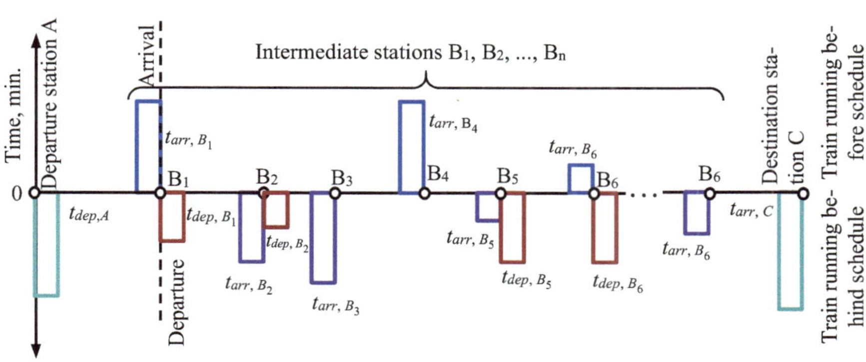


**Fig. 5.4** Various available combinations of train keeping to schedule on the route

the *General rules of passenger transportation* [9]. The train manager or some other person with the train management responsibilities can decide the departure or arrival time of a train on the CQI (e.g. a senior attendant).

The value of the CQI $K$, depending on $K_B$, which describes the organization and technology of travel by train, and the variable $x_{B1}$ of the product $\overline{Q}_{B1} \cdot x_{B1}$, showing how the train keeps to schedule when arriving at or departing from the stations, is calculated by the formula:

$$x_{B1} = 1 - \frac{num}{\beta(2+b)t_{arr,p}^2}, \tag{5.2}$$

The numerator ($num$) of the formula (5.2) is calculated by the equation:

$$num = \gamma \cdot \left[\left(t_{dep,p} + t_{dep,A}\right)^2 - t_{dep,p}^2 + \left(t_{arr,p} + t_{arr,C}\right)^2 - t_{arr,p}^2\right] + (1-\gamma)\sum_{e=1}^{b}\left[\left(t_{arr,p} + t_{arr,B}\right)^2 - t_{arr,p}^2 + \left(t_{dep,p} + t_{dep,B}\right)^2 - t_{dep,p}^2\right]_e,$$

where $t_{dep,A}$ is the delay time of the train's departure from station A, min; $t_{arr,C}$ is the delay time of the train's arrival at the destination station C, min; $t_{arr,B}$ is the delay time of the train's arrival at intermediate station $B_e$, min; $t_{dep,B}$ is the delay time of the train's departure from intermediate station $B_e$, min; $t_{arr,p} = t_{dep,p}$ is the tolerance delay time of the train's arrival to or departure from the departure, destination or intermediate station (taken as 5 min because this delay time does not require any report writing); $b$ is the number of intermediate stations on the route, $e = 1, 2, \ldots, b$; $\beta$ is a coefficient estimating the number of intermediate stations on route $b$ (Table 5.2 and Fig. 5.6); $\gamma$ is a coefficient indicating the degree (level) of "fill-rate" of a train with passengers boarding at intermediate stations $B_e$ on the journey from stations A to C.

The coefficient $\gamma$ is calculated by dividing the number of passengers $N_{A+C}$ who boarded at the departure station A and travelled to the arrival station C by the total number of passengers $N_{total}$, according to formula (5.3):

$$\gamma = \frac{N_{A+C}}{N_{total}} = \frac{N_{A+C}}{N_A + \sum_{e=1}^{b} N'_{Be}} = \frac{N_{A+C}}{N_{A+C} + \sum_{e=1}^{b} N_{Be}}. \tag{5.3}$$

**Table 5.2** The results of calculating the coefficient $\beta$, when there is a certain number of intermediate stations $b$ on the route and $\gamma = 1$ (the average route quality is obtained)

| Number of intermediate station $b$ | 0 | 1 | 2 | 3 | 4 | 5 | 6 | 7 | 8 | 9 | 10 | 11 |
|---|---|---|---|---|---|---|---|---|---|---|---|---|
| The coefficient $\beta$ | 6 | 4 | 3 | 2.4 | 2 | 1.71 | 1.5 | 1.33 | 1.2 | 1.09 | 1 | 0.92 |

The total number of passengers ($N_{total}$) consists of passengers boarding at the starting station A and intermediate stations $B_e$ ($N_A + \sum_{e=1}^{b} N_{Be}$), or of passengers travelling from station A to station C (the number of passengers $N_{A+B}$) and passengers disembarking in the intermediate stations $B_e$ (the number of passengers $\sum_{e=1}^{b} N_{Be}$) (Fig. 5.5). $N_{total}$ is not equal to the sum of the number of embarkations and disembarkations.

Applying the principle of a sustainable balance between the number of embarking and disembarking passengers, the following options are used to calculate the coefficient $\gamma$ for the "fill-rate" of a train with passengers on a journey:

1. The train is empty (without passengers) before boarding commences at the starting station A. On arrival at the terminus station C, all passengers on the train are disembarked, i.e. the train is left empty. The number of passengers embarking and disembarking at all stations is equal.

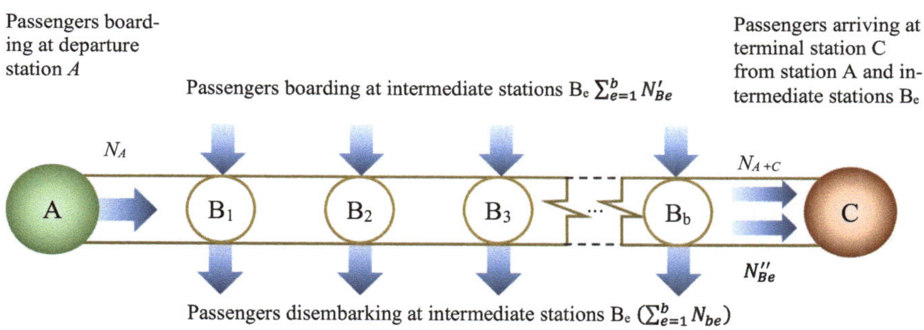


**Fig. 5.5** A model for the establishment and distribution of train passenger flows on a journey

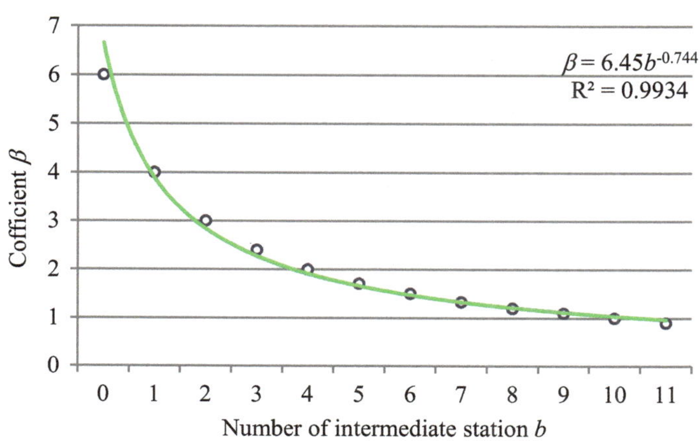


**Fig. 5.6** The dependence of the coefficient $\beta$ on the number of intermediate stations on the route $b$

2. All passengers (the number is $N_{total}$) holding tickets from the departure station A to the arrival at the terminus station C, and passengers embarking and disembarking at the intermediate stations $B_e$ shall be considered as having travelled on the train.
3. The number of passengers on board is not physically measured (counted), but is recorded through the booking of travel tickets. The number of passengers on the train is recorded in the carriage passenger record sheet, which is filled in by carriage attendants. Illegal passengers are not accounted for.
4. The maximum number of passengers that can travel on a train at any one time is equal to the total number of seats in all its carriages. Along the entire route from stations A to C, the train can carry not only fewer but also more passengers than the number of seats available on the train, with a large number of passengers embarking and disembarking at intermediate stations $B_e$.
5. Only passengers departing from a starting station A and travelling to terminus station C (their number $N_{A+C}$) are affected by timetable adherence at both stations. For passengers embarking and disembarking at intermediate stations $B_e$, the timetable at station A is irrelevant. For some passengers who boarded at intermediate stations $B_e$ and travelled to the terminus station C (their number $N''_{Be}$), the arrival delay is also of importance, however, it is not considered in the model (5.2).

The coefficient can vary from 1 to 0.

When = 1, it indicates that all passengers boarding at the starting station A travelled to the terminus station C, and no passengers boarded at the intermediate stations $B_e$.

For train passengers, adherence to the timetable (delays and early arrivals) is therefore irrelevant when passing through intermediate stations $B_e$.

When the coefficient = 0.5, it indicates that half (50%) of the passengers travelled from the starting station A to the terminus station C, while the other half (50%) of the passengers disembarked at the intermediate stations $B_e$, and that they are sensitive to the adherence of the timetable at intermediate stations along the route.

When = 0, it indicates that all passengers boarded and disembarked the train only at intermediate stations $B_e$, therefore the adherence to the timetable for the departure of the train at station A and the arrival at station C is not important for them and has no impact on the quality of their journey.

The permissible train delay time (delay tolerance) $t_{arr.p}$ and $t_{dep.p}$ may vary from country to country or from route to route and may not necessarily equal 5 min.

It depends on the volume of train traffic on the route and the compatibility of train timetables with the timetables of other public transport modes (road, air).

The calculated values of the coefficient $\beta$ that depend on the number of intermediate stations $b$ along the train route are given in Table 5.2.

As the number of intermediate stations $b$ on the route increases, the coefficient $\beta$ decreases according to the exponential relationship (Fig. 5.6). The coefficient $\beta$ has been calculated in a way that its value allows to ensure (have, guarantee) an average level of passenger transport quality according to the variable $x_{B1} = 0.5$ of sub-criterion

B1. It is assumed that $x_{B1} = 0.5$ when the train delays its arrival or departure by 5 min, as much as the tolerance $t_{arr,p} = t_{dep,p}$ allows. The graph (Fig. 5.6) shows the case when $\gamma = 1$, i.e. when all passengers travel from the starting (departure) station A to the final (arrival) station C.

### 5.1.2 Delivering Meals to Passengers

Meals are often included in the first-class carriage fare for international trains. Breakfast is sometimes served in the sleeping carriages of night trains. However, not all trains in the world are catering for passengers smoothly. Vending machines are used not only to sell tickets, but also food, drinks and souvenirs.

Food quality, food delivery and prices are often an issue on Indian trains. Even on trains that are considered to be better, passengers are sometimes given outdated, overpriced food, or they don't get food when they want it. Indian Railways is in the process of addressing this issue. Vending machines are planned to be installed to dispense ready-to-eat food to customers 24 h a day, 7 days a week. The vending machine will offer passengers the products they want at maximum retail price. It will have two payment options—cash and card. The scheme is to install vending machines in every third carriage. In addition to ready-to-eat food, other food options will be available from the machine. A brief description of the selected dish and its price will be displayed. The vending machine will sell both vegetarian and conventional food [10].

Passengers in different carriages may be served differently when it comes to meals on board. The model therefore assesses the quality of the food service on each carriage.

Two quality levels have been used to assess this sub-criterion: 0 means that train carriage meals are not included in the ticket price for travel in the train carriage and, therefore, is not provided to passengers, 1 means that meals are included in the ticket price for travel in this train carriage and is provided to passengers. Its mean rank (priority) 14 ($\overline{R}_{B2} = 14$) out of the 19 sub-criteria of the key criterion B signals average significance (Table 3.8). The variable $x_{B2}$ of the CQI $K_B$, evaluating the significance of the sub-criterion, describing this service, is calculated by the formula:

$$x_{B2} = \frac{\sum_{c=1}^{n_{car}} \frac{PP_{fc}}{PP_{max}}}{n_{car}}, \tag{5.4}$$

where: $x_{B2}$ is the variable of the second sub-criterion of key criterion B, based on which the actual sub-criterion value, ranging from 0 to 1, is calculated; $PP_{fc}$ is the actual evaluation in points (0 or 1) of delivery rations provision for passengers in the $c$-th car ($c = 1, 2, \ldots, n_{car}$); $PP_{max}$ is the maximum number of points (1) given to rations provision; $n_{car}$ is the number of railway cars in the train.

### 5.1.3 Provision, Layout, and Condition of Bed Linen

Since 2010, the number of overnight passenger trains in Europe has decreased significantly. The "ideal" night train can run non-stop from 22:00 to 8:00 h and allow a passenger to sleep for 8 h or more, but this is rare. On overnight trains, extra service such as berth bedding is available [11]. Sleeping carriages can be of various designs. The berths are used as seats during the day and are laid out for the night by a passenger or carriage attendant. The carriage bunks are generally padded.

Each passenger is provided with a blanket and a pillow for sleeping, along with a set of bed linen. The quality of a train passenger's sleep depends largely on the comfort of the berth (mattress) and the cleanliness and quality of the bedding. The higher the quality of the provision, laying and condition of the bed linen, the higher the score for this sub-criterion, ranging from 0 to 10.

Each carriage on the train is assessed. Score 0 is given when there is no bed linen in the carriage. Score 1 is given when: a passenger is presented with bed linen that is not wrapped in special sealed packaging, so there is no guarantee that it has not been used; a passenger has to pay for the bed linen in cash; there is no obligation for the carriage attendant to lay it out. Score 2 is awarded when: a passenger is presented with bed linen that is wrapped in a special sealed package, so there is a guarantee that the bed linen has not been used; a passenger has to pay for the bed linen in cash; the attendant is not obliged to lay it out for the passenger. Score 3 is awarded when bed linen is included in the ticket fare and it is packaged in a special hermetically sealed container, and the carriage attendant is not obliged to lay it out for the passenger. A carriage receives score 4 when bed linen is included in the ticket fare and packed in a special hermetically sealed container, and the carriage attendant is obliged to lay out the bed linen at the request of the passenger. A carriage is assessed with a score 5 when the passenger is provided with bedding included in the ticket fare, the lower berths of the carriage are laid, but the quality of the bedding is questionable, and the attendant is required to lay out the upper berths at the request of the passenger. A carriage is awarded a score 6 when the passenger is provided with bed linen included in the ticket fare, the lower berths of the carriage are laid out, the bed linen is of good quality (clean and ironed), and the attendant lays out the upper berths on the passenger's request. A carriage is scored 7 if the passenger is provided with bed linen included in the ticket fare, the lower and upper berths of the carriage are laid out, but the quality of the bed linen is questionable. A carriage is scored 8 when the passenger is provided with the bedding included in the ticket fare, all berths in the carriage are made up and the quality of the bedding is good. A carriage receives a score 9 when the passenger is provided with bed linen included in the ticket fare, all berths in the deluxe carriage are made up, but the quality of the bed linen is questionable. A maximum of 10 points is awarded for a carriage where the passenger is provided with bedding included in the ticket price, all berths in the deluxe carriage are made up and the quality of the bedding is high.

Its mean rank (priority) is $\overline{R}_{B3} = 8$ (Table 3.8) indicates the average significance of this sub-criterion across the other sub-criteria of key criterion B.

The variable $x_{B3}$ of the CQI $K_B$, evaluating the delivery of bedclothes to passengers, their state and making the bed for passengers is calculated by the formula:

$$x_{B3} = \frac{\sum_{c=1}^{n_{car}} \frac{BL_{fc}}{BL_{max}}}{n_{car}}, \tag{5.5}$$

where $x_{B3}$ is the variable of the third criterion of key criterion B, used for calculating the actual sub-criterion value ranging from 0 to 1; $BL_{fc}$ is the actual estimate in points (0, 1, 2, 3, 4, 5, 6, 7, 8, 9 or 10 points) delivery of bedclothes, their condition and making up the bed for passengers; $BL_{max}$ is the highest possible estimate of the considered service (10 points); $n_{car}$ is the number of railway cars in the train ($c = 1$, 2, …, $n_{car}$).

### 5.1.4 Ordering Food and Drinks from the Dining-Car and Delivering Them to the Compartment by Carriage Attendant

First-class passengers have high expectations of the carriage attendant. Attendants are expected to make beds, carry luggage, explain train procedures and deliver refreshments and newspapers. Long-distance train attendants often take orders from carriage passengers and bring them meals from the train restaurant car [12].

The variable $x_{B4}$ of this sub-criterion is calculated for the whole train, since the service can be requested by passengers of all carriages.

The actual value $FD_f$ for sub-criterion B4 varies from 0 to 3: 0 means that this service is not available on the train and passengers should go to the dining-car themselves, 1 shows that only passengers of two-seat compartments of the first-class cars can order food and drinks to be delivered from the dining-car, 2 means that only passengers of two- or four-seat compartments of the first-class cars can order food and drinks to be delivered from the dining-car, 3 means that passengers of all railway cars can order food and drinks to be delivered from the dining-car. Its mean rank (priority) is $\overline{R}_{B4} = 13$ across the 19 sub-criteria that make up key criterion B. The variable $x_{B4}$ of the CQI $K_B$, evaluating the considered service (criterion), is calculated as follows:

$$x_{B4} = \frac{FD_f}{FD_{max}}, \tag{5.6}$$

where $x_{B4}$ is the variable of the fourth sub-criterion of key criterion B, used for calculating the actual sub-criterion value ranging from 0 to 1; $FD_f$ is the actual estimate in points (0, 1, 2 or 3 points) possibility of ordering food and drinks from the dining car to be delivered to the compartment (via car attendant); $FD_{max}$ is the highest possible estimate of the considered service in points (3 points).

### 5.1.5 On-Board Delivery of Popular Newspapers and Magazines

Until the early 1900s, many people travelled to work by train, commuting from the suburbs and villages to the cities. Arnold Bennett, a popular novelist of the time, encouraged people to take advantage and make the best use of this new and unfamiliar "our time". The train was a great place to sit and read. It is rumoured that Allen Lane, a book publisher, came up with the idea of publishing best-selling novels in paperback and offering them for sale in railway stations, grocery and department stores while waiting for a train at Exeter station in 1934. He was upset that the station had no books for sale. The average British commuter spends 1 h 40 min on the train every day. Taken over the entire year, that's a lot of time spent on the road doing nothing.

No wonder many people do read [13]. For people with busy lives and time pressures, travelling by train is a great opportunity to read. Many of the world's railway stations and trains have libraries and distribute periodicals.

Passengers on all carriages of the train can make use of this service, which is rated from 0 to 3.

0 means that popular newspapers and magazines are not available on the train, 1 shows that passengers can buy popular newspapers and magazines on the train, 2 shows that some free-of-charge and paid newspapers are available on the train, 3 means that all newspapers and magazines on the train are free of charge. Its mean rank (priority) of significance is $\overline{R}_{B5} = 15$ (Table 3.8). The variable $x_{B5}$ of the CQI $K_B$, evaluating the considered service, is calculated by the formula:

$$x_{B5} = \frac{NM_f}{NM_{max}}, \tag{5.7}$$

where $x_{B5}$ is the variable of the fifth sub-criterion of key criterion B, used for calculating the actual sub-criterion value, ranging from 0 to 1; $NM_f$ is the actual service estimate in points (0, 1, 2 or 3 points) distribution of popular newspaper and magazines on the train; $NM_{max}$ is the highest possible service estimate (3 points).

### 5.1.6 On-Board Internet Access

Internet access on a train is one of the most often discussed issues in the modern railway industry. Its mean rank (priority) of significance is $\overline{R}_{B6} = 11$ (Table 3. 8). More and more railway companies (especially those which operate suburban trains) provide a free wireless network (WiFi) service. In this way, these companies ensure the satisfaction of their existing customers during the trips and attract even more customers. The implementation of similar services is often difficult due to ineffectiveness of the employed technologies and incompatibility with the rules of the Internet connection use in the public space. Passenger expectations are changing which raises the issues like: what level of service they can expect during their journey and what they can do during that time. In the Netherlands, for example, WiFi is installed on almost all trains. Most passengers have internet access on their mobile phones, making it less required on the train. Power and USB sockets are increasingly important. Passengers who use a computer on board a train often tend to work on more than one device [14].

Internet access (IA) on a train is assessed as follows: 0 means the absence of the IA on a train, 1 means the presence of paid IA on a train, 2 means the presence of free IA on a train. The variable $x_{A6}$ of the CQI $K_B$, evaluating the considered service on a train, is calculated by the formula:

$$x_{B6} = \frac{IA_f}{IA_{max}}, \tag{5.8}$$

where $x_{B6}$ is the variable of the sixth sub-criterion of key criterion B, the value of which is used for calculating the actual sub-criterion value, ranging from 0 to 1; $IA_f$ is the actual estimate in points (0, 1 or 2 points) possibility to access the internet; $IA_{max}$ is the highest possible service estimate (2 points).

### 5.1.7 On-Board Ticket Purchase Option

Transport operators provide information on ticketing, and passengers choose the most favourable way of purchasing their ticket. Train tickets can usually be bought at the ticket office in the station, from a ticket machine or on the Internet. Railway companies are working to make the ticketing procedure easily accessible to passengers.

The Chinese government has introduced information and communications technology (ICT) to facilitate ticket purchases during The Spring Festival Travel Rush (SFTR). The study developed a structural equation model to investigate the impact of the online ticket booking system, socio-economic status and perceived control on passengers' well-being during SFTR. The findings show that the online booking system has a direct positive impact on passengers' well-being. The online ticket booking system has made it easier for people from lower socio-economic backgrounds to buy tickets [15].

Ticketing locations and methods have to be adapted to the needs of different passengers. The Office of Rail Regulation (ORR) in Great Britain has undertaken a review of rail ticket retailing and commissioned research into the rail passenger market to understand passenger attitudes towards rail ticketing. Some surveyed passengers expressed a preference for non-traditional ticketing locations such as newsagents and supermarkets [16]. Tickets can often be purchased on board the train (from the train manager or conductor), but the service charge usually makes tickets more expensive.

Not every passenger who wants to travel by train buys a ticket in advance at a ticket office or online. This is due to subjective reasons. If a passenger boards a train without a ticket, he or she can only travel legitimately after buying a ticket from the train manager, if this service is available.

The availability of this service is scored: 0 means that this service is not provided on the train, 1 means that the considered service is available in an ordinary sleeping (sitting) car, 2 shows that the considered service is provided in the sleeping car with four-seat compartments, 3 means that the service is provided in an ordinary sleeping or sitting car and in the sleeping car with four-seat compartments, 4 means that the service is available in an ordinary sleeping (sitting) car, a sleeping car with four-seat compartments and a first-class car with two-seat compartments, 5 shows that the considered service is provided in all types of cars.

The mean rank (priority) $\overline{R}_{B7} = 6$ of the significance of this sub-criterion indicates that the service is of considerable importance to passengers (Table 3.8). The variable $x_{B7}$ of the CQI $K_B$, evaluating the considered service (sub-criterion), is calculated by the formula:

$$x_{B7} = \frac{BT_f}{BT_{max}}, \tag{5.9}$$

where $x_{B7}$ is the variable of the seventh sub-criterion of key criterion *B*, used for calculating the actual sub-criterion value, ranging from 0 to 1; $BT_f$ is the actual estimate in points (0, 1, 2, 3, 4 or 5 points) possibility of buying a ticket on the train (from the train manager); $BT_{max}$ is the highest possible service estimate (5 points).

### 5.1.8 Dining-Car Reservations

In most cases, the dining-car has fewer seats compared to the number of passengers on the train. Therefore, on long train journeys, and especially when the journey starts in the evening and continues into the night, passengers wishing to dine out want a better guarantee of access to the dining-car.

In Switzerland and Europe, for a comfortable train journey, reservations for seats in the carriage and in the dining-car should be made in advance of the journey. For some countries, reservations are mandatory or automatically included in the ticket price. Depending on the country and the railway company, it may be possible to reserve seats in the dining-car for individuals or groups. Reservations may be invalidated if it becomes necessary to change rolling stock shortly before the departure of a train [17]. On some trains, reservations can be made through the carriage attendant.

Reservation of a seat in a dining-car is not significant for the quality of the journey (priority $\overline{R}_{B8} = 17$) (Table 3.8).

This option is scored: 0 means that this service is not provided on the train, 1 denotes that tables in the dining-car may be reserved only for passengers of the first-class cars with two-seat compartments, 2 means that tables in the dining-car may be reserved for passengers of the first-class cars with two- and four-compartments, 3 shows that tables in the dining-car may be reserved for passengers of all types of railway cars.

The variable $x_{B8}$ of the CQI $K_B$, evaluating the considered service (sub-criterion), is calculated by the formula:

$$x_{B8} = \frac{DCR_f}{DCR_{max}}, \tag{5.10}$$

where $x_{B8}$ is the variable of the eighth sub-criterion of key criterion B, used for calculating the actual sub-criterion value, ranging from 0 to 1; $DCR_f$ is the actual estimate in points (0, 1, 2 or 3 points) possibility to reserve a table in the dining-carriage; $DCR_{max}$ is the highest possible service estimate (3 points).

### 5.1.9 Taxi Call Facility

Taxi call for a train passenger option provided by train and carriage attendant. In Europe, taxi services account for 5% of public transport services. European travellers appreciate the reliability and consistency of travel conditions. A taxi service is perfectly tailored to meet these expectations. Taxi companies offer flexible and customer-oriented door-to-door services 24 h a day. Taxis are comfortable, safe, and easily accessible. Taxi companies offer personalised, high-quality customer service. Taxi drivers can come to pick up passengers from airports, train stations, etc. Flexible services provided by taxi companies as well as by companies offering other modes of transport, play a key role in sustainable human mobility [18]. For example, transport infotainment platform provider "Gomedia" has partnered with shared travel specialist "Moses Mobility" to provide rail passengers with a tool to book a taxi while on the train. "Moses Mobility" is partnering with several ride-hailing services, including "Uber" and "Gett", to ensure sustainable trips. Passengers travelling to nearby destinations are identified. They are offered a taxi ride together with other passengers from the station to their destination. The company says the same technological solution can be applied in the event of rail disruption to make passenger transport easier, more reliable, and more cost-effective. Transport management is becoming smarter as technology improves [19].

Calling a taxi from the train is rated as follows: 0 means that the service is not provided on the train, 1 shows that there is a possibility to call a taxi for a passenger at the terminal (train formation) station, but it is a paid service, 2 means that there is a possibility to call a taxi for a passenger at the terminal (train formation) station and this service is provided free of charge, 3 denotes that there is a possibility to call a taxi for a passenger at the terminal (train formation or destination) station, but it is a paid service, 4 means that it is possible to call a taxi for a passenger at the terminal (train formation or destination) station and this service is provided free of charge, 5 shows that the considered service can be provided at any station, where the train stops, but it is a paid service, 6 means that the service can be provided at any station, where the train stops, and it is free of charge.

The availability of a taxi is irrelevant for the quality of the train journey, as the priority (rank) of this sub-criterion is $\overline{R}_{B9} = 18$ when compared to the other sub-criteria of the key criterion B (Table 3.8). The variable $x_{B9}$ of the CQI $K_B$, evaluating the service under consideration (sub-criterion), is calculated by the formula:

$$x_{B9} = \frac{TCF_f}{TCF_{max}}, \tag{5.11}$$

where $x_{B9}$ is the variable of the ninth sub-criterion of key criterion B, used for calculating the actual sub-criterion value, ranging from 0 to 1; $TCF_f$ is the actual estimate in points (0, 1, 2, 3, 4, 5 or 6 points) possibility of ordering a taxi; $TCF_{max}$ is means its highest possible service estimate (6 points).

### *5.1.10 On-Board Card Payment Option*

Option to pay on board using a card. The introduction of the *MoKa* mobile payment system on long-distance trains has enabled German train passengers visiting a restaurant car to pay not only in cash. Previously, passengers had to pay for their food and beverages in cash or by credit card at the stationary counter. Since mid-March 2020, all old tills have been replaced by the *MoKa* system. Now, railway staff carry a card reader. They no longer need to go to a stationary till in the restaurant to arrange payments. Passengers have the option to pay using other payment methods and not just cash. Staff can offer food and beverages and take orders from a passenger when standing at the passenger seat. Service staff do not need to walk around unnecessarily. This system is regularly being updated, as a train restaurant is very different from the one that operates in a stationary building. On land, problems rarely arise when using a fixed or wireless network to connect to your own servers or those of payment service providers. However, on a train with speeds of up to 320 km/h, it is practically impossible to maintain a constant connection. There are many reasons for this, e.g., the transition between radio cells that the train passes through during its journey, poor network coverage along the route, border crossings or sometimes just the weather [20].

The option to pay for services on board by card is valued in the following way: 0 means that this service is not provided on the train, 1 denotes that the service is provided on the train, but a passenger should go with the train conductor to the dining car, where a scanner of electronic cards is installed, 2 means that the considered service is provided, while the scanner of electronic cards is located at the train conductor's workplace, 3 shows that the considered service is provided on the train, with the scanner equipped in every railway car. The variable $x_{B10}$ of the CQI $K_B$, evaluating the service under consideration (sub-criterion), is calculated by the formula:

$$x_{B10} = \frac{PC_f}{PC_{max}}, \tag{5.12}$$

where $x_{B10}$ is the variable of the tenth sub-criterion of key criterion B, used for calculating the real sub-criterion level, ranging from 0 to 1; $PC_f$ is the actual estimate in points (0, 1, 2 or 3 points) possibility of setting for onboard services by payment card; $PC_{max}$ is the highest possible service estimate (3 points).

### 5.1.11 Souvenir Shopping on the Train

Souvenirs that embody the spirit and the culture of the country visited can last a lifetime and bring back memories. However, sometimes people fail to find the right souvenir, or are unable to buy one for other reasons. The on-board souvenir shop would then be very beneficial for a train passenger. Souvenirs and souvenir retailing is for the most part an untapped field of research [21, 22].

When travelling by train from one country to another, it's nice to be able to buy souvenirs that remind you of your journey or have a practical use. Respondents and experts rank this option as the least important sub-criterion. Its priority across the 19 sub-criteria of the key criterion B is $\overline{R}_{B11} = 19$ (Table 3.8). The impact of this service on the quality of the journey is assessed by a score.

The following scores apply to souvenir shopping on the train: 0 means that souvenirs are not sold on the train; 1 shows that souvenirs are sold on the train. The variable $x_{B11}$ of the CQI $K_B$, evaluating this service (sub-criterion), is calculated by the formula:

$$x_{B11} = \frac{SS_f}{SS_{max}}, \tag{5.13}$$

where $x_{B11}$ is the variable of the eleventh sub-criterion of key criterion B, used for determining the real sub-criterion level, ranging from 0 to 1; $SS_f$ is the actual estimate in points (0 or 1 points) sale of souvenirs on the train; $SS_{max}$ is the highest possible service estimate (1 point).

### 5.1.12 Meeting Passenger Preferences for On-Board Music and In-Train Information

Evaluation of the radio music broadcast and in-train information with reference to the feedback from passengers. On the train, passengers can be adversely affected by the noise from the rail-wheel interaction, music or loud talking from fellow passengers. In addition to its negative effects on hearing, noise also impairs the functioning of the human central nervous system. Noisy environments are annoying, cause fatigue and weaken concentration. A study was conducted to evaluate the impact of music tempo and cleanliness on train for passengers on Dutch railways (NS: Dutch Railways) [23]. The study focused on the music tempo (slow and fast) and cleanliness (dirty compartment and clean compartment) with respect to the passenger perception of the environment and time. The experiment took place in a simulated train compartment with fast or slow music. After a simulated train journey, 123 respondents answered questions about their perception of time (subjective time and time estimation) and their experience about the environment (pleasure, excitement, dominance, appreciation of the journey, upcoming intentions). The study found that with slow music and a clean room, passengers had a feeling that the journey was shorter. They also experienced less (negative) agitation. In the dirty compartment of the train, they failed to feel the rhythm of the music and they lost track of the journey time.

The priority (mean rank) of this sub-criterion is $\overline{R}_{B12} = 16$. It is the fourth from the bottom and the one that shows low importance for the quality of passenger traffic (Table 3.8). In the questionnaire that was delivered to passengers on the train, passengers had to rate the music broadcast and in-train information with reference to the following points: 0 very poor, 1 poor, 2 satisfactory, 3 very good, 4 excellent. The variable $x_{B12}$ of the CQI $K_B$, evaluating this service (sub-criterion), is calculated by the formula:

$$x_{B12} = \frac{\sum_{p=1}^{n_P} \frac{RM_{fp}}{RM_{max}}}{n_P}, \quad (5.14)$$

where $x_{B12}$ is the variable of the twelfth sub-criterion of key criterion B, used for calculating the actual sub-criterion level, ranging from 0 to 1; $RM_{fp}$ is the actual estimate in points (0, 1, 2, 3 or 4 points) music broadcast and information in conformity with passenger requests; $RM_{max}$ is the highest possible service estimate (4 points); $n_P$ is the number of respondents (passengers), evaluating the considered service in points ($p = 1, 2, \ldots, n_P$).

### *5.1.13 Securing Passenger Luggage and Belongings on Board the Train*

It is generally accepted that it is passenger responsibility when it comes to the security of their luggage and belongings. However, in one case, a 74-year-old passenger on an Indian train was finally compensated for the loss of her valuables after a 14-year legal battle with the railways. The Delhi State Consumer Commission (DSCC) considered that the railways had a duty to take care of the train passenger. The passenger must arrive safely at the destination with their luggage and belongings. And rail staff are responsible for this [24].

Passengers usually carry personal documents, money, belongings and luggage on board the train, the proper protection of which is a key factor in the quality of the journey. The priority of this sub-criterion B13 is $\overline{R}_{B13} = 2$ (Table 3.8). Security of these items can be provided by the installation of lockers or it is passenger responsibility to safeguard them, i.e. there is no possibility of storing items securely in the train carriages.

This service is graded by score: 0 means that the passenger's luggage, documents and other things are not protected on the train. A passenger himself should be responsible for their protection, 1 shows that the luggage is not protected, while the documents and other valuable things on the train are protected for the specified payment, 2 means that the passenger's luggage is not protected, while documents and other valuable things on the train are protected free of charge, 3 means that the passenger's luggage, documents and other things on the train are protected for the specified payment, 4 denotes that there are locked containers for passenger's luggage on the train, therefore, his/her luggage and other things are protected. The variable $x_{B13}$ of the CQI $K_B$, evaluating this service (sub-criterion), is calculated by the formula:

$$x_{B13} = \frac{PL_f}{PL_{max}}, \tag{5.15}$$

where $x_{B13}$ is the variable of the thirteenth sub-criterion of key criterion B, used for calculating the actual sub-criterion level, ranging from 0 to 1; $PL_f$ is the actual estimate in points (0, 1, 2, 3 or 4 points) protection of passenger luggage and other valuable things on the train; $PL_{max}$ is the highest possible service estimate (4 points).

### 5.1.14 Purchasing Overseas Medical Insurance on Board the Train

On international trains, purchasing an overseas medical insurance is usually not available or necessary. A person intending to travel abroad obtains insurance at the time the visa is issued. However, passengers and company staff give this option a fairly high priority (rank) $\overline{R}_{B14} = 7$ (Table 3.8).

This option has been scored: 0 means that the service is not provided on the train, 1 means that the service is provided on the train, 2 denotes that there is no need to get a foreign health insurance policy because the health of a passenger going abroad is insured when he/she gets a visa. The variable $x_{B14}$ of the CQI $K_B$, evaluating this service (sub-criterion), is calculated by the formula:

$$x_{B14} = \frac{HIP_f}{HIP_{max}}, \tag{5.16}$$

where $x_{B14}$ is the variable of the fourteenth sub-criterion of key criterion B, used for calculating the actual sub-criterion level, ranging from 0 to 1; $HIP_f$ is the actual estimate in points (0, 1 or 2 points) possibility to get a foreign health insurance policy on the train; $HIP_{max}$ is the highest possible service estimate (2 points).

### 5.1.15 Obtaining a Visa at the Frontier Point

When travelling on an international train outside the Schengen Agreement countries, a visa is required in a passenger passport, the terms and conditions of which are established by the countries. A passenger with no visa cannot cross the state border and must get off the train, as in most cases a visa is not issued at the frontier. This fact does not contribute to the quality of international train travel. Among the 19 sub-criteria of the key criterion B, the availability to obtain a visa at the frontier point is highly rated by respondents and experts: priority $\overline{R}_{B15} = 3$ (Table 3.8).

This service is scored like this: 0 means that there is no such possibility during the trip by train, 1 means that the considered service may be provided, but a passenger should interrupt the trip and, after getting a visa, take another train to continue travel, 2 implies that a passenger can get a visa and continue the travel on the same train. The variable $x_{B15}$ of the CQI $K_B$, evaluating this service (sub-criterion), is calculated by the formula:

$$x_{B15} = \frac{GV_f}{GV_{max}}, \tag{5.17}$$

where $x_{B15}$ is the variable of the fifteenth sub-criterion of key criterion B, used for calculating the actual sub-criterion level, ranging from 0 to 1; $GV_f$ is the actual service estimate in points (0, 1 or 2 points) possibility to get a visa at the frontier post; $GV_{max}$ is the highest possible service estimate (2 points).

### 5.1.16 The Appearance of the Train Staff

On the first trains, train attendants were professionals and had to be dressed in a way that made them feel they were going on a trip of a lifetime. Since the beginning of rail transport, railway employees have impressed the public with their appearance. Since 1840, railway staff have worn exquisite clothes. This gave rise to the rail staff uniform and nameplates. Train attendants were the employees who interacted the most with passengers. They were the railway company representatives to the people travelling on the train. They had to be dressed up to match the glamour of the carriage they were in charge of. From 1870 onwards, railway companies began to issue official uniforms to all train crews except the engine crew. Railway uniforms have changed dramatically since the early days. They are now tailored according to contemporary fashion trends [25]. The uniform of the train staff of today is to be practical: suitable for a wide range of working conditions and temperature changes, resistant to frequent washing and cleaning. Clothing must be safe and fit the needs of the staff member wearing it. It must also be clean, tidy and in style, and represent the rail company vision.

The appearance of train staff is rated by respondents and experts as of moderate importance, the B16 sub-criterion priority (rank) is $\overline{R}_{B16} = 9$ (Table 3.8).

The uniform, shoes, hairstyle, nameplate, the opinions of passengers should be considered. For this purpose, a 20 special questionnaire was prepared, where a passenger evaluating this sub-criterion could give it the following points: 0 means very poor, 1 indicates poor, 2 means satisfactory, 3 means very good, 4 denotes excellent. The variable $x_{B16}$ of the CQI $K_B$, evaluating this service (sub-criterion), is calculated by the formula:

$$x_{B16} = \frac{\sum_{p=1}^{n_P} \frac{ATS_{fp}}{ATS_{max}}}{n_P}, \tag{5.18}$$

where $x_{B16}$ is the variable of the sixteenth sub-criterion of key criterion B, used for calculating the actual sub-criterion level, ranging from 0 to 1; $ATS_{fp}$ is the actual estimate in points (0, 1, 2, 3 or 4 points) exterior the appearance of TS members; $ATS_{max}$ is the highest possible estimate (4 points); $n_P$ is the number of respondents (P), evaluating the considered sub-criterion in points.

### *5.1.17 Communication Culture of Train Service Staff with Passengers and Other Members of Staff*

One's inner culture affects the way one interacts with other people. People with similar cultural values find it easier to communicate. The staff member serving the train cannot choose co-workers or passengers. Trains transport people of different characters and cultures. They are not always well-disposed. A passenger may be emotionally agitated, scared, angry or have a disability. The quality of a passenger's journey depends to a large extent on the communication, conflict management and decision-making skills of the train staff. Sometimes situations may be unconventional. For example, the kindness and creativity of the train manager on International Passenger Train No 5 facilitated the successful arrival to the destination of a passenger carrying 20 cages for cats each containing two or three cats (up to two animals can be transported in one individual compartment) [26]. Developing a culture of communication within the company is essential to increase employee job satisfaction and loyalty to the company.

The culture of communication between employees influences the quality of teamwork, productivity, efficiency and the psychological microclimate.

Providing various services to passengers, a car attendant should conform to the rules of communication with passengers: to meet passengers in a friendly way, to listen to them attentively and to answer their questions, to be always polite and tactful, to speak to the passengers calmly and express his/her ideas concisely and clearly [27]. The car attendant should not be disrespectful to colleagues and superiors, use profanity and make inappropriate remarks [28]. Evaluating the communication of the service staff with passengers and each other, the opinions of passengers should be considered. The priority (mean rank) $\overline{R}_{B17} = 4$ of this sub-criterion indicates its very high importance across the 19 sub-criteria of Key Criterion B (Table 3.8). For this purpose, a special questionnaire was designed, where passengers give the following points to the considered issue: 0 means very poor, 1 means poor, 2 denotes satisfactory, 3 means very good and 4 means excellent. The variable $x_{B17}$ of the CQI $K_B$, evaluating this sub-criterion, is calculated by the formula:

$$x_{B17} = \frac{\sum_{p=1}^{n_P} \frac{CCTS_{fp}}{CCTS_{max}}}{n_P}, \tag{5.19}$$

where $x_{B17}$ is the variable of the seventeenth sub-criterion of key criterion B, used for calculating the actual sub-criterion level, ranging from 0 to 1; $CCTS_{fp}$ is the actual estimate in points (0, 1, 2, 3 or 4 points) communication of the train service staff with passengers and other members of staff; $CCTS_{max}$ is the highest possible sub-criterion estimate (4 points); $n_P$ is the number of respondents (passengers), evaluating the considered sub-criterion in points.

### 5.1.18 *Proficiency in Foreign Languages of the Train Staff*

In line with the EU language policy, mother tongue is for identity, English is for communication, and to be a true European you need to know at least one or two other languages. Employers who acknowledge the importance of emotional intelligence appreciate the benefits of multilingualism. Especially those who need to recruit staff prepared to work internationally. It has been proven that people who speak several languages have a better understanding and appreciation of other countries and cultures.

They are much more tolerant [29].

The foreign language proficiency of TS is of medium significance: priority of sub-criteria B18 is $\overline{R}_{B18} = 10$ (Table 3.8).

The impact of the language proficiency of TS on the quality of the train journey is scored as follows: 0 means that a person does not know any foreign language, 1 shows that a person knows (to some extent) a foreign language, allowing him/her to communicate with passengers, 2 means that a person knows (to some extent) two foreign languages, 3 means that a person knows (to some extent) three foreign languages, 4 means that a person knows (to some extent) four and more foreign languages. The variable $x_{B18}$ of the CQI $K_B$, evaluating this sub-criterion, is calculated by the formula:

$$x_{B18} = \frac{\sum_{a=1}^{n_{TS}} \frac{FLTS_{fa}}{FLTS_{max}}}{n_{TS}}, \tag{5.20}$$

where $x_{B18}$ is the variable of the eighteenth sub-criterion of key criterion B, used for calculating the actual sub-criterion level, ranging from 0 to 1; $FLTS_{fa}$ is the actual estimate in points (0, 1, 2, 3 or 4 points) knowledge of foreign languages by the TS; $FLTS_{max}$ is the highest possible estimate (4 points); $n_{TS}$ is the number of experts (TS) $(a = 1, 2, \ldots, n_{TS})$.

### *5.1.19 Competence, Objectivity and Communication Skills of Customs Officers and Frontier Guards*

Improvement of international services must take centre stage. The frontier procedure itself is complicated due to the necessity of identity checks, customs formalities, etc.

It is also challenging from an operational perspective due to differences in "safety" languages, signalling systems, power supply, national regulations, etc. [30]. Customs officers and frontier guards are often the first people to represent the country. They should be competent in answering the questions of passengers, as well as polite, tactful, and respectful. In terms of its priority, this sub-criterion is very important for passengers. Its priority is and is $\overline{R}_{B19} = 5$ (Table 3.8).

Evaluating this sub-criterion, the estimates of passengers should be considered, which may be expressed as follows: 0 means very poor, 1 means poor, 2 denotes satisfactory, 3 means very good, 4 means excellent. The variable $x_{B19}$ of the CQI $K_B$, evaluating this sub-criterion, is calculated by the formula:

$$x_{B19} = \frac{\sum_{p=1}^{n_P} \frac{COFG_{fp}}{COFG_{max}}}{n_P}, \tag{5.21}$$

where $x_{B19}$ is the variable of the nineteenth sub-criterion of key criterion B, used for calculating the actual sub-criterion level, ranging from 0 to 1; $COFG_{fp}$ is the actual estimate in points (0, 1, 2, 3 or 4 points) competence, objectivity and communication skills of customs officers and frontier guards; $COFG_{max}$ is the highest possible sub-criterion estimate (4 points); $n_P$ is the number of passengers $(p = 1, 2, \ldots, n_P)$.

## 5.2 Additive Model for the Sub-criteria of the Key Criterion B

The corrected formula for calculating the comprehensive quality index $K_B$, evaluating the sub-criteria describing organization and technology of travel by train, is expressed as follows:

$$
\begin{aligned}
K_B = \overline{Z}_B^* \cdot \Bigg[ & \overline{Q}_{B1} \cdot \left(1 - \frac{num}{\beta(2+b)t_{arr,p}^2}\right) + \overline{Q}_{B2} \cdot \frac{\sum_{c=1}^{n_{car}} \frac{PP_{fc}}{PP_{max}}}{n_{car}} \\
& + \overline{Q}_{B3} \cdot \frac{\sum_{c=1}^{n_{car}} \frac{BL_{fc}}{BL_{max}}}{n_{car}} + \overline{Q}_{B4} \cdot \frac{FD_f}{FD_{max}} + \overline{Q}_{B5} \cdot \frac{NM_f}{NM_{max}} \\
& + \overline{Q}_{B6} \cdot \frac{IA_f}{IA_{max}} + \overline{Q}_{B7} \cdot \frac{BT_f}{BT_{max}} + \overline{Q}_{B8} \cdot \frac{DCR_f}{DCR_{max}} \\
& + \overline{Q}_{B9} \cdot \frac{TCF_f}{TCF_{max}} + \overline{Q}_{B10} \cdot \frac{PC_f}{PC_{max}} + \overline{Q}_{B11} \cdot \frac{SS_f}{SS_{max}} \\
& + \overline{Q}_{B12} \cdot \frac{\sum_{p=1}^{n_P} \frac{RM_{fp}}{RM_{max}}}{n_P} + \overline{Q}_{B13} \cdot \frac{PL_f}{PL_{max}} + \overline{Q}_{B14} \cdot \frac{HIP_f}{HIP_{max}} \\
& + \overline{Q}_{B15} \cdot \frac{GV_f}{GV_{max}} + \overline{Q}_{B16} \cdot \frac{\sum_{p=1}^{n_P} \frac{ATS_{fp}}{ATS_{max}}}{n_P} \\
& + \overline{Q}_{B17} \cdot \frac{\sum_{p=1}^{n_P} \frac{CCTS_{fp}}{CCTS_{max}}}{n_P} + \overline{Q}_{B18} \cdot \frac{\sum_{a=1}^{n_{TS}} \frac{FLTS_{fa}}{FLTS_{max}}}{n_{TS}} \\
& + \overline{Q}_{B19} \cdot \frac{\sum_{p=1}^{n_P} \frac{COFG_{fp}}{COFG_{max}}}{n_P} \Bigg].
\end{aligned}
\tag{5.22}
$$

The all symbols of model (5.22) are given in Sect. 5.1 for the analysis of each sub-criterion from B1 to B19.

Using the considered model as well as real research data and the allowable or best values of each sub-criterion, the significance of the key criterion B for evaluating the quality of passenger transportation by train, expressed by a single value, may be obtained.

## 5.3 Practical Application and Adequacy Assessment of the Key Criterion B

Using the formulas suggested in the paper, all CQI components (19 sub-criteria) and their significance may be calculated based on the real data taken from standards, technical specifications, and other reliable sources. The testing was performed on the international passenger train (IPT) No. 6.

***Departure and arrival of trains at the scheduled time***. The test was performed, based on the schedule of the IPT No. 6. According to schedule, the travel time of the train from the departure station A to the terminal (arrival) station C is 14 h 58 min (898 min). At the departure station in Vilnius, 91 passengers took the train. Seventy-seven of them travelled to the terminal station C. In general, there were 102 passengers on the train. The coefficient obtained for the delay of departure of the train from the initial station A and the arrival to the terminal station C was $g = 0.75$. There are 8 intermediate stations $B_e$ on the route.

The train departed from Vilnius (A) on time and arrived at $B_1$ frontier station according to schedule but departed from it 2 min behind schedule (staying there for 37 min) (Table 5.3). The train was 2 min late to arrive at $B_2$, stayed there for 38 min (2 min less than provided for in the schedule) and, therefore, departed from it on time. The train arrived at $B_3$ and $B_4$ on time and departed from them on time as well. It was 3 min late to arrive at $B_5$ and 3 min late to depart from it. The train arrived at $B_6$ and $B_7$ on time and departed from them on time as well. However, it was 5 min late to arrive at $B_8$, stayed there for a shorter time (18 min) than specified and, therefore, departed from it on time. The train arrived at the terminal station, C, according to schedule.

Evaluating the travel of the train, formula (5.2) was used to determine the value of variable $x_{B1}$ of the CQI $K_B$. The denominator of formula (5.2) is $(2+b)t^2_{arr,p} = 1.2(2+8)\cdot 5^2 = 300$; its numerator is $num = 0.75\ 0 + (1-0.75)\cdot(24+24+0+0+78+0+0+75) = 50.25$. Their ratio (quotient) is 0.1675. The final value of

**Table 5.3** Timetable of the international train IPT No. 6 and data on arrivals at and departures from stations during the study period

| Timetable of IPT No. 6 | | | | Difference in departure time (± min) | Difference in arrival time (± min) | The train was stationary (min) |
|---|---|---|---|---|---|---|
| Arrival time (hh:mm) | Standing time (min) | Departure time (hh:mm) | Stations | | | |
| - | - | 16:00 | Departure A | 0 | – | - |
| 16:31 | 35 | 17:06 | Intermediate $B_1$ | – 2 | 0 | 37 |
| 18:25 | 40 | 19:05 | Intermediate $B_2$ | 0 | – 2 | 38 |
| 19:35 | 2 | 19:37 | Intermediate $B_3$ | 0 | 0 | 2 |
| 20:08 | 3 | 20:11 | Intermediate $B_4$ | 0 | 0 | 3 |
| 21:40 | 29 | 22:09 | Intermediate $B_5$ | + 3 | – 3 | 29 |
| 00:32 | 17 | 00:49 | Intermediate $B_6$ | 0 | 0 | 17 |
| Time zone (GMT) | | | | | | |
| 03:08 | 4 | 03:12 | Intermediate $B_7$ | 0 | 0 | 4 |
| 05:07 | 23 | 05:30 | Intermediate $B_8$ | 0 | – 5 | 18 |
| 08:58 | - | - | Arriving C | – | 0 | – |

*Note* A minus sign (–) indicates that the train was delayed in arriving or departing, while a plus sign (+) indicates that the train departed or arrived before schedule the variable $x_{B1}$ is $1 - 0.1675 = 0.8325$. Since the train departed from the departure station A on time and the delay time of arrival at or departure from some intermediate stations $B_e$ was not large, the result close to unity, i.e. $x_{B1} = 0.833$, was obtained. When the train meets the schedule, $x_{B1} = 1$ is obtained as an ideal case. We assume that the worst case of the train's delay is the situation, when it departs from the initial station A, then stops immediately and stays there for the whole time of the trip to the terminal station C. In this case, a new train may be formed, while the first run is considered void. The quality level of the sub-criteria of key criterion B and the respective estimate of the variable $x_{Bi}$, based on real data, are presented in Table 5.4.

***Delivering meals to passengers.*** IPT No. 6 has six cars (not including the dining-car) (Fig. 4.8). The assessment of this sub-criterion considers the fact that passengers in the first-class carriages of the IPT of the Joint-Stock Company "LTG Link" are served meals included in the ticket price. The variable of the CQI $K_B$ was calculated by formula (5.4) and $x_{B2} = 0.167$ was obtained.

***Provision, layout, and condition of bed linen.*** The evaluation of this sub-criterion (service) is carried out with the consideration that on international trains bedlinen (two sheets, a pillowcase, and a towel) are included in the ticket price and provided to passengers on all types of train carriages. Bedlinen must be safely sealed. The carriage attendant must lay out the berth at the request of a passenger. In the first-class carriages with two-seat compartments, the berths are made when the car is being prepared for the journey and passengers have not yet boarded the carriage. The train IPT No. 6 has six carriages (not including the dining-car). The variable of the CQI $K_B$ was calculated by formula (5.5) and $x_{B3} = 0.50$ was obtained.

***Ordering food and drinks from the dining-car and delivering them to the compartment by carriage attendant.*** Every carriage of the train has a menu. Carriage attendant should deliver food and drinks from the restaurant car to the compartment of passenger, who has made an order. The train IPT No. 6 has six cars (not including the dining-car). The variable of the CQI $K_B$ was calculated by formula (5.6) and the maximum $x_{B4} = 1$ was obtained.

***On-board delivery of popular newspapers and magazines.*** Newspapers on IPT No6 of Joint-Stock company "LTG Link" are priced at the same price as on news-stands. There are also some free newspapers for train passengers. The variable of the CQI $K_B$ was calculated by formula (5.7) and $x_{B5} = 0.667$ was obtained.

***On-board internet access.*** The IPT No 6 of Joint-Stock Company "LTG Link" does not provide this service. The variable of the CQI $K_B$ was calculated by formula (5.8) and the critically low $x_{B6} = 0$ was obtained.

***On-board ticket purchase option (from the train manager).*** A passenger, who had no time to buy a ticket for the IPT No. 6 at a booking office, can buy it from the train manager, who has $5 \div 8$ seats reserved in the sixth car (a sleeping car with four-seat compartments). The service of selling a ticket on the train costs 1.50 Euro (on the local routes this payment is not taken from passengers, getting on the train at the station, where a booking office does not work). The variable of the CQI $K_B$ was calculated by formula (5.9) and $x_{B7} = 0.400$ was obtained.

***Dining-car reservations.*** A passenger can reserve a table in restaurant cars of the international passenger trains of the Joint-Stock Company "LTG Link". Passengers

**Table 5.4** Quality level of the sub-criteria of the key criterion B and the respective value of variables $x_{Bi}$ based on real data

| Variable $x_{Bi}$ of the key criterion B (formula) | Real (actual) value of the component | The calculated value $x_{Bi}$ |
|---|---|---|
| $x_{B1}$ (5.2) | $b = 8$; $\beta = 1.2$; $N_{A+C} = 77$; $N_{total} = 102$; $\gamma = 0.75$; $t_{dep,p} = 5$; $t_{arr,p} = 5$; $t_{dep,A} = 0$; $t_{arr,C} = 0$; $t_{arr,B1} = 0$; $t_{dep,B1} = -2$; $t_{arr,B2} = -2$; $t_{dep,B2} = 0$; $t_{arr,B3} = 0$; $t_{dep,B3} = 0$; $t_{arr,B4} = 0$; $t_{dep,B4} = 0$; $t_{arr,B5} = -3$; $t_{dep,B5} = 3$; $t_{arr,B6} = 0$; $t_{dep,B6} = 0$; $t_{arr,B7} = 0$; $t_{dep,B7} = 0$; $t_{arr,B8} = -5$; $t_{dep,B8} = 0$ | 0.833 |
| $x_{B2}$ (5.4) | $n_{car} = 6$; $PP_{f1} = 0$; $PP_{f2} = 0$; $PP_{f4} = 0$; $PP_{f5} = 0$; $PP_{f6} = 0$; $PP_{f7} = 1$; $PP_{max} = 1$ | 0.167 |
| $x_{B3}$ (5.5) | $n_{car} = 6$; $BL_{f1} = 4$; $BL_{f2} = 4$; $BL_{f4} = 4$; $BL_{f5} = 4$; $BL_{f6} = 4$; $BL_{f7} = 10$; $BL_{max} = 10$ | 0.500 |
| $x_{B4}$ (5.6) | $FD_f = 3$; $FD_{max} = 3$ | 1 |
| $x_{B5}$ (5.7) | $NM_f = 2$; $NM_{max} = 3$ | 0.667 |
| $x_{B6}$ (5.8) | $IA_f = 0$; $IA_{max} = 2$ | 0 |
| $x_{B7}$ (5.9) | $BT_f = 2$; $BT_{max} = 5$ | 0.400 |
| $x_{B8}$ (5.10) | $DCR_f = 3$; $DCR_{max} = 3$ | 1 |
| $x_{B9}$ (5.11) | $TCF_f = 2$; $TCF_{max} = 6$ | 0.333 |
| $x_{B10}$ (5.12) | $PC_f = 1$; $PC_{max} = 3$ | 0.333 |
| $x_{B11}$ (5.13) | $SS_f = 0$; $SS_{max} = 1$ | 0 |
| $x_{B12}$ (5.14) | $n_P = 20$; $RM_{f5} = 1$; $RM_{f,1,2,6–14,16,17} = 2$; $RM_{f,3,4,15,18,20} = 3$; $RM_{f,19} = 4$; $RM_{max} = 4$ | 0.575 |
| $x_{B13}$ (5.15) | $PL_f = 0$; $PL_{max} = 4$ | 0 |
| $x_{B14}$ (5.16) | $HIP_f = 2$; $HIP_{max} = 2$ | 1 |
| $x_{B15}$ (5.17) | $GV_f = 0$; $GV_{max} = 2$ | 0 |
| $x_{B16}$ (5.18) | $n_P = 20$; $ATS_{f,1–3,5,7–10,13,14,16–20} = 3$; $ATS_{f,4,6,11,12,15} = 4$; $ATS_{max} = 4$ | 0.813 |
| $x_{B17}$ (5.19) | $n_P = 20$; $CCTS_{f,1,7–10,14–16,18} = 3$; $CCTS_{f,2–6,11–13,17,19,20} = 4$; $CCTS_{max} = 4$ | 0.888 |
| $x_{B18}$ (5.20) | $n_{TS} = 11$; $FLTS_{f,3–5,8,11} = 1$; $FLTS_{f,1,2,6,9,10} = 2$; $FLTS_{f,7} = 3$; $FLTS_{max} = 4$ | 0.409 |
| $x_{B19}$ (5.21) | $n_P = 20$; $COFG_{f,4,7,9,10,14–16,18} = 2$; $COFG_{f,1–3,8,13,20} = 3$; $COFG_{f,5,6,11,12,17,19} = 4$; $COFG_{max} = 4$ | 0.725 |

rarely use this option. A passenger from any car on the IPT No. 6 can reserve the table in the dining-car. The variable of the CQI $K_B$ was calculated by formula (5.10) and the maximum $x_{B8} = 1$ was obtained.

***Taxi call facility.*** The sub-criterion for this service considers the fact that a passenger on an IPT No. 6 can request the carriage attendant or the train manager to call a taxi for him prior to the arrival of the train at the terminus station. This service is free of charge. The variable oh the CQI $K_B$ was calculated by formula (5.11) and $x_{B9} = 0.333$ was obtained.

***On-board card payment option.*** Passengers can pay with a payment card for the provided services, goods, and tickets in the international trains. The scanner of electronic cards is in the restaurant car of the train. The train manager should accompany a passenger, who wishes to pay with a payment card, to the attendant-barman and perform the required operation for the passenger. Only cash is accepted for the services provided on the local trains. A passenger can pay with a payment card, but, in this case, he / she should go to the dining-car, where a scanner of electronic cards is found (1 point). The variable of the CQI $K_B$ was calculated by formula (5.12) and $x_{B10} = 0.333$ was obtained.

***Souvenir shopping on the train.*** Souvenirs are not sold on the trains of the Joint-Stock Company "LTG Link". The variable of the CQI $K_B$ was calculated by formula (5.13) and the critically low $x_{B11} = 0$ was obtained.

***Meeting passenger preferences for on-board music and in-train information.*** Radio shows, music, and in-train information (about the trip duration, intermediate stations, the provided services, etc.) are broadcast on all trains of the Joint-Stock Company "LTG Link". This information is regulated by special rules. The test for evaluating this sub-criterion on the IPT No. 6 was performed based on the data obtained from 20 questionnaires filled in by passengers. The variable of the CQI $K_B$ was calculated by formula (5.14) and $x_{B12} = 0.575$ was obtained.

***Securing passenger luggage and belongings on board the train.*** Passengers are responsible for the protection of their luggage, documents, and other valuable things during the trip on the IPT No. 6. The variable of the CQI $K_B$ was calculated by formula (5.15) and the critically low $x_{B13} = 0$ was obtained.

***Purchasing overseas medical insurance on board the train.*** There is no need to get a foreign health insurance policy on the international trains of the Joint-Stock Company "LTG Link" because the health of a passenger going abroad is insured when he/she gets a visa. The variable of the CQI $K_B$ was calculated by formula (5.16) and the maximum $x_{B14} = 1$ was obtained.

***Obtaining a visa at the frontier point.*** Visas are not issued at the frontier points. Passengers without a visa or having an invalid visa should get off the train at the frontier point. He/she should go to the office, issuing visas, and having got a visa can continue travel on another train. It takes too much time and causes serious problems for passengers. The test was performed for the IPT No. 6, when there was no possibility for passengers to get visas on the frontier post. The variable of the CQI $K_B$ was calculated by formula (5.17) and the critically low $x_{B15} = 0$ was obtained.

***The appearance of the train staff (uniform, shoes, hairstyle, nameplate).*** All members of the TS should wear uniforms during the trip and follow the rules, titled as the *The rules of wearing the uniform by the service staff members* [31]. The uniform details, its delivery, period of wearing, maintenance, and replacement, as well as wearing of the identification badge and nameplate, are described in these rules, which pay great attention to clothing and accessories of the TS. The test for evaluating this criterion on the IPT No. 6 was performed based on the data obtained from 20 questionnaires filled in by passengers. The variable of the CQI $K_B$ was calculated by formula (5.18) and $x_{B16} = 0.813$ is obtained.

***Communication culture of train service staff with passengers and other members of staff.*** The test for evaluating this sub-criterion on the IPT No. 6 was performed based on the data obtained from 20 questionnaires filled in by passengers. The variable of the CQI $K_B$ was calculated by formula (5.19) and $x_{B17} = 0.888$ was obtained.

***Proficiency in foreign languages of the train staff.*** The train conductor (senior attendant) of an international train should speak a foreign language so that he / she could communicate with passengers and administration of the railway, on which the train runs [32]. Car attendant must speak at least one foreign language. The data were obtained from the survey of 11 service staff members (one train conductor and 10 car attendants, questioned by him) of the IPT No. 6. The variable of the CQI $K_B$ was calculated by formula (5.20) and $x_{B18} = 0.409$ was obtained.

***Competence, objectivity and communication skills of customs officers and frontier guards.*** The test was performed on the IPT No. 6 based on the data of 20 questionnaires filled in by passengers. The variable of the CQI $K_B$ was calculated by formula (5.21) and $x_{B19} = 0.725$ was obtained.

The quality level of the key criterion B and the respective $x_{Bi}$ value, based on the real data, are presented in Table 5.4. The quality of train, represented by the estimate of the variable $x_{Bi}$, when based on the highest values of the data it is equal to 1, when based on the average values is equal to 0.5, while based on the lowest values of the data, is equal to 0.

Using the formula (5.22), the comprehensive quality index $K_B$ of the key criterion B describing organization and technology of travel by train, is calculated as follows:

$$\begin{aligned}
K_B = 0.228(&0.1128 \times 0.833 + 0.0279 \times 0.167 + 0.0532 \times 0.500 \\
&+ 0.0323 \times 1 + 0.0273 \times 0.667 + 0.0444 \times 0 + 0.0604 \times 0.400 \\
&+ 0.0227 \times 1 + 0.0212 \times 0.333 + 0.0429 \times 0.333 + 0.0180 \times 0 \\
&+ 0.0250 \times 0.575 + 0.0965 \times 0 + 0.0540 \times 1 + 0.0959 \times 0 \\
&+ 0.053 \times 0.813 + 0.0832 \times 0.888 + 0.0498 \times 0.409 \\
&+ 0.0795 \times 0.725) = 0.228 \times 0.5074 = 0.1157.
\end{aligned}$$

The obtained index of the set of sub-criteria describing organization and technology of travel by train is $K_B = 0.1157$. The significance of the other key criteria (C, D) for the CQI component $K$ will be determined by using other additive models in Chaps. 6 and 7.

The highest quality level, determined by using the real data, was obtained for the sub-criteria B4 ($x_{B4} = 1$), B8 ($x_{B8} = 1$) and B14 ($x_{B14} = 1$). The lowest quality level, determined by using the real data, was obtained for the criteria B6 ($x_{B6} = 0$), B11 ($x_{B11} = 0$), B13 ($x_{B13} = 0$) and B15 ($x_{B15} = 0$). The average quality level was determined, based on the real data, for B3 ($x_{B3} = 0.50$). First, the real values of the sub-criteria, whose variables $x_{Bi} = 0$ or are about zero, should be improved.

The sum $\sum_{i=1}^{19} \overline{Q}_{Bi} \times x_{Bi} = 0.5074$ of local weights $\overline{Q}_{Bi}$ of the 19 sub-criteria of the key criterion B (Table 3.8 and Fig. 3.7b) and the products $\overline{Q}_{Bi} \times x_{Bi}$ of the variables $x_{Bi}$ (Table 5.4) indicates that the level of the journey on the IPT No. 6 is average in terms of process organization and technology.

The computed product (Fig. 5.7) of the weight $\overline{Q}_{Bi}$ of each of the nineteen sub-criteria of this key criterion and the variable $x_{Bi}$ of the train under consideration varies from 0 (sub-criteria B6, B11, B13, B15) to 0.0940 (sub-criteria B1). The following sub-criteria have a higher than average impact (average $\overline{Q}_{Bi} \times x_{Bi} = 0.0267$) on improving the quality of travel: departure and arrival of the train at the scheduled time (B1), communication culture of rail staff with passengers and other members of staff (B17), competence, objectivity and communication skills of customs officers and frontier guards (B19), purchasing overseas medical insurance on-board the train (B14), the appearance of the TS (uniform, shoes, hairstyle, identification card) (B16) and ordering food and drinks from the dining-car and delivering them to the compartment by carriage attendant (B4). The sub-criteria B6, B11, B13 and B15 are the most detrimental to the quality of travel on IPT No. 6, as their products $\overline{Q}_{Bi} \times x_{Bi}$ are equal to 0. On this train, the availability of bed linen in the carriage, its condition, and the layout of the berth for passengers (B3), on-board ticket purchase option (from the train manager) (B7), dining-car reservation (B8) were close to the average level.

Considering the weight of the key criterion B ($\overline{Z}_B^* = 0.228$) and the sum 0.5074 of the products $\overline{Q}_{Bi} \times x_{Bi}$ of all 19 sub-criteria, the effect of this key criterion on the complex quality indicator $K$ for the train journey was computed (formula 5.1). The impact on the quality of the train travel process organisation and the technology of the train under consideration is indicated by the comprehensive quality indicator $K_B = 0.11570$. The other part $K$ of the Quality Level Indicator (QLI) consists of the

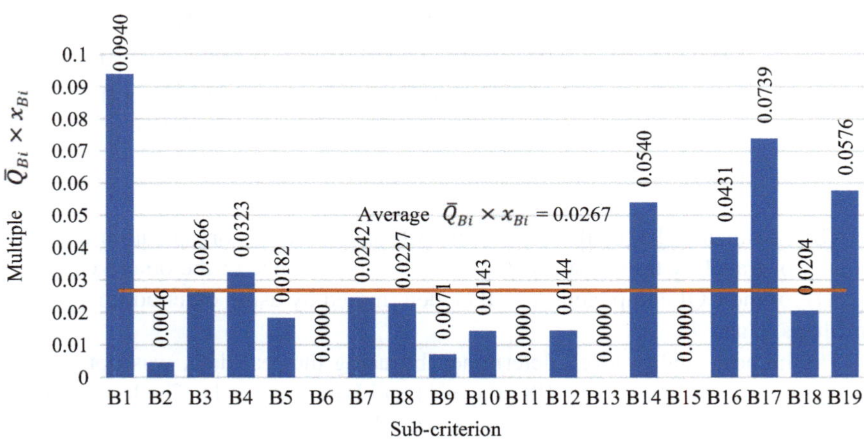


**Fig. 5.7** Findings of the calculations of the impact of the sub-criteria of the main criterion B on the quality of the trip of the IPT No. 6

computed CQIs $K_A$, $K_C$ and $K_D$ of the key criteria A, C and D, which have been tested for the IPT No. 6 and are presented in the following chapters of this monograph.

The resolved example shows that the additive mathematical model (5.22) is correct as it allows the application of the normalised subjective weights $\overline{Q}_{Bi}$ of the sub-criteria given by the experts and the respondents and the actual experimental parameters of the passenger train under consideration to obtain a calculation result $K_B$ to be between 0 and $\overline{Z}_B^*$. The methodology and mathematical model can therefore be used in practice to assess the quality of train travel.

The developed original additive mathematical model is used for calculating the values of each of the 19 sub-criteria based on the criterion's normalized weight multiplied by its variable, showing the correspondence of the sub-criterion variable to specified, critical, the highest, permissible or the best values. Comparing the real values of the sub-criterion variable with the specified, critical, the highest, permissible and the best values, the value, showing the closeness of a particular sub-criterion describing the trip by train to the best level sought (when its value is equal to one) and the worst inadmissible level (when its value is about 0), is calculated.

The use of a multicriteria model allows for evaluating the quality of travel by train objectively and expressing it in a single value. The value of $K_B$, calculated based on the values of the real parameters, allows the authors to state that the suggested mathematical model is valid and can be used in practice. The significance of the sub-criteria, describing organization and technology of travel by train for the comprehensive quality index $K_B$, may be evaluated by determining only the variables $x_{Bi}$ and using them in the suggested mathematical model.

## References


1. Sivilevičius H, Maskeliūnaitė L, Petkevičienė B, Petkevičius K (2012) The model of evaluating the criteria, describing the quality of organization and technology of travel by international train. Transport 27(3):307–319. https://doi.org/10.3846/16484142.2012.724448
2. Jiang F, Yu D-b, Ni S-q (2017) An objective train timetabling quality evaluation method. Math Prob Eng 3047963:1–10. https://doi.org/10.1155/2017/3047963
3. Tiong KY, Ma Z, Palmqvist C-W (2023) Analyzing factors contributing to real-time train arrival delays using seemingly unrelated regression models. Transp Res Part A Policy Pract 174:103751
4. Spanninger T, Trivella A, Büchel B, Corman F (2022) A review of train delay prediction approaches. J Rail Transp Plan Manage 22(2022):100312. https://doi.org/10.1016/j.jrtpm.2022.100312
5. Wang P, Zhang Q-P (2019) Train delay analysis and prediction based on big data fusion. Transp Saf Environ 1(1):79–88. https://doi.org/10.1093/tse/tdy001
6. Huang Z, Niu H, Gao R, Fan H, Liu C (2021) Optimizing train timetable based on departure time preference of passengers for high-speed rails. J Adv Transp. https://doi.org/10.1155/2021/6611289
7. Liu Y, Shi T, Lv X, Zan R, Wang L, Zhang Y, Ye X (2016). Research on train reception and departure system based on intelligent video analysis. In: International Conference on Service Operations and Logistics, and Informatics (SOLI) 10–12 July 2016, Beijing, China, pp 234–238. IEEE. https://doi.org/10.1109/SOLI.2016.7551693

8. Techninio geležinkelių naudojimo nuostatai (1996) Lietuvos Respublikos susisiekimo ministerija. Patvirtinta Lietuvos Respublikos susisiekimo ministro 1996 m. rugsėjo 20 d. įsakymu Nr.297. Vilnius: Informacijos ir leidybos centras, 126 p (in Lithuanian)
9. Bendrosios keleivių vežimo taisyklės (2011) Patvirtinta AB „Lietuvos geležinkeliai" generalinio direktoriaus 2011 m. gruodžio 13 d. įsakymu Nr. Į-966. Vilnius: AB „Lietuvos geležinkeliai" (in Lithuanian)
10. Prashar S (2017) Now vending machines will serve you food on trains. Retail, From the Economic Times. https://retail.economictimes.indiatimes.com/news/food-entertainment/food-services/now-vending-machines-will-serve-you-food-on-trains/59367155
11. Bird G, Collins J, Da Settimo N, Dunmore D, Ellis S, Khan M, Kwok M, Leach T, Preti A, Ranghetti D, Vollath C (2017) Research for Tran Committee—passenger night trains in Europe: the end of the line? Study. Directorate-General for Internal Policies, Policy Department for Structural and Cohesion Policies. European Parliament. European Union, p 228. https://www.europarl.europa.eu/RegData/etudes/STUD/2017/601977/IPOL_STU(2017)601977_EN.pdf
12. Sperandeo A (2006) The people who work on trains. Trains. https://www.trains.com/trn/train-basics/abcs-of-railroading/the-people-who-work-on-trains/
13. Why do People Read on Trains? 6 Reasons... (2023) London North Eastern Railway (LNER). https://www.lner.co.uk/our-destinations/travel-inspiration/why-do-people-read-on-trains-6-reasons/
14. Burroughs D (2020) Capacity vs comfort: balancing the interior design. International Railway Journal. https://www.railjournal.com/in_depth/capacity-vs-comfort-balancing-the-interior-design
15. Guan Y, Wu B, Jia J (2020) Does online ticket booking system make people better off? An empirical study on railway service. Transp Res F Traff Psychol Behav 73(3):143–154. https://doi.org/10.1016/j.trf.2020.03.014
16. Rail ticket retailing: the passenger perspective. Market research findings. For the Office of Rail Regulation, by BDRC Continental (2015) BDRC continental, 33 p. file:///C:/Users/40507/Downloads/Rail-ticket-retailing-passenger-perspective-report170615%20(2).pdf
17. Seat reservations in trains and dining cars (2023) SBB. https://www.sbb.ch/en/travelcards-and-tickets/tickets-for-switzerland/seat-reservations.html
18. "Did you know?" Facts and Figures on Taxi Transport in Europe (2009) International Road Transport Union (IRU). Brussels, Belgium, pp 1–12. https://www.iru.org/sites/default/files/2016-01/en-dyk-taxi.pdf
19. Dyson M (2019) Partnership gives rail passengers onboard taxi-booking service. Technology. BTN. Business travel news Europe. https://www.businesstravelnewseurope.com/Technology/Partnership-gives-rail-passengers-onboard-taxi-booking-service
20. Mobile checkout system on long distance transport. Working together to offer more payment options (2020) Deutsche Bahn. https://www.dbsystel.de/dbsystel-en/Digital-Stories-en/Working-together-to-offer-more-payment-options-6175452
21. Héroux L, Church NJ (2014) Marketing strategies of gift and souvenir shops in Canada and the United States. A comparative exploratory study. Téoros 33(2):1042438ar. https://doi.org/10.7202/1042438ar
22. Peter S, Anandkumar V (2022) Souvenir shopping: a value chain approach. Int J Res Anal Rev (IJAR) 9(2):619–627
23. Gostelie FH (2017) Clean, lean and mean: a fast train is a clean train. The effect of music tempo on the travel experience of ns passengers in the train under various cleanliness conditions. Student Theses. University of Twente, pp 1–64. https://essay.utwente.nl/72046/1/Master
24. Jain A (2017) Railways liable for safety of passengers and luggage, says consumer forum. The Hindu. https://www.thehindu.com/news/cities/Delhi/railways-liable-for-safety-of-passengers-luggage-says-consumer-forum/article17688589.ece
25. Lambrecht J (2020) Railroad uniforms and accessories. Railpage. News—Railway History. http://www.railpage.com.au/news/s/railroad-uniforms-and-accessories
26. Fedaravičius T (2014) Į kelionę—šunys, katės, begemotas. lžinios.lt (in Lithuanian)

27. Keleivinio vagono palydovo pareiginė instrukcija (1999) (VL/120) Patvirtinta Komercijos direktoriaus 1999 m. rugsėjo 3 d. Informacijos ir leidybos centras, Vilnius, 20 p (in Lithuanian)
28. AB „Lietuvos geležinkeliai" įmonių grupės darbuotojų etikos kodeksas (2021) Patvirtinta AB „Lietuvos geležinkeliai" generalinio direktoriaus 2021 m. AB „Lietuvos geležinkeliai", Vilnius. https://www.litrail.lt/documents/10279/0/Darbuotoj%C5%B3%20etikos+kodeksas+2021-02-03 (in Lithuanian)
29. Kairiūnaitė A (2016) Docentė: užsienio kalbų mokėjimas gerina net vairavimo ir matematikos įgūdžius. LRT Klasikos laida „Manasis aš", Delfi. https://www.delfi.lt/news/daily/education/docente-uzsienio-kalbu-mokejimas-gerina-net-vairavimo-ir-matematinius-igudzius.d?id=72151018 (in Lithuanian)
30. Cerny P (2021) 10 questions about night trains that you always wanted, but never dared, to ask. Heinrich Böll Stiftung, Brussels, European Union. https://eu.boell.org/en/10-questions-night-trains
31. Keleivius aptarnaujančio personalo uniformos dėvėjimo taisyklės (2003) Patvirtinta 2003 m. gegužės 7 d. įsakymu Nr. 116. AB „Lietuvos geležinkeliai",Vilnius (in Lithuanian)
32. Tarptautinio susisiekimo keleivinio traukinio viršininko instrukcija (2003) Gelspa, Vilnius, 16 p (in Lithuanian)

# Chapter 6
# Rail Travel Costs and Quality

*Railway ticket excites more hope than a lottery.*
Paul Morand

## 6.1 Impacts of Rail Travel Costs on the Quality of the Journey—Elements of a Model

Rail transport is a cost-effective way of transporting passengers. As with any passenger traffic operator, pricing and capacity management are the two key tools to ensure the profitability of a rail passenger service provider: pricing affects the demand for the service, while capacity management determines the availability of the service. The study by Kamandanipour et al. [1] developed an expert system as a decision-making tool for an integrated pricing and capacity management issue of a passenger rail service provider, which is of great importance for the success of the service provider.

The principal (main) goal of railway transport is to attract a maximum number of customers and load on planned or existing lines with minimum of investment cost, personnel, equipment, energy consumption, operating and maintenance costs [2]. De Oña et al. [3] proposes an analysis of rail service quality (SQ) conducted on the basis of user's perceptions, expressed in terms of satisfaction and importance assigned to the various service characteristics.

The present chapter aims to suggest an original multicriteria mathematical model for evaluating the weights and variables of six sub-criteria, describing the expenses of the trip and the provided services (ETPS) by international passenger train (IPT), as well as their influence on the comprehensive quality index (CQI), and to validate the results through the case study of the numerical CQI calculations made for the IPT No. 6.



H. Sivilevičius and L. Maskeliūnaitė, *Assessment of the Quality of Passenger Transportation by Train Using Multiple Criteria Decision Making Methods*, Lecture Notes in Intelligent Transportation and Infrastructure,
https://doi.org/10.1007/978-3-031-85877-2_6

The local weights of six sub-criteria, belonging to key criterion C and describing the ETPS by international train, were obtained from the survey of experts, when three categories of respondents, such as passengers (P) and experts, including the train staff (TS) and the administration staff (AS), provided their estimates (Table 3. 9 and Fig. 3.7). The mean weight, $\overline{Z}_C^*$, shows the significance of the key criterion C (with a different number of the respondents and experts in each category). This weight was calculated as follows (3.35).

The significance estimates (in weights) provided to the key criterion referring to all groups P, TS and the AS are presented in Table 3.16.

The quality of passenger transportation by any train can be determined only roughly, subjectively and intuitively. Therefore, to evaluate it more accurately, a qualitative method and the CQI $K$, allowing for the quality of travel on a particular route to be expressed by a single number, were developed ($K = K_A + K_B + K_C + K_D = 0\ldots1$).

The model for calculating the significance of the key criterion C factors on the quality of international train travel is given in the formula [4]:

$$K_c = \overline{Z}_C^* \cdot \sum_{i=1}^{m} \left(\overline{Q}_{Ci} \cdot x_{Ci}\right) = \overline{Z}_C^* \cdot \left(\overline{Q}_{C1} \cdot x_{C1} + \overline{Q}_{C2} \cdot x_{C2} + \cdots + \overline{Q}_{C6} \cdot x_{C6}\right). \quad (6.1)$$

where $K_C$ is the component of the CQI of the international train, depending on the ETPS (and ranging from 0 to $\overline{Z}_C^*$); $\overline{Z}_C^*$ is the mean weight, showing the estimates of the respondents and experts assigned to the weights of the key criterion C is calculated by the formula (3.35) (Table 3.16 $\overline{Z}_C^* = 0.2580$); $\overline{Q}_{C1}, \ldots, \cdot\overline{Q}_{Cm}$ (Table 3.9 and Fig. 3. 7) denote the local weight values of $i$-th sub-criterion of the key criterion C ($i = 1, 2, \ldots, m$) determined by the expert evaluation method; $x_{C1}, \ldots, \cdot x_{Cm}$ are the variables of $i$-th sub-criterion of the key criterion C ($i = 1, 2, \ldots, m$), whose estimates are used for determining the real sub-criterion value, ranging from 0 to 1.

To calculate the CQI, evaluating the significance of the trip by an international train, depending on the ETPS, the variables of any sub-criterion in the set $x_{C1}, \ldots, \cdot x_{C6}$, serving as a basis for calculating the real sub-criterion value ranging from 0 to 1, should be determined.

### 6.1.1 Ticket Fare

It is not only transport modes that are in competition with each other in the provision of passenger transport services. Companies also compete. In Sweden, there are several railway companies that compete and try to be innovative and offer new services to their passengers. Competition has been a key factor behind the current growth in passenger numbers in Sweden since 2010 (+ 45%). This is one of the highest rates in Europe. Traditional sleeping carriages satisfy passengers, but to be profitable and sustainable they require a much higher ticket fare. The need to pay more for tickets (for a traditional sleeping compartment), at least for the moment, is too small. The focus is on improving passenger comfort with new and more comfortable mattresses, blankets, and pillows. For privacy, customers can book a separate compartment [5].

China's railway reform has been increasing competition between rail passenger transport and other modes of transport. In this situation, a uniform and unchanging tariff structure is gradually becoming a preventive measure for the growth of railway revenues and the development of the railway system. A new approach to fare optimisation strategy based on revenue management theory is considered by Zheng and Liu [6]. Based on a separate pricing strategy, a fare optimisation model has been developed. The new approach has been tested on the Beijing-Shanghai HSR line. The study shows that a pricing strategy can not only increase revenues but can also be useful in regulating existing demand and stimulating potential demand.

The proposed innovative model is designed to optimise price and seat allocation in HSR. The findings of the study show that the model has the flexibility to adjust the ticket price and seat allocation for the corresponding sales period in line with passenger demand, and without increasing the train capacity raise the total expected revenue by 5.92% [7]. A new method for estimating generalised travel costs and the factors that influence them is proposed in the analysis by Chen and Gao [8] on the characteristics of rail passenger transport.

Of the six sub-criteria of key criterion C, ticket fare (TF) has the greatest impact on the quality of the train journey: its rank is $\overline{R}_{C1} = 1$, local weight is $\overline{Q}_{C1} = 0.3324$ (Table 3.9).

When the highest ($TF_{max}$) and the lowest ($TF_{min}$) fares of the ticket to a sleeping (S), compartment (C) or luxury (L) car, are known, the value of the CQI component, depending on the expenses $K_C$, of the trip by train, the product of $\overline{Q}_{C1} \cdot x_{C1}$ and the variable, $x_{C1}$, which shows the influence of the ticket fare on the quality of the trip, can be calculated by the formula:

$$x_{C1} = \frac{\sum_{y=1}^{3} n_y \cdot \left(1 - \frac{TF_f - TF_{min}}{TF_{max} - TF_{min}}\right)}{n_{car}}, \tag{6.2}$$

where $x_{C1}$ is the variable of the first sub-criterion of the key criterion C; $n_y$ is the number of cars of $y$-th category (S, C or L) in the train; $TF_f$, $TF_{min}$ and $TF_{max}$ denote the factual, the lowest and the highest ticket fare, respectively, for a car of $y$-th category, €; $n_{car}$ is the number of cars in the train ($n_{car} = n_S + n_C + n_L$).

### 6.1.2 *Price of Meals Served in the Dining-Car and Their Quality*

For long and medium distance journeys, it is necessary to offer a catering service for passengers. A long train journey can be an exceptional experience for a passenger, depending on the quality of the food in the restaurant car, the value for money, the variety of the offer, the communication skills of the staff and the appearance of the staff.

On the trains of European railway companies, there is usually a restaurant car, which is open to all passengers, regardless of the class of carriage they are travelling in. For example, one of the countries with high quality and reasonably priced meals on trains is the Czech Republic. Most Swiss trains have restaurants serving high quality cuisine, including national dishes, however, these are quite expensive. The German railway operator DB focuses on the use of local ingredients in its restaurant cars. The menu on the train aims to offer healthy and nutritious meals. One disadvantage of DB is that most meals are heated (cooked) using microwaves. The comfortable restaurant cars are a distinctive advantage. UK train routes are known for their high-quality food, prepared by top chefs. There is a trend towards stricter food standards in the rail industry. Most rail operators, such as Thalys, Alleo, Trenitalia, SNCF, and Eurostar, are sourcing local products and they also offer vegan food [9].

An online survey was conducted in Indonesia. Long-distance train passengers were surveyed on the impact of service and food quality upon their satisfaction with their journey. Also, for food delivery to the passenger. Bivariate and multivariate statistical tests were used to analyse the data. This study showed that overall service quality was positively correlating with customer satisfaction regarding rail catering and overall food quality. It was also shown that service and food quality simultaneously explained about 17.8% of the variability in customer satisfaction. The main aspect of service and food quality that influenced customer satisfaction was the delivery of the food. The study found that service and food quality had a beneficial effect on customer satisfaction with rail food [10].

With the rapid development of high-speed railways in China, passengers are increasingly demanding higher levels of on-board catering services. Studies on train catering systems have led to the improvement of ordering and delivery efficiency, the level of on-board catering services, and passengers' access to on-board restaurants [11].

The dishes served in the dining car can be evaluated based on the four criteria as follows:

- price of meals (PM);
- food (dish) quality (FQ);
- assortment abundance (AA);
- meal change cycle (MCC).

In the absence of the data on passengers' estimates of the significance of each of these sub-criteria, the weight 1. The weight is taken the same whilst variable is calculated as follows:

$$x_{C2} = \frac{x_{C2,PM} + x_{C2,FQ} + x_{C2,AA} + x_{C2,MCC}}{4}, \tag{6.3}$$

where $x_{C2}$ is the variable of the second sub-criterion of the key criterion C; $x_{C2,PM}$ is the variable of the second key criterion C, showing the price of meals; $x_{C2,FQ}$ is the variable of the second key criterion C, showing the food quality; $x_{C2,AA}$ is the variable of the second key criterion C, showing the assortment abundance; $x_{C2,MCC}$ is the variable of the second key criterion C, showing the meal change cycle.

More accurate evaluation of the significance of these sub-criteria can be performed by conducting a questionnaire survey. The variable summand of the CQI component, $K_C$, determining the PM, FQ, AA and MCC, is calculated by the formula:

$$x_{C2} = \omega_{PM} \cdot x_{C2,PM} + \omega_{FQ} \cdot x_{C2,FQ} + \omega_{AA} \cdot x_{C2,AA} + \omega_{MCC} \cdot x_{C2,MCC}, \tag{6.4}$$

where $\omega_{PM}$ is the weight price of meals; $\omega_{FQ}$ is the weight of the food quality; $\omega_{AA}$ is the weight of the assortment abundance; $\omega_{MCC}$ is the weight of the meal change cycle.

In this study, the weights, $\omega_{PM}, \omega_{FQ}, \omega_{AA}$ and $\omega_{MCC}$, of the sub-criteria, describing the services provided in the dining-car, were determined based on the judgements of 12 passengers ($j = 1, 2, \ldots, n_P$) about the provided services ($i = 1, 2, \ldots, n_{ps}$). They gave the ranks $R_{ij}$ to the considered four sub-criteria ($n_{ps} = 4$) (Table 6.1). The number of the respondents, $n_P = 12$, was three times as large as the number of the sub-criteria describing services, $n_{ps} = 4$, and, therefore, was sufficient for evaluation.

The estimate made by the group of respondents (P) is taken as the solution to the problem (or its result) only if it is Consistent. This estimate, expressed by the coefficient $W$ (Kendall's coefficient of concordance), was calculated by the formula (2.21):

$$W = \frac{12S}{n_P^2\left(n_{ps}^3 - n_{ps}\right)} = \frac{12 \cdot 618}{12^2(4^3 - 4)} = 0.858,$$

where $S$ is the sum of the squared deviations of the sum of ranks $\sum_{j=1}^{n_P} R_{ij}$ (assigned by each of 12 passengers to the services provided in the dining-car of the train) from the mean rank value $\overline{R} = \frac{n_P(n_{ps}+1)}{2} = \frac{12(4+1)}{2} = 30$:

$$S = \sum_{i=1}^{n_{ps}} \left[ \sum_{j=1}^{n_P} - \frac{n_P\left(n_{ps} + 1\right)}{2} \right]^2 = 4 + 289 + 1 + 324 = 618.$$

The Pearson chi-squared test was used to determine if the judgements of twelve respondents were in agreement:

$$\chi^2 = n_P\left(n_{ps} - 1\right) \cdot W = 12(4 - 1) \cdot 0.858 = 30.89.$$

The number of the degrees of freedom $\nu = n_{ps} - 1 = 4 - 1 = 3$ was calculated for the number of respondents, $n_P = 12$, and the number of the compared quality factors, $n_{ps} = 4$, while a relatively high significance level, $\alpha = 0.01$, was selected. The critical value, $\chi^2_{\nu,\alpha}$, corresponding to the number of degrees of freedom and the selected significance level, $\chi^2_{\nu,\alpha}$, equal to 11.34 (which is much smaller than the

**Table 6.1** Ranks $R_{ij}$ for service indicators given by passengers who used the dining-car catering services

| Passenger $j = 1, 2, \ldots, n_P$ | Provided service characteristic $i = 1, 2, \ldots, n_{ps}$ | | | |
|---|---|---|---|---|
| | PM | FQ | AA | MCC |
| E1 | 2 | 1 | 3 | 4 |
| E2 | 2 | 1 | 3 | 4 |
| E3 | 1 | 2 | 3 | 4 |
| E4 | 3 | 1 | 2 | 4 |
| E5 | 3 | 1 | 2 | 4 |
| E6 | 3 | 1 | 2 | 4 |
| E7 | 2 | 1 | 3 | 4 |
| E8 | 2 | 1 | 3 | 4 |
| E9 | 3 | 1 | 2 | 4 |
| E10 | 3 | 1 | 2 | 4 |
| E11 | 2 | 1 | 3 | 4 |
| E12 | 2 | 1 | 3 | 4 |
| $\sum_{j=1}^{n_P} R_{ij} = R_i$ | 28 | 13 | 31 | 48 |
| $\overline{R}_i = \sum_{j=1}^{n_P} R_{ij}/n_P$ | 2.333 | 1.083 | 2.583 | 4.000 |
| $\sum_{j=1}^{n_P} R_{ij} - \frac{n_P(n_{ps}+1)}{2}$ | –2 | –17 | 1 | 18 |
| $\left[\sum_{j=1}^{n_P} R_{ij} - \frac{n_P(n_{ps}+1)}{2}\right]^2$ | 4 | 289 | 1 | 324 |

calculated value ($\chi^2 = 30.89$), was selected from the statistical Appendix Table 8 [12].

The lowest value of the concordance coefficient, $W_{min}$, was calculated from formula (2.24), with the significance level $\alpha = 0.01$ and the number of the degrees of freedom $\nu = 3$. Therefore, it could be stated that the respondents' estimates were still in concordance:

$$W_{min} = \frac{\chi^2_{\nu,\alpha}}{n_P(n_{ps} - 1)} = \frac{11.34}{12(4 - 1)} = 0.315 < 0.858.$$

The significances (normalized weights) of the sub-criteria, describing services provided to passengers in the dining-car of the train, were determined based on the method ARTIW-L developed by the author. The weight of first sub-criterion PM equal (formula 2.17):

$$\omega_i = \frac{(n_{ps}+1) - \overline{R}_i}{\sum_{i=1}^{n_{ps}} \overline{R}_i} = \frac{(4+1)2.333}{10} = 0.2667.$$

The weights of these sub-criteria, which were calculated by formula (2.17), based on the average ranks $\overline{R}_i$, assigned by 12 respondents to four criteria describing the services provided in the dining-car, are given in Table 6.2.

The influence of the dish price on the quality of the trip by train can be expressed by formula (6.5), showing the variable value of this sub-criterion:

$$x_{C2,PM} = \frac{PM_{min}}{PM_{average}} = \frac{PM_{min}}{\sum_{d=1}^{n_C} \frac{PM_d \cdot D_d}{D}}, \qquad (6.5)$$

where $PM_{min}$ is the lowest price of the dish served in the dining-car of the train, €; $PM_{average}$ is the average price of the dish, €; $n_C$ is the number of the categories of dishes served in the dining-car ($d = 1, 2, \ldots, n_C$); $PM_d$ is the price of the dish of $d$-th category, €; $D_d$ is the rate of the $d$-th dish ordering, i. e. the number of the served (sold) dishes of this kind; $D$ is the rate of ordering (serving) the dishes of all categories offered in the dining-car, i. e. the total number of all served dishes, ($D = D_1 + D_2 + \cdots + D_{n_C}$).

The quality of the food served in the dining-car of the train can be expressed by formula (6.6), when the variable is calculated:

$$x_{C2,FQ} = \frac{FQ_0 \cdot m_0 + FQ_1 \cdot m_1 + \cdots + FQ_4 \cdot m_4}{FQ_{max}(m_0 + m_1 + \cdots + m_4)}, \qquad (6.6)$$

where $x_{C2,FQ}$ is the variable showing the food quality; $FQ_0$, $FQ_1$, …, $FQ_4$ denote the estimate of a food expressed in points as follows: $FQ_0 = 0$ (very poor), $FQ_1 = 1$ (poor), $FQ_2 = 2$ (satisfactory), $FQ_3 = 3$ (good) and $FQ_4 = 4$ (very good, the highest possible estimate in points), $FQ_{max} = 4$; $m_0, m_1, \ldots, m_4$ denote the number of points assigned by the respondents, evaluating the food ($m_0 + m_1 + m_2 + m_3 + m_4 = m_{FQ}$ denotes the number of all respondents (P), who estimated the quality of the served dishes).

The assortment abundance was described by the ratio of the real number of dishes available in the dining-car in the investigated period, $n_{AAf}$, to the largest amount of

**Table 6.2** The average ranks, $\overline{R}_i$, awarded by passengers to the quality of services provided in the dining-car, and their weights, $\omega_i$

| Significance (weight) | The criteria, describing the quality of services provided in the dining-car | | | |
|---|---|---|---|---|
| | PM | FQ | AA | MCC |
| $\overline{R}_i$ | 2.333 | 1.083 | 2.583 | 4.000 |
| $\omega_i$ | 0.267 | 0.391 | 0.242 | 0.100 |
| Priority order | 2 | 1 | 3 | 4 |

dishes ever offered in international trains, $n_{AAmax}$:

$$x_{C2,AA} = \frac{n_{AAf}}{n_{AAmax}}, \tag{6.7}$$

where $x_{C2,AA}$ is the variable of the second sub-criterion of the key criterion C, determining the assortment abundance; $n_{AAf}$ is the factual number of dishes offered in the dining-car during the investigated trip; $n_{AAmax}$ is the maximum number of dishes ever offered in the dining-car.

The meal change cycle was evaluated in points as follows: 0 means that dishes are not changed, 1 denotes that dishes are changed once a month or more often, 2 means that dishes are changed once in two weeks, 3 denotes that dishes are changed once a week, and 4 means that dishes are changed in every trip. The cycle was expressed by the ratio of the estimate in points given to the real meal change cycle, $MCC_f$, and the estimate in points given to the shortest cycle, $MCC_{max}$:

$$x_{C2,MCC} = \frac{MCC_f}{MCC_{max}}, \tag{6.8}$$

where $MCC_f$ denotes the real estimate in points awarded to the meal change cycle; $MCC_{max}$ is the highest estimate (4 points) given to the cycle.

### 6.1.3 Price of Newspapers and Magazines

The price of newspapers and magazines sold in the train was evaluated in points as follows: 0 means that the prices of newspapers and magazines offered in the train are higher than their prices in the news-stalls, 1 means that the prices of newspapers and magazines offered in the train and their prices in news-stalls are the same, 2 shows that the prices of newspapers and magazines offered in the train are lower than their prices in the news-stalls.

The variable $x_{C3}$ of the CQI $K_C$, assessing the price of newspapers and magazines in the train, was calculated by the formula:

$$x_{C3} = \frac{PNM_f}{PNM_{max}}, \tag{6.9}$$

where $x_{C3}$ is the variable of the third sub-criterion of the key criterion C; $PNM_f$ shows the estimate of the criterion describing the price of newspapers and magazines offered in the train in points (0, 1 or 2); $PNM_{max}$ is the highest possible estimate (2 points) given to the price of newspapers and magazines in the train.

### 6.1.4 Price of the Health Insurance Card Valid Abroad

Health insurance is required in the case, when a passenger of the train requires medical help because of an acute disease. The required medical help is usually provided either in an outpatient department or in a hospital abroad.

The variable $x_{C4}$ of the CQI $K_C$, assessing health insurance price abroad, was calculated by the formula:

$$x_{C4} = \left(1 - \frac{PH_f - PH_{min}}{PH_{max} - PH_{min}}\right), \tag{6.10}$$

where $x_{C4}$ is the variable of the fourth sub-criterion of the key criterion C; $PH_f$, $PH_{min}$, $PH_{max}$ denote the real, the lowest and the highest price of health insurance abroad (€).

### 6.1.5 Visa Fees

Visas are an important tool used by States to regulate the possible entry of nationals of other States into the territory concerned. Visa fees can vary considerably. The paper [13] presents a novel dataset based on a manual collection of visa fees for travel between a global set of country pairs in seven different categories (tourist, work, student, family reunification, business, transit, and other). This analysis reveals a significant global divergence in visa fees. This is unfair in terms of the right to travel for people living in different parts of the world. Europeans find travelling easier. Visa fees in Africa and South Asia can often amount to several weeks' average income. There is a fundamental paradox: the richer the country, the less its nationals pay for visas to go abroad (both in absolute terms and in relation to their income). Poisson pseudo-maximum likelihood regression analyses reveals that a variety of factors influence the costs of visas between countries: reciprocal treatment, processing costs, historical-cultural ties, geographic proximity, and regional, religious, economic, and political discrimination.

The visa fee sub-criterion C5 is ranked second ($\overline{R}_{C5} = 2$) in terms of significance in key criterion C. European Union (EU) citizens do not need a visa when travelling in Schengen area countries. A visa is required for Belarusian and Russian citizens travelling to an EU country, as well as for EU citizens travelling to Belarus or Russia.

Once the maximum and minimum fee of a visa is known for each country through which and to which an international passenger train goes, the variable $x_{C5}$, which assesses the impact of the actual cost of a visa on the Comprehensive Quality of Travel Index (CQI) for a train journey, $K_C$, is calculated from the formula:

$$x_{C5} = \frac{\sum_{v=1}^{n_v} \cdot \left(1 - \frac{VF_f - VF_{min}}{VF_{max} - VF_{min}}\right)}{n_v}, \tag{6.11}$$

where $x_{C5}$ is the variable of the fifth sub-criterion of the key criterion C; $n_v$ is the number of visas required for the trip, ($v = 1, 2, \ldots, n_v$); $VF_f$, $VF_{min}$, $VF_{max}$ denote the real, the lowest and the highest prices of $v$-th visa's fees.

### 6.1.6 *Delivery of Meals (Included in the Ticket Price) to Passengers Travelling in the First-Class Double Compartment*

The price of the delivery of meals included in the ticket price to passengers was calculated in points. In this case, 0 means that the delivery of meals is not included in the ticket price and, therefore, this service is not provided to the passengers in the car; 1 denotes that the delivery of supper or breakfast is included in the ticket price; 2 means that the delivery of both breakfast and supper are included in the ticket price.

The variable $x_{C6}$ of the CQI $K_C$, determining the price of meal delivery included in the ticket for passengers travelling in the first-class double compartment, was calculated by the formula:

$$x_{C6} = \frac{DM_f}{DM_{max}}, \tag{6.12}$$

where $x_{C6}$ is the variable of the sixth sub-criterion of the key criterion C; $DM_f$ denotes the real estimate (in points) of the delivery of meals (or its absence), included (not included) in the ticket price of the passengers travelling in the first-class double compartment (0, 1 or 2 points); $DM_{max}$ is the highest possible estimate (2 points) of the delivery of meals included in the ticket price.

## 6.2 Additive Model for the Sub-criteria of the Key Criterion C

The comprehensive quality index $K_C$, describing the criteria associated with the expenses of the trip by an international train and the provided services, was calculated as follows:

$$K_C = \overline{Z}_C^* \cdot \left[ \overline{Q}_{C1} \cdot \frac{\sum_{y=1}^{3} n_y \cdot \left(1 - \frac{TF_f - TF_{min}}{TF_{max} - TF_{min}}\right)_y}{n_{car}} \right.$$

$$+ \overline{Q}_{C2} \cdot \left( \omega_{PM} \cdot \frac{PM_{min}}{\sum_{d=1}^{n_C} \frac{PM_d \cdot D_d}{D}} \right.$$

$$+ \omega_{FQ} \cdot \frac{FQ_0 \cdot m_0 + FQ_1 \cdot m_1 + \cdots + FQ_4 \cdot m_4}{\mathrm{FQ}_{max}(m_0 + m_1 + \cdots + m_4)}$$

$$+ \omega_{AA} \cdot \frac{n_{AAf}}{n_{AAmax}} + \omega_{MCC} \cdot \frac{MCC_f}{MCC_{max}}\Bigg)$$

$$+ \overline{Q}_{C3} \cdot \frac{PNM_f}{PNM_{max}} + \overline{Q}_{C4} \cdot \left(1 - \frac{PH_f - PH_{min}}{PH_{max} - PH_{min}}\right)$$

$$+ \overline{Q}_{C5} \cdot \frac{\sum_{v=1}^{n_v} \cdot \left(1 - \frac{VF_f - VF_{min}}{VF_{max} - VF_{min}}\right)_v}{n_v} + \overline{Q}_{C6} \cdot \frac{DM_{\bar{f}}}{DM_{max}}\Bigg]. \qquad (6.13)$$

Due to the application of the considered model, as well as the real research data and the allowable or best values of each sub-criterion, the significances of the key criterion C for evaluating the quality of passenger transportation by train could be described by a single value.

As $K_C$ increases from zero to $\overline{Z}_C^*$, the impact of the cost of international train travel on the quality of train travel increases.

## 6.3 Practical Application and Adequacy Assessment of the Key Criterion C Model

Using the formulas suggested in the paper for determining CQI components (six sub-criteria), the authors could calculate their significances based on the real data taken from the surveys (on the quality of dishes) and other reliable sources. The testing was performed in the train IPT No. 6.

***Ticket fare***. Basic fares for local transportation services provided by JSC "LTG Link" were determined, depending on the class of services offered in passenger cars (2nd or 3rd class) and the speed of the train (fast or ordinary trains).

The price of passenger transportation does not depend on the type of traction and railway lines. The IPT No. 6 has one sleeping (S) car ($n_S = 1$), four compartment (C) cars ($n_C = 4$) and one luxury (L) car ($n_L = 1$) (Fig. 4.8). Seeking higher occupancy of passenger cars in various days of the week, as well as higher profit from the trips by the "LTG Link" trains, special coefficients (depending on the days of the week) were introduced for fares for travelling in various berths in the cars. The upper berth fare is higher than that of a lower berth. The lowest fares are on May 8–9, while the highest fares are on April 28–29, June 8–12, July 13, November 2–5 and December 29–30.

The train has six passenger cars of S, C and L categories (Fig. 4.8). To calculate the CQI summand, the expanded formula (6.2), which is given below, was used for various car categories of the IPT No. 6:

$$x_{C1} = \frac{n_S\left(1 - \frac{TF_{fS} - TF_{minS}}{TF_{maxS} - TF_{minS}}\right) + n_C\left(1 - \frac{TF_{fC} - TF_{minC}}{TF_{maxC} - TF_{minC}}\right) + n_L\left(1 - \frac{TF_{fL} - TF_{minL}}{TF_{maxL} - TF_{minL}}\right)}{n_{car}}, \tag{6.14}$$

where $n_{car}$ is the number of cars in the train ($n_{car} = n_S + n_C + n_L$); $n_S$ is the number of ordinary second-class sleepers, $n_C$ is the number of the second-class 4-berth compartment cars and $n_L$ is the number of the first-class luxury two-berth cars; $TF_{fS}$, $TF_{minS}$, $TF_{maxS}$ denote the regular, the lowest and the highest fare in the car of S category, €; $TF_{fC}$, $TF_{minC}$, $TF_{maxC}$ denote the regular, the lowest and the highest fare in the car of C category, €; $TF_{fL}$, $TF_{minL}$, $TF_{maxL}$ denote the factual, the lowest and the highest fare in the car of L category, €, respectively.

The fares for travelling in the cars of S and C categories used in formula (6.14) were assumed to be the mean values of the fares for travelling in the lower and upper berths, €, while for the cars of L category, only the fare for travelling in a lower berth, which was the same in this case, was considered. The fares varied depending on particular seasons. Thus, on May 11, regular fares were used, while on May 9 they were the lowest, and, on June 8, the fares were the highest. The value of $x_{C1} = 0.287$ was obtained for the IPT No. 6.

***Price of meals served in the dining-car and their quality***. Formulas (6.5)–(6.8) were used for calculating the CQI variable $x_{C2}$. The price of the meal served in the dining-car was calculated by formula (6.5). The lowest price of a dish was $PM_{min} = 1.16$ €, the average price of the dish $P_{average} = 3.42$. Ten various dishes were served in the dining-car ($n_C = 10$), while their prices ($PM_d$) were as follows. The total number of the served dishes was $D = 97$ (Table 6.3). The variable of the quality sub-criterion, C2, was, $x_{C2,PM} = 0.339$.

The oral survey of the 60 passengers of the train was made (by the train manager), which had shown very positive evaluation (3 points) of the quality of the dishes served by the majority of passengers. Formula (6.6) was used for evaluating the quality of meals (dishes). For $FQ_0 = 0$, $FQ_1 = 1$, $FQ_2 = 2$, $FQ_3 = 3$, $FQ_4 = FQ_{max} = 4$ and $m_0 = 0$, $m_1 = 5$, $m_2 = 3$, $m_3 = 23$, $m_4 = 29$, ($m_0 + m_1 + \ldots + m_4 = 60$) (Table 6.4), it was found that $x_{C2,FQ} = 0.817$. On the day of the study, the manager of IPT No. 6 counted the number of meals $n_{AAf}$ served to passengers in the dining-car and compared it with the highest number of meals $n_{AAmax}$ ever served in the dining-car of this train. Formula (6.7) was used to assess the assortment abundance. When $n_{AAf} = 14$ and $n_{AAmax} = 24$, $x_{C2,AA} = 0.583$. During the period under consideration, the meals in the IPT No 6 dining-car are changed every month. The meal change

**Table 6.3** Price of ten dishes and number of portions sold

| Indicator | Dish number ($d = 1, 2, \ldots, n_C$) | | | | | | | | | | Average |
|---|---|---|---|---|---|---|---|---|---|---|---|
| | 1 | 2 | 3 | 4 | 5 | 6 | 7 | 8 | 9 | 10 | Total |
| $PM_d$, € | 1.16 | 1.74 | 1.88 | 2.32 | 2.61 | 4.92 | 5.21 | 5.50 | 5.79 | 7.24 | 3.42 |
| $D_d$, number | 22 | 5 | 9 | 10 | 9 | 5 | 15 | 12 | 5 | 5 | 97 |

cycle was evaluated by formula (6.8) as follows: when $MCC_f = 1$ and $MCC_{max} = 4$, $x_{C2,MCC} = 0.25$.

Formula (6.4) calculates the variable $x_{C2}$, which is:

$$\begin{aligned} x_{C2} &= \omega_{PM} \cdot x_{C2,PM} + \omega_{FQ} \cdot x_{C2,FQ} + \omega_{AA} \cdot x_{C2,AA} + \omega_{MCC} \cdot x_{C2,MCC} \\ &= 0.267 \cdot 0.339 + 0.391 \cdot 0.817 + 0.242 \cdot 0.583 + 0.100 \cdot 0.250 \\ &= 0.576. \end{aligned}$$

***Price of newspapers and magazines***. The prices of newspapers and magazines in the train did not differ from those in the news stalls ($PNM_f = 1$, $PNM_{max} = 2$). The CQI summand was calculated by formula (6.9), and, for a general case, $x_{C3} = 0.50$, was found.

***Price of health insurance card valid abroad***. Lithuanian citizens going to another country (e.g. by the IPT No. 6) should have health insurance valid abroad. The insurance is valid for 1–3 days and price $PH_f = 2.32$ €, while the lowest price of the required medical service guaranteed by it is $PH_{min} = 1.74$ € and the highest price is $PH_{max} = 3.48$ €. The variable summand, $x_{C4}$, of the CQI component was calculated by formula (6.10), which yielded $x_{C4} = 0.667$.

***Visa fees***. Lithuanian passengers travelling by the train IPT No. 6 should have two visas ($n_v = 2$), including a transit visa and a visa country of final station. Visas can be of various types (e.g. visas for tourism and commerce) and can be valid for a specified period of time (e.g. for one, two or more visits). Let us assume that a passenger needs a tourist visa for one visit only (the period of up to 30 days) to final country and a transit visa for one visit. It usually takes about 6–7 working days to get these visas. It is assumed that for the considered period, a transit single entry visa costs $VF_{fB} = 20.27$ €, with a minimum price of $VF_{minB} = 18.83$ € and a maximum price of $VF_{maxB} = 24.62$ €. It is assumed that for the considered period, a transit single entry visa for final country costs $VF_{fR} = 63.72$ €, with a minimum price of $VF_{minR} = 57.92$ € and a maximum price of $VF_{maxR} = 98.47$ €. Formula (6.11) was used to calculate the variable summand, $x_{C5}$, of CQI, which yielded $x_{C5} = 0.804$ (for the real data).

***Delivery of meals (included in the ticket price) to passengers travelling in the first-class double compartment***. At present, breakfast and supper are included only in the price of the tickets to the L category passenger cars of the international passenger trains No 6 "LTG Link". To evaluate the quality of the delivery of meals (the variable

**Table 6.4** Factual data from the evaluation of the quality of food in the IPT No. 6 dining-car

| Linguistic assessment of food quality | Very poor | Poor | Satisfactory | Good | Very good | Total |
|---|---|---|---|---|---|---|
| Meal quality assessment in the form of scores | $FQ_0 = 0$ | $FQ_1 = 1$ | $FQ_2 = 2$ | $FQ_3 = 3$ | $FQ_4 = 4$ | – |
| Number of passengers who rated the quality of meals | $m_0 = 0$ | $m_1 = 5$ | $m_2 = 3$ | $m_3 = 23$ | $m_4 = 29$ | $m_{FQ} = 60$ |

summand, $x_{C6}$, of CQI) included in the ticket price, formula (6.12) was used, which yielded $x_{C6} = 1$, as $DM_f = 2$ and $DM_{max} = 2$.

The value of the CQI component, depending on the ETPS. The quality of the items and services described by the sub-criteria of key criterion C and the respective $x_{Ci}$ value, based on the real data, are presented in Table 6.5. The quality level of train shown by the estimate of the variable $x_{Ci}$, based on the highest values of the real data, was equal to 1, but based on the average values, it was equal to 0.5.

The comprehensive quality index $K_C$ of the sub-criteria describing the ETPS was calculated by formula (6.13) as follows:

$$\begin{aligned} K_C &= 0.258(0.3324 \times 0.287 + 0.1030 \times 0.576 \\ &\quad + 0.0547 \times 0.500 + 0.1509 \times \\ &\quad \times 0.667 + 0.2722 \times 0.804 + 0.0868 \times 1) \\ &= 0.1518. \end{aligned}$$

The index referring to the key criterion, describing the expenses of the trip by international train and the provided services, was found to be $K_C = 0.152$. It is approximately 40 percent less than the maximum (indicating the best quality) $K_{Cmax} = 0.2580$. Average quality will be when $K_C = \frac{1}{2} \cdot \overline{Z}_C^* = \frac{1}{2} \cdot 0.258 = 0.129$. The poor quality will be when $K_C = 0$, and excellent quality $K_C = \overline{Z}_C^* = 0.258$.

**Table 6.5** Quality level of the sub-criteria of the key criterion C and the respective value of variables $x_{Ci}$ based on real data

| Variable $x_{Ci}$ of the key criterion C (formula) | Real (actual) value of the variable component | The calculated value $x_{Ci}$ |
|---|---|---|
| $x_{Ci}$ (6.2, 6.14) | $n_S = 1$; $n_C = 4$; $n_L = 1$; $TF_{fS} = 69.87$ €; $TF_{minS} = 58.94$ €; $TF_{maxS} = 74.24$ €; $TF_{fC} = 101.86$ €; $TF_{minC} = 81.06$ €; $TF_{maxC} = 110.17$ €; $TF_{fL} = 196.21$ €; $TF_{minL} = 154.67$ €; $TF_{maxL} = 212.81$ € | 0.287 |
| $x_{C2}$ (6.5–6.8) | $n_C = 10$; $D = 97$; $PM_{min} = 1.16$ €; $PM_{average} = 3.42$ €; $PM_d$ and $D_d$ (in Table 6.3) $x_{C2,PM} = 0.339$ $FQ_0 = 0$; $FQ_1 = 1$; $FQ_2 = 2$; $FQ_3 = 03$; $FQ_4 = 4$; $m_0 = 0$; $m_1 = 5$; $m_2 = 3$; $m_3 = 23$; $m_4 = 29$ (Table 6.4); $x_{C2,FQ} = 0.817$ $n_{AAf} = 14$; $n_{AAmax} = 24$; $x_{C2,AA} = 0.583$ $MCC_f = 1$; $MCC_{max} = 4$; $x_{C2,MCC} = 0.250$ | 0.576 |
| $x_{C3}$ (6.9) | $PNM_f = 1$; $PNM_{max} = 2$ | 0.500 |
| $x_{C4}$ (6.10) | $PH_f = 2.32$ €; $PH_{max} = 1.74$ €; $PH_{max} = 3.48$ € | 0.667 |
| $x_{C5}$ (6.11) | $n_v = 2$; $VF_{fB} = 20.27$ €; $VF_{minB} = 18.83$ €; $VF_{maxB} = 24.62$ €; $VF_{fR} = 63.72$ €; $VF_{minR} = 57.92$ €; $VF_{maxR} = 98.47$ € | 0.804 |
| $x_{C6}$ (6.12) | $DM_f = 2$; $DM_{max} = 2$ | 1 |

The CQI calculated according to this methodology can be used by decision makers in passenger rail transport companies. They can evaluate the suitability of individual trains to carry passengers.

The sum $\sum_{i=1}^{6} \overline{Q}_{Ci} \times x_{Ci} = 0.5883$ of the products $\overline{Q}_{Ci} \times x_{Ci}$ of the normalised local weights.

$\overline{Q}_{Ci}$ of the six sub-criteria of the key criterion C (Table 3.9 and Fig. 3.7c) and the variables $x_{Ci}$ (Table 6.5) indicates that the quality of travel on IPT No. 6 in terms of travel costs and services provided is slightly above average.

The product (Fig. 6.1) of each of the local weight $\overline{Q}_{Ci}$ of six sub-criteria of this key criterion, and of the variable $x_{Ci}$ for the IPT No. 6 varies from 0.0274 (C3—price of newspapers and magazines) to 0.2188 (C5—visa fee), which is much higher than the average (average $\overline{Q}_{Ci} \times x_{Ci} = 0.0981$). Sub-criteria C1 (ticket fare), C4 (cost of on-board health insurance card) and C6 (delivery of meals included in the ticket fare for passengers travelling in a first-class double compartment) have a close to average impact on the quality of the train journey. The impact of sub-criterion C2 (price of meals served in the dining-car and their quality) on the quality of the trip is about half of the average impact.

Considering the weight of the key criterion C ($\overline{Z}_C^* = 0.258$) and the sum of the products $\overline{Q}_{Ci} \times x_{Ci}$ of its 6 sub-criteria amounting to 0.5883, the impact of this key criterion on the CQI $K_C$ was calculated (formula 6.1). The impact of train travel costs and service factors on the quality of the international passenger train under study is comprehensively reflected by the CQI $K_C = 0.1518$. The other part of the Quality Level Indicator (QLI) $K$ consists of the calculated CQI $K_A$, $K_B$ and $K_D$ for key criteria A, B and D, which have been studied in the IPT No. 6, and are presented in other chapters of the monograph.

The solved example shows that the additive mathematical model (6.13) is correct. It allows, by applying normalised subjective local weights $\overline{Q}_{Ci}$ for the sub-criteria provided by the raters (experts and respondents) and the experimentally determined

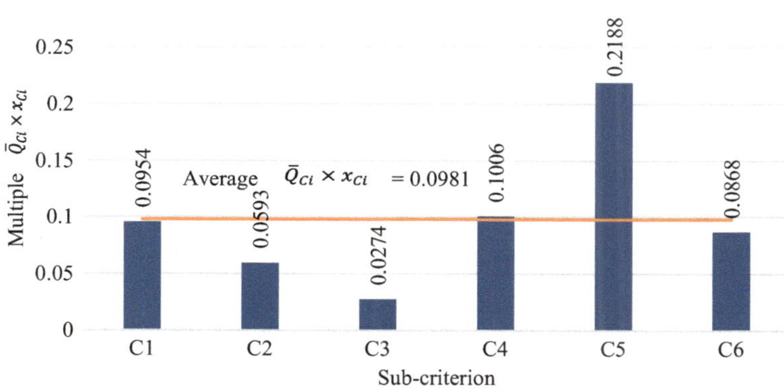


**Fig. 6.1** Findings of the calculations of the impact of the sub-criteria of the main criterion C on the quality of the trip of the IPT No. 6

factual parameters of the international passenger train under study, to obtain a computational result between 0 and $\overline{Z}_C^* = 0.228$. The new methodology and mathematical model can therefore be used in practice to assess the quality of train travel.

The normalized local weights of six sub-criteria of the key criterion C also differed considerably (Table 3.9). Thus, the respondents and experts of all three categories, including P and the TS and AS, evaluated the sub-criteria, such as the ticket price (0.3324) and the price of a visa (0.2722) as the most important sub-criteria considered. The sub-criteria of the price of newspapers and magazines (0.0547), as well as the delivery of meals (included into the ticket price), were evaluated by the respondents (particularly, by the passengers travelling in the first-class double compartments) as the least important criteria (0.0868).

The price of meals served in the dining-car and their quality (C2) was the fourth most significant of the key criterion C (with the local weight of 0.1030). It consisted of the PM, FQ, AA and MCC. The Kendall rank correlation method and the equations suggested by the authors yielded the following weights of the sub-criteria, describing the quality of services provided in the dining-car of the train (which were obtained based on the ranks) (Table 6.2): $\omega_{PM}$ (0.267), $\omega_{FQ}$ (0.391), $\omega_{AA}$ (0.242) and $\omega_{MCC}$ (0.100).

The accuracy of the calculated quality index $K_C$ of the mathematical model (formula 6.13), used by the authors to express the weights of the sub-criteria (key criterion C), describing the ETPS in a single value, was checked based on the real data on the trip by the IPT No. 6. The calculated CQI, $K_C$, referring to this key criterion C, was equal to 0.152. This value is between 0 (the lowest quality) and 0.258 (the highest quality) and makes almost 60% of the ideal value for the main criterion C. The results obtained in the study allow the authors to conclude that the quality of the trip by the considered train is higher than the average quality of the trips by train.

## References


1. Kamandanipour K, Nasiri MM, Konur D, Yakhchali SH (2020) Stochastic data-driven optimization for multi-class dynamic pricing and capacity allocation in the passenger railroad transportation. Expert Syst Appl 158(113568):1–16. https://doi.org/10.1016/j.eswa.2020.113568
2. Hansen IA (2010) Railway network timetabling and dynamic traffic management. Int J Civil Eng 8(1):19–32
3. De Oña R, Eboli L, Mazzulla G (2014) Key factors affecting rail service quality in the northern Italy: a decision tree approach. Transport 29(1):75–83. https://doi.org/10.3846/16484142.2014.898216
4. Sivilevičius H, Maskeliūnaitė L (2019) The model assessing the impact of price and provided services on the quality of the trip by train: MCDM approach. Bus Admin Manage 22(2):51–67. https://doi.org/10.15240/tul/001/2019-2-004
5. Cerny P (2021) 10 questions about night trains that you always wanted, but never dared, to ask. Heinrich Böll Stiftung, Brussels, European Union. https://eu.boell.org/en/10-questions-night-trains
6. Zheng J, Liu J (2016) The research on ticket fare optimization for China's high-speed train. Math Probl Eng 2016:1–8. https://doi.org/10.1155/2016/5073053

7. Qin J, Zeng Y, Yang X, He Y, Wu X, Qu W (2019) Time-dependent pricing for high-speed railway in China based on revenue management. Sustainability 11(16):1–18. https://doi.org/10.3390/su11164272
8. Chen J-H, Gao Z-Y (2005) Optimal railway passenger-ticket pricing. J China Railway Soc 27:16–19
9. Brandhuber S (2021) Discover the best train food services. Rail Europe. https://www.raileurope.com/en-us/blog/best-train-food-services
10. Palupi IR, Darmawan NA, Prawiningdyah Y (2020) The effect of service and food quality on customer satisfaction in Indonesian train catering. Int J Acad Res Bus Social Sci 10(13):88–97. https://doi.org/10.6007/IJARBSS/v10-i13/6891
11. Yan J, Hu L (2021) Research on intelligent design of train catering service system based on rail train commuting volume. In: International Conference on Control Science and Electric Power Systems (CSEPS), 28–30 May 2021, pp 157–160. IEEE, Shanghai. https://doi.org/10.1109/CSEPS53726.2021.00038
12. Kendall M, Gibbons JD (1990) Rank correlation methods, 5th edn. Oxford University Press, New York, NY
13. Recchi E, Deutschmann E, Gabrielli L, Kholmatova N (2021) The global visa cost divide: how and why the price for travel permits varies worldwide. Polit Geogr 86(102350):1–14. https://doi.org/10.1016/j.polgeo.2021.102350

# Chapter 7
# Intrinsic Safety on Train Journeys

*When the train driver is looking for new ways—the train derails.*
Stanislaw Jerzy Lec

## 7.1 The Impact of Intrinsic Safety Elements on the Quality of Train Travel

Rail transport is focused on passenger and freight safety. Safety requires the application of new technologies to prevent fire risks, enhance communication between train staff and passengers, ensure the efficient functioning of the train itself. The results of an in-depth interview of a railway expert with 115 passengers, and the analysis of passenger rail safety statistics data show that passenger trains have changed significantly. The aesthetics of trains has improved, they are more comfortable and safer. Travellers believe that passenger transport has undergone a major metamorphosis, yet many of the safety-related changes are not visible to passengers. This might be linked to insufficient communication by rail operators on innovations, and consequently to building a positive image of the company and rail transport as such [1].

More and more people in mega cities are choosing to travel by public transportation due to its convenience and punctuality. It is widely acknowledged that there may be some potential associations between passengers. Their travel behavior may be working together, shopping together, or even some abnormal behaviors, such as stealing or begging. Thus, analyzing association between passengers is very important for management departments. It is very helpful to make operational plans, provide better services to passengers and ensure public transport safety [2].

Severe accidents on the railway system require the evacuation of large numbers of people. This may often be carried out in a hurry, under imminent threat and in difficult circumstances, such as in the event of a fire or if the evacuation takes place in a tunnel or on a bridge. Such events are rare, but the consequences can be severe.



H. Sivilevičius and L. Maskeliūnaitė, *Assessment of the Quality of Passenger Transportation by Train Using Multiple Criteria Decision Making Methods*, Lecture Notes in Intelligent Transportation and Infrastructure,
https://doi.org/10.1007/978-3-031-85877-2_7

Mass evacuations are a common occurrence when one or more trains are stopped due to a track hazard, a vehicle accident, or a traffic management problem. The evacuation of passengers and staff is subject to the risk of electrocution or being hit by passing trains. In situations where there is no immediate threat to the people on board, it may take a long time to evacuate them. Over time, poor environmental conditions, such as high temperatures, can cause discomfort and inconvenience for passengers. Passengers may have the urge to evacuate spontaneously.

The study by Kecklund et al. [3] aimed to better understand different evacuation situations and passenger behaviour. It also identified potential areas for improvement regarding evacuation. Another study aimed to investigate the effect of crew actions during the evacuation of two HSTs under different fire scenarios [4]. The analysis focuses on an evacuation drill carried out by the Spanish Railroad Administration, RENFE Operadora. This analysis is used to obtain modelling inputs. It also analyses the impact of the actions of the train crew on the early evacuation of passengers (when a fire is detected, and the train is still moving). The achieved results contribute to railway safety. They demonstrate that there are qualitative and quantitative advantages when the evacuation is controlled by the train crew rather than the passengers themselves using a self-rescue strategy. Based on the findings of the study, recommendations for emergency management are outlined.

In rail transport, the location and timing of an accident cannot be predicted and determined in advance, but the likelihood and consequences can be minimised by taking appropriate safety measures.

In the event of a train accident (collision), the number of casualties can be high, which is why passive safety is very crucial. Passive safety of passengers in a car and on a train is compared. Based on a comprehensive literature review, the state of the art of passive safety protection in cars and trains is analysed from three dimensions: interior, human characteristics and occupant posture. The method of protection in cars is mainly restraint, while in trains it is separation. In terms of posture, the research related to car occupants mostly considers posture in reality and in future automated vehicles, while the research related to posture of train passengers focuses more on lateral and standing occupants. The safety issues in cars and trains are different. However, to ensure the passive safety of train occupants, the interior design of cars can be used as a reference to reduce occupant injuries. In addition, human characteristics and posture have a significant impact on passenger injuries, which is a factor that would be relevant for passive safety in trains [5].

The safety of freight and passenger transportation by the railway network has a significant effect on growth in its attractiveness and competitiveness. Safety is one of the most important "cornerstones" of passenger and freight transportation [6]. With an increasing demand for high-performance and cost-effective railway transport, it is required that railway systems correspond two main criteriafault tolerance and safety for freight and passenger traffic [7].

Since railroads play a crucial role in transport systems to ensure efficient and safe operation, risk assessment of the derailment of vehicles using reliable criteria is an important topic. Since risk assessment of derailment is a critical issue, the researchers attempted to identify and evaluate an effective derailment criterion. In the process of assessing the risk of derailment, the Nadal coefficient was widely adopted [8].

Rail traffic safety is a set of railway traffic characteristics showing how road users and rolling stock are protected from accidents and their consequences. Railway safety can be divided into two categories: external and internal [9]. This work assesses the internal safety of the railway trip, i.e., creating favourable driving conditions for passengers to avoid accidents on the train (in the carriage) or making the consequences of the accident minimal.

The model for calculating the significance of the key criterion D factors which characterise the quality of intrinsic safety on international train travel is given in the formula:

$$K_D = \overline{Z}_D^* \cdot \sum_{i=1}^{m} \left(\overline{Q}_{Di} \cdot x_{Di}\right) = \overline{Z}_D^* \cdot \left(\overline{Q}_{D1} \cdot x_{D1} + \overline{Q}_{D2} \cdot x_{D2} + \cdots + \overline{Q}_{D8} \cdot x_{D8}\right) \quad (7.1)$$

where $K_D$ is a comprehensive quality index (CQI) related to the internal safety (ranking from 0 to $\overline{Z}_D^*$); $\overline{Z}_D^*$ is the mean weight, reflecting experts and respondents' estimates of the significance of the key criterion D (Table 3.16; $\overline{Z}_D^* = 0.2860$); $\overline{Q}_{D1}, \ldots, \overline{Q}_{Dm}$ denote the local weight values of $i$-th sub-criterion of D key criterion ($i = 1, 2, \ldots, m$) determined the expert evaluation method (Table 3.10 and Fig. 3. 7); $x_{D1}, \ldots, x_{Dn}$ are the variables of $i$-th sub-criterions of key criterion D, whose estimates are used the real sub-criterion value, ranking 0 to 1.

The weights of 8 sub-criterion belonging to key criterion D, which describe the internal safety trip by train, are found from the survey assessors, when three categories of respondents P and experts TS and AS provided their judgements. The mean weight $\overline{Z}_D^*$ shows the significance of all sub-criteria of key criterion D (when the number of respondents and experts in each assessors category differs). This weight is calculated by the formula (3.36).

### *7.1.1 Number and Condition of Fire Extinguishing Equipment*

Fire is a major hazard due to the risk of death or injury. It can also cause a lot of material damage. With the proper use of fire-fighting equipment, most fires can be extinguished in the early stages of fire progression. From the start of a fire, it can take just a few minutes for the consequences to be severe. Responding quickly and correctly is vital. Effective training is needed to ensure that the public, who are not experts in firefighting, can use a fire extinguisher properly. This may help prevent fires. Fire prevention professionals have become interested in virtual reality (VR) as an intelligent, safe, and effective training method [10].

The European Standard [11] specifies the requirements for the runnability under fire conditions of railway rolling stock carrying passengers. The standard specifies fire conditions: for which it is not necessary to specify running capacity requirements because there is no significant potential for serious injury or risk to life; or it is reasonable to expect that trains will continue to run under control; or it is not reasonably possible to impose requirements due to the exceptional nature of the fire.

Fire detection and suppression solutions are becoming a common safety instrument on newly designed and manufactured trains in Europe and other parts of the world International and national legislation often sets minimum safety requirements for different modes of transport. However, it does not provide any guidance on how to assess the performance of such systems. The legislation addresses the need for the system but does not provide a scheme for testing its operation. To compensate for this lack of guidelines and standards, European manufacturers of rolling stock fire safety systems, along with certification bodies, have developed a series of guidelines to define the performance of the system and the success criteria for firefighting. The focus is on the experience gained from the use of these guidelines. The testing methods are described, and the choice of these methods is justified [12]. The document [13] provides requirements for systems to detect, control or manage the effects of a fire in a railway vehicle to establish a safe area until the evacuation of passengers and staff is complete.

Modern trains are equipped with fire containment and control systems (FCCS). Rolling stock is made of materials that are persistent to the ignition: they are supposed to burn slowly and emit less smoke and toxic products [14]. Fire containment, control and extinguishing measures shall be used in accordance with the standards and technical requirements in passenger, baggage, mail, and baggage carriages with video equipment compartments and in carriages with diesel engines. In addition to the portable fire extinguisher (PFE), an automatic powder extinguisher (APE) or aerosol fire extinguishing equipment (AFE) with similar technical characteristics shall be installed at the electrical equipment control panel of each carriage (Fig. 7.1).

Each passenger train shall be equipped with industrial personal protective equipment (PPE), at least 3 pieces per train set. Currently, it is installed in every carriage of the train set (one unit per carriage).

New passenger carriages, as well as refurbished and modernized carriages, shall be equipped with an automatic fire alarm system (FAS) designed to detect and report the outbreak of a fire in a controlled space (carriage). The facility consists of fire detectors, a fire detection and control unit (FDCU) and communication between the detectors and the FDCU [15]. An AFE shall be fitted on passenger carriages in newly manufactured or overhauled versions, allowing the use of water reserves and water supply systems [15].

The score for the quantity and condition of fire extinguishers on each passenger train carriage shall be (0, 1 or 2): 0—extinguishers are present but the number of extinguishers in the carriage is insufficient as they do not comply with the standards for the provision of fire extinguishers for international passenger trains, or the extinguishers are not sealed (at least one of them), or the seal is broken; 1—the number of fire extinguishers in the carriage meets the standards for the provision of fire extinguishers for international passenger trains, but not all extinguishers are sealed or the seal is broken; 2—the number of fire extinguishers complies with the standards

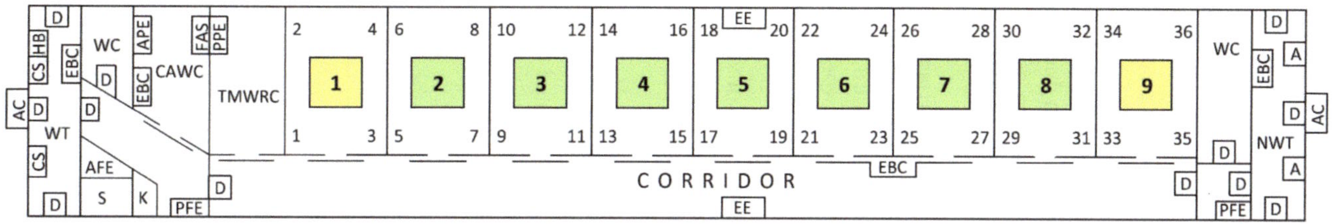


**Fig. 7.1** IPT No. 6/5, diagram of a 4-berth sleeping compartment car: *AC* automatic coupler; *WT* work tambour; *CS* coal stack; *D* door; *HB* hand brake; *EBC* emergency brake cock; *WC* sanitary unit; *CAWC* car attendant work compartment; *APE* automatic powder extinguisher; *AFE* aerosol fire extinguisher equipment; *FAS* fas alarm system; *S* boiler (boiler dome); *K* kettle; *PFE* portable fire extinguisher; *TMWRC* train manager work and rest compartment; *PPE* personal protective equipment; *EE* emergency exit; *NWT* non-working tambour; *A* ashtray; *1–9* passenger compartments (compartment 1 is the carriage attendant's lounge compartment; compartment 9 is for the electro-mechanic of the train and restaurant carriage staff. Tickets for these are not available. The carriage seats 24 passengers); the odd numbers (1, 3, 5, …) indicate the lower berths; the even numbers (2, 4, 6,…) indicate the upper ones

for the provision of fire extinguishers for international passenger carriages and all extinguishers are sealed.

The number and condition of APEs to extinguish fires at the control panel of the electrical installations on each carriage of a passenger train shall be evaluated by the following scores (0, 1 or 2): 0—no APE on board; 1—fire extinguisher on board, but it is technically defective, either because it has expired or because the extinguisher is mechanically damaged (cracked casing), its positioning in the electrical control panel is not in accordance with the requirements of the construction documentation; 2—APE on board, it is technically sound, the condition checks have been carried out on time, the extinguisher is not damaged, and the extinguisher is positioned in accordance with the requirements of the construction documentation.

The following scores (0, 1 or 2) shall be used to assess the provision of PPE for each passenger carriage: 0—no PPE onboard; 1—PPE onboard but not in good condition; 2—PPE onboard and in good condition.

The presence and condition of the FAS in each carriage of a passenger train shall be assessed by the following scores (0, 1 or 2): 0—FAS is not part of the carriage design and is therefore not present; 1—FAS is on board, but is inoperative because it is out of order; 2—the carriage is equipped with a working FAS that can send out audible (acoustic) and visual (optical) signals of ignition, indicating the location of the fire inside the carriage.

The presence and condition of an AFE for the use of water from the water supply system of the carriage shall be assessed with the following scores (0, 1 or 2): 0—no AFE on board; 1—AFE on board, but not in working condition (defective); 2—AFE on board and in working condition.

The value of the CQI $K$, depending on $K_D$, which describes the internal safety trip by train, and the variable $x_{D1}$ of the product $\overline{Q}_{D1} \cdot x_{D1}$, showing number of fire extinguishing equipment and its condition, is calculated by the formula:

$$x_{D1} = \frac{\sum_{c=1}^{n_{car}} \sum_{d=1}^{n_c} \left( \frac{FED_f}{FED_{max}} \right)_c}{n_{car} \cdot n_{ex}} = \frac{\sum_{c=1}^{n_{car}} \left( \frac{PFE_f}{PFE_{max}} + \frac{APE_f}{APE_{max}} + \frac{PPE_f}{PPE_{max}} + \frac{FAS_f}{FAS_{max}} + \frac{AFE_f}{AFE_{max}} + \right)_c}{n_{car} \cdot n_{ex}}, \quad (7.2)$$

where $x_{D1}$ is the variable of the first sub-criterion of the key criterion D; $FED_f$ $FED_{max}$—factual and maximum scores of the types of fair extinguishing devices in the carriage $c$; $PFE_f$—the factual score of number and condition of portable fire extinguishers in carriage $c$ (0, 1 or 2 points); $PFE_{max}$—the score of maximum necessary number and condition of portable fire extinguishers in each carriage $c$ (2 points); $APE_f$—the score of factual number and condition of automatic powder extinguishers in carriage $c$ (0, 1 or 2 points); $APE_{max}$—the maximum possible score of number and condition of automatic powder extinguishers in each $c$ carriage (2 points); $PPE_f$—the score of factual supplying carriage $c$ with manufactured (industry) personal protective equipment and its condition (0, 1 or 2 points); $PPE_{max}$—the maximum possible score of supplying each carriage $c$ with manufactured personal protective equipment and its condition (2 points); $FAS_f$—the score of the factual availability and condition of the automatic fire alarm system in carriage $c$ (0, 1 or 2 points); $FAS_{max}$—the highest possible score of the availability and condition of the automatic fire alarm system in the carriage $c$ (2 points); $AFE_f$—the factual score of the availability and condition of the aerosol fire extinguishing equipment in carriage $c$ (0, 1 or 2 points); $AFE_{max}$—the maximum possible score of the availability and condition of the aerosol fire extinguishing equipment in each carriage $c$ (2 points); $n_{car}$—the number of carriages in passenger train ($c = 1, 2, \ldots, n_{car}$); $n_{ex}$—the number of the types of fire extinguishing devices in the carriage ($d = 1, 2, \ldots, n_{ex}$).

### 7.1.2 First Aid Kits

Passengers travelling on international trains are entitled to first aid when necessary. With the disruption of production and logistics due to the war in Ukraine, the market is facing a shortage of re-stocked first aid kits. In this context, the Ministry of Health of the Republic of Lithuania has imposed an obligation on drivers to restock their first aid kits by 1 July 2023. The need to restock the first aid kit is because its stock has remained unchanged for 18 years. The principles of administering first aid have changed over time. The renewed first aid kit for cars and rail transport includes a pressure bandage, ice packs, tourniquets to stop arterial bleeding and a disposable face mask for resuscitation (artificial respiration) [16]. There are no pharmaceuticals in the first aid kit on board, except for anti-burn medication. The first aid kit is with the train manager. There is also a stretcher in the "staff" carriage.

The composition of the first aid kit of the train has been analysed [17]. The study employed content analysis, comparative and logical analysis, and expert interviews. A survey of 453 experts, including train managers and carriage attendants, was used. The respondents were specialists from the regional branches of Ukrainian Railways. First aid kits on trains from Ukraine, Russia, the Republic of Kazakhstan, and the Republic of Belarus were analysed. The feedback from the questionnaires showed that all respondents were convinced of the need for a first aid kit on board the train. However, many specialists were unaware of the presence and location of first aid kits on board the train. 75.7% of respondents think that there should be two types of first aid kits on-board—in the train manager's carriage and in the carriage of a train attendant. The availability of an on-board first aid kit is well known to only 53.7% of experts. Many professionals believe that its composition needs to be upgraded. Staff should regularly attend first aid courses and be certified. Based on the content analysis, it was established that the carriage should have one first aid kit and one additional first aid kit for emergency use. The advantages and disadvantages of first aid kits have been identified. Ways to further improve them have also been proposed. When choosing a first aid kit, it is recommended to look at in-car first aid kits.

The availability of first aid kits on board the train is scored between 0 and 2 points. Score 0 is when first aid kit is not available on board. When the first aid kit is on board, but its stock of bandages and essential medication is insufficient in an emergency, such as a train crash or a terrorist attack, the score is 1. Score 2 indicates the presence of first aid kits on board the train, where the first aid kit contains enough bandages and medicines to meet an emergency.

The value of the CQI $K$, developing on $K_D$, which describes the internal safety trip by train, and the variable $x_{D2}$ of the product $\overline{Q}_{D2} \cdot x_{D2}$, showing availability of first medical aid measures, is calculated by the formula:

$$x_{D2} = \frac{FAK_f}{FAK_{max}}, \tag{7.3}$$

where $x_{D2}$ is the variable of the 2th sub-criterion of the key criterion D; $FAK_f$ is the factual assessment of the first aid kits on board the train (0, 1 or 2 points); $FAK_{max}$ is the maximum possible score given to the assessment of the availability of first aid kits on-board the train (2 points).

### *7.1.3 Ambulance Call Facility*

Travelling by train, especially over long distances, can sometimes cause health problems for passengers (due to climate change, food poisoning, heart disease). Children and the elderly tend to be affected more often. It is essential that first aid is available on board the train and that train staff are trained to provide it.

Cardiovascular diseases (CVD) are the leading cause of death among international travellers. It is not known whether CVD is a deterrent to international travel. The study [18] aimed to describe the travel experiences of individuals with CVD. The objectives were to find out what travellers with CVD consider to be the biggest barriers that they face in travelling, and then work out recommendations for CVD travellers, doctors, and travel organisers. Semi-structured interviews were conducted with CVD patients who participated in a CVD prevention or cardiac rehabilitation programme. Demographic and clinical data such as gender, age and types of cardiovascular conditions were recorded. Twelve patients (eight men), mean age 68 $\pm$ 7.58 years, consented to semi-structured interviews (26–78 min). The principal themes that evolved from the interviews included altered perceptions of travel, access to medical care abroad, problems with medicines, problems with medical equipment at airports, restricted leisure travel activities and optimal self-care. All respondents considered travel to be beneficial for health and did not consider CVD to be a significant barrier to international travel. Specific cardiovascular diseases have led to increased travel anxiety. These findings highlight the unique CVD patient experience when travelling abroad. Optimisation of cardiovascular health and responsible health behaviour during travel would facilitate medically unfeasible travel abroad. The findings of the study can provide useful information regarding pre-travel health issues for travellers with CVD.

There are also Passenger Assist services for elderly and disabled passengers provided by some train companies. The railway company can organise assistance for passengers to get on and off the train, reserve a seat for a wheelchair, assist travellers with animals or pets.

An ambulance is called on board when the first aid facilities on board are insufficient or the passenger requires medical attention. It can be summoned by the train manager via radio through the train driver at the nearest station. The radio station on board the train shall be at the disposal of the train manager or the electrician-mechanic and, during their time off, of the carriage attendant in charge of the train [19].

The ability to call an ambulance to a train (not to each carriage) is scored 0, 1 or 2 points. Where there is no option to call an ambulance to the train, it is scored 0 points. Where the call for an ambulance to the train can be made by radio via the train driver, this option is scored 1 point. When an ambulance can be called on board the train directly via the nearest railway station (GSM-R radio system), the option is scored 2 points.

The value of the CQI $K$, depending on $K_D$, which describes the internal safety trip by train, and the variable $x_{D3}$ of the product $\overline{Q}_{D3} \cdot x_{D3}$, showing availability of challenge medical emergency in train, is calculated by the formula:

$$x_{D3} = \frac{ACF_f}{ACF_{max}}, \tag{7.4}$$

where $x_{D3}$ is the variable of the 3th sub-criterion of the key criterion D; $ACF_f$ is the factual assessment of the emergency ambulance call-out to trains option (0, 1 or 2 points); $ACF_{max}$ is the maximum possible score of the emergency ambulance call-out to trains option (2 points).

### 7.1.4 *Operational Status of Axle Box Overheating and the Fire Alarm System*

Rolling stock axle boxes are among the most relevant components of railway rolling stock. The axle box is the bearing casing that supports the axle load of the vehicle and insulates the lubricant needed to keep the bearings running. The axle box is therefore central to both rolling stock safety and maintenance. To improve the safety and reliability of the axle box, diagnostic systems have been introduced in recent years to detect malfunctions or bearing damage. This is done with on-board or roadside monitoring systems that react to vibrations or heat. The performance of these systems depends on the algorithms employed to predict the condition of the bearings [20].

The prediction of axle temperature in advance and the assessment of the working condition of axle bearings on a high-speed train is vital for enhancing the safety of train operation and reducing the risk of accidents. For this, the method of monitoring the temperature of the train axle was applied. The results showed that the proposed method is sufficiently accurate and can be a reliable tool for axle temperature prediction [21].

A three-dimensional vehicle and track dynamics model has been developed that accurately accounts for the dynamic performance of the high-speed train when in operation. The temperature and vibration characteristics of the axle box bearing can be used for the assessment, design, and maintenance of the bearing [22].

The axle box overheating control signalling is designed to improve train safety. It provides continuous monitoring of the temperature of the axle boxes (while the train is running or stationary) and helps prevent train accidents when the axle boxes get too hot, which can lead to the breakage of their roller bearings. When the bearing gets hot, the alarm light on the control panel of the carriage lights up and an audible alarm is given. The system for controlling the heating of the axle boxes is controlled by a switch or button on the control panel. It is strictly forbidden to switch it off while on the move.

The maximum permissible values $T_{lim}$ of the axle-box and axle-neck (ABAN) heating depend on the ambient air temperature [23] (Table 7.1). The $T_{lim}$ values may change as the bearing design improves, but the principle of ABAN temperature assessment is intended to be the same.

These values $T_{lim}$ are compared with the actual ABAN temperature $T_{ABANf}$. During the inspection (after the train has stopped) they are used to assess the heat hazard by means of a non-contact remote thermometer. If the hot axle-box detection system (HABD) detects that the technical condition of a running train is dangerous and poses a threat to traffic safety, the axle of the carriage shall be replaced and repaired in the designated siding.

If any of the ABAN temperatures of a train carriage exceeds the limit value (Table 7.1), not only is there a risk of breakage of the wheel neck and derailment, but also of the whole train crashing. It is not logical to assess the quality of a train on averaged actual ABAN temperature of all the axles in accordance with the limit temperature. The safety of the whole passenger train in terms of axle box temperature shall be assessed by taking the worst (highest) ABAN measured value $T_{ABANfmax}$ and not by calculating an arithmetic or other (geometric, weighted) average temperature.

**Table 7.1** The limiting heating values for the ABAN of passenger cars, recorded by automatic control devices

| The temperature of ambient environment air ($T_{air}$), °C | The limiting heating values for the ABAN ($T_{lim}$), °C | Difference $T_{lim} - T_{air}$ |
|---|---|---|
| Below − 30, between − 30 and − 20 | 20 | 50–40 |
| Between − 20 to − 10 | 30 | 50–40 |
| Between − 10 to − 1 | 40 | 50–41 |
| Between 0 and + 10 | 50 | 50–40 |
| Between + 11 and + 20 | 60 | 49–40 |
| + 21 and higher | 70 | 49–40 |

The worst axle box temperature value $T_{ABANfmax}$ is the one closest to the permissible value $T_{lim}$ given in Table 7.1.

The value of the CQI $K$, depending on $K_D$, which describes the internal safety trip by passenger train, and the variable $x_{D4}$ of the product $\overline{Q}_{D4} \cdot x_{D4}$, showing operational state of axlebox overheating and the firealarm system, is calculated by the formula:

$$x_{D4} = \frac{T_{lim} - T_{ABANf}}{T_{lim} - T_{air}}, \quad (7.5)$$

where $x_{D4}$ is the variable of 4th sub-criterion of the key criterion D; $T_{ABANf}$ is the factual temperature of the measured ABAN, °C; $T_{air}$ is the temperature of environment (ambient) air on the measured site (station), °C; $T_{lim}$ is the limiting heating temperature value for ABAN of passenger cars, °C (Table 7.1).

The correlation of the ultimate temperature of the passenger train carriage axle box with the ambient air temperature is shown by the linear regression equation calculated on the data in Table 7.1:

$$T_{lim} = 0.9211T_{air} + 45, \quad (7.6)$$

with coefficient of determination $R^2 = 0.9211$. This and the straight line (Fig. 7.2) shows that as the ambient air temperature increases, the ABAN ultimate temperature also increases $T_{lim}$ as the efficiency of ABAN natural cooling using warmer ambient air decreases. In any case, the temperature of the axle box must not exceed $T_{lim} =$ 70°C during the warm summer period.

As the actual ABAN temperature $T_{ABANf}$ increases from the ideal temperature$T_{\mathrm{air}}$ towards the ultimate temperature $T_{lim}$, this dangerous trend is likely to continue. In most cases, the cause of the temperature approach towards the limit temperature is not known and is likely to reach or exceed the limit temperature due to an ABAN fault (Fig. 7.3).

In most cases, the actual temperature of the ABAN is higher than the ambient air temperature. The data presented in Table 7.1 and Fig. 7.2 show that the maximum permissible temperature difference $\Delta T_{ABAN} = T_{lim} - T_{air}$ is between 40 °C and 50 °C. The lower the ambient air temperature, the more intensive the cooling of the

**Fig. 7.2** The influence of the outdoor temperature $T_{air}$ variation on the limiting ABAN temperature $T_{lim}$

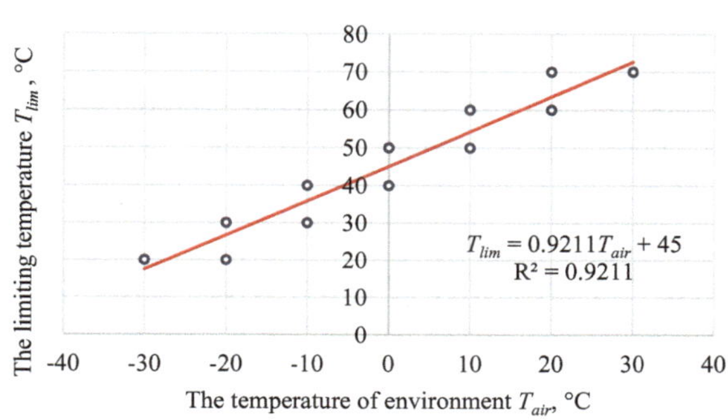

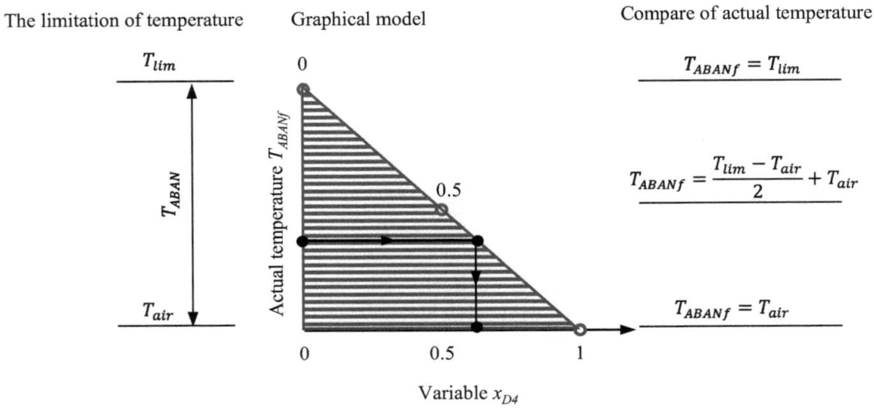


**Fig. 7.3** Graphical model of the influence of the passenger train ABAN temperature on its variable $x_{D4}$

ABAN and the less likely it is to overheat and consequently fail. The axle box of the train carriage with the greatest difference $\Delta T_{maxl} = T_{ABANf} - T_{air}$ between the actual temperature $T_{ABANf}$ and the ambient air temperature$T_{air}$ measured at the 1st railway siding ($l = 1, 2, \ldots, n_{stage}$) shall be considered as the most dangerous for safety. Given the maximum difference between these temperatures $\Delta T_{maxl}$, the values $T_{ABANf}$ and $T_{air}$ are also taken, which allow the variable$x_{D4}$ to be calculated from formula (7.5).

A graphical model of the effect of the actual temperature $T_{ABANf}$ of the carriage ABAN on the variable $x_{D4}$ (Fig. 7.3) shows that the value of the variable $x_{D4}$ decreases in a proportional manner as $T_{ABANf}$ approaches from $T_{air}$ to $T_{lim}$.

The model for calculating the value $x_{D4}$ of a variable (Fig. 7.3) shows that the passenger train with the highest quality $x_{D4} = 1$ according to the ABAN actual temperature $T_{ABANf}$ is the one with the hottest axle box temperature $T_{ABANfmax} = T_{air}$ measured on the $l$ st intermediate stage. A train in which the temperature of the hottest axle box $x_{D4} = 0.5$ is equal to the temperature calculated according to the same formula is of medium quality:

$$T_{ABANfmax} = \frac{T_{lim} - T_{air}}{2} + T_{air}. \tag{7.7}$$

The train with the worst quality in terms of overheating ($x_{D4} = 0$) is the train with the hottest axle box temperature $T_{ABANfmax}$ which equals $T_{lim}$ or is higher than this.

### 7.1.5 *Availability of Emergency Exits On-Board*

Most passenger train accidents are resolved quickly, and passengers do not need to be evacuated from the train. On the contrary, it is safer for passengers to stay on the train.

Nevertheless, emergencies still occur, and they sometimes result in fatalities. Delayed evacuation of passengers and crew could increase the number of casualties. The purpose of regulations issued by the US Federal Railroad Administration (FRA), US Department of Transportation (DOT) is to ensure the safety of passengers and crew on intercity and commuter railways, as well as safe, timely and effective evacuation in emergencies. The FRA Research and Development Office supports an ongoing research programme that investigates how to improve the safety of passengers and crew on passenger trains in the event of an emergency. One of the FRA studies focuses on the evaluation of evacuation times. Exiting a passenger train after an accident is a complex process which depends on dynamic factors and conditions. There are a few variables that affect the time required for a passenger to disembark from a passenger train in an abnormal or emergency. There is no one methodology to assess the emergency exit system for passenger train carriages, or the impact of failures in this system on the time taken to exit the train. Several methods have been used to determine passenger exit times, however, none of the exit models have been validated using actual passenger data from a railway carriage. The time it takes to evacuate a railway carriage is heavily influenced by individual efficiency and physical obstacles. The location of the carriage and emergency exits are critical [24].

New and overhauled passenger carriages must have additional emergency exits. The number of these emergency exits shall be at least two per carriage and their location shall allow the evacuation of passengers to be completed before the fire hazardous factors reach the permitted limit values [15].

The availability (number and condition) of emergency exits on each carriage of a passenger train is scored 0, 1, 2, 3 or 4 points. Where there are no emergency exits in the carriage, the score is 0. Where there is only one emergency exit, and it is not in good condition (no fire safety memos, no evacuation signs, no special hammer for breaking glass), the carriage is awarded 1 point. Where a carriage has just one proper or two improper emergency exits, it scores 2 points. Score 3 is awarded for a carriage with two emergency exits, with only one in good working order. Two or more properly maintained emergency exits indicates the highest carriage rating, which is scored 4 points.

The value of the CQI $K$, depending on $K_D$, which describes the internal safety trip by passenger train, and the variable $x_{D5}$ of the product $\overline{Q}_{D5} \cdot x_{D5}$, showing availability of emergency exits in carriage, is calculated by the formula:

$$x_{D5} = \frac{\sum_{c=1}^{n_{car}} \left( \frac{EE_f}{EE_{max}} \right)_c}{n_{car}}, \tag{7.8}$$

where $x_{D5}$ is the variable of the 5-th sub-criterion of the key criterion D; $EE_f$ is the factual score of emergency exits in carriage $c$ (0, 1, 2, 3 or 4 points); $EE_{max}$ is the maximum possible score of emergency exits in carriage $c$ (4 points); $n_{car}$ is the number of carriages in the passenger train ($c = 1, 2, \ldots, n_{car}$).

### *7.1.6 Maintaining Railings, Stairs, Gangways, Doors, and Locks*

The maintenance of railings, stairs, gangways, doors, and locks is of paramount importance for the safety of passengers and train staff. Train door systems often cause injury to passengers. They are the cause of most incidents leading to train delays. Noonan [25] reviews the design and performance aspects of train doors in relation to injuries and other incidents. Modern train door systems are equipped with safety mechanisms that control the force, speed, and direction (opening or closing) of door movement. The mechanisms and controls that operate the doors must comply with strict reliability standards. In addition to the design and maintenance elements of the door systems, there are controls in place for the train staff to stop the train in the event of a trouble. When door structures and controls work as intended, injuries and other accidents can be avoided. The forensic examination of the train door function examines design documents, maintenance records and operating procedures. Professionals in the design and operation of railway systems can identify the causes of an incident.

Reducing the cost of running the railways and improving the reliability of services, especially in terms of punctuality, is essential. Train doors are a key element in maintenance costs and the causes of train delays. Door drive defects are the most common; especially on older trains, where door failures account for 25–50% of rolling stock defects [26].

Defective or uncomfortable train seats, doors, stairs, railings, and gangways often cause incidents on the railway system. A new set of safety signs for young children is being developed to prevent accidents on trains [27]. Troubleshooting of the train door system is carried out using the motor current signal that operates the door. The test rig is designed to investigate different failure modes under extreme conditions, as well as natural and artificial wear of critical components [28].

New trains are constantly joining Britain's network. Their design complies with European rules on train floor height. However, there is often a gap between the body of the carriage and the platform. For some passengers, boarding the carriage is a challenge, especially for passengers with small children, women, and elderly people. A train staircase system is being investigated to facilitate the boarding and disembarking of passengers on a low platform [29, 30]. The impact of vertical step height on the boarding and disembarking time of train passengers is analysed [31]. The likelihood of injuries to standing passengers with different handrails in a metro train collision is assessed [32]. A pilot study on the location of vertical handrails to facilitate boarding and disembarking of wheelchair users at metro stations is being conducted [33].

Before or during a train journey, elements of the carriage that affect the safety of passengers may fail. Railings, stairs, gangways, doors, locks (RSGDL) and other elements during the journey are serviced by a train electrician/mechanic. The score (0, 1, 2, 3 or 4) is awarded to assess the condition of the five elements RSGDL of a passenger carriage. Where all RSGDLs on a carriage are not technically sound, the score is 0. Score 1 is awarded if the railings are sound but the state of the stairs, gangways, doors, and locks is not technically sound. A carriage with sound railings and stairs, but with technically faulty gangways, doors and locks is of medium quality according to this sub-criterion with a score of 2 points. Where the railings, stairs and gangways of a carriage are sound, but the doors and locks are technically defective, a score of 3 points is awarded. The best passenger carriage is the one with all RSGDLs in good working order, which gives it a score of 4 points.

The value of TQI $K$, depending on $K_D$, which describes the internal safety trip by passenger train, and the variable $x_{D6}$ of the product $\overline{Q}_{D6} \cdot x_{D6}$, showing efficiency of railings, stairs, gangways, doors and locks, is calculated by the formula:

$$x_{D6} = \frac{\sum_{c=1}^{n_{car}} \left( \frac{RSGDL_f}{RSGDL_{max}} \right)_c}{n_{car}}, \tag{7.9}$$

where $x_{D6}$ is the variable of the 6-th sub-criterion of the key criterion D; $RSGDL_f$ is the factual score of railings, stairs, gangways, doors and locks in carriage $c$ (0, 1, 2, 3 or 4 points); $RSGDL_{max}$ is the maximum score of railings, stairs, gangways, doors and locks in carriage $c$ (4 points); $n_{car}$ is the number of carriages in the train ($c = 1, 2, \ldots, n_{car}$).

### *7.1.7 Handbrake Status and Function, Condition of Breaks*

Rolling stock is required to have automatic brakes, whereas passenger carriages and locomotives additionally need electric air brakes. Automatic brakes on rolling stock must be maintained to specified standards, easy to operate and reliable under a wide range of conditions of use, guaranteeing smooth braking and stopping of the train in the event of brake pipework disconnection, loss of air supply or pull of an emergency brake (emergency brake cock or lever) [34]. Emergency brake cocks are installed, sealed, and labelled in the opening of a passenger carriage body (one in each opening) and in the carriages themselves—one in the corridor and one in the carriage attendant's compartment. The passenger carriage is equipped with 4 emergency brakes. There is a hand brake in the passenger carriage (where passengers board and disembark) (Fig. 7.1).

Braking systems on rail vehicles ensure road safety. The main factors affecting the performance of the braking system are the braking force, the mass and speed of the vehicles, the stopping or braking distance, the condition of the railway and environmental factors. The work [35] considers such braking systems as disc and tread brakes, dynamic brakes, aerodynamic brakes, vacuum brakes, and electro-pneumatic brakes. This review study, with its comprehensive analysis of scientific and applied research, can contribute to the productivity of institutions and organisations working on braking systems for railway vehicles.

There are two types of emergency braking facilities on a passenger train carriage: the handbrake (HB) and 4 emergency brake cock (EBC).

The HB is in the working opening of the carriage. It is used to stop the wheels of the carriage chassis by a train attendant after an alarm signal from the train driver is received. When the HB wheel is turned, the brake pads of the carriage are pressed against the wheelset until the train comes to a stop. The 4 EBCs are installed in well visible and easily accessible places on the carriage. They shall be fitted in the neutral position and bear proper seals. In case of the necessity to stop the train, the train staff or, exclusively, a passenger can break the seal of any of the EBC four handles to lock the wheelsets of the carriage.

The status and functioning of the HB on the carriage is scored 0, 1 or 2 points. Where the HB lever is missing and inoperative, the score is 0. Where the HB lever is present but not working, or where the HB lever is absent, but the brake mechanism is working, the score for the carriage shall be 1 point. A carriage with a perfectly functioning HB is awarded 2 points.

The condition of the EBC shall be scored 0, 1, 2, 3, 4 for each train carriage. When all four EBCs on a carriage are not sealed, the score is 0. When three EBCs on a carriage do not have seals, the carriage shall be awarded 1 point and when two EBCs on a carriage do not have seals, the carriage shall be awarded 2 points. A carriage with just one EBC without a seal is scored 3 points. When all the carriage EBCs are sealed, the carriage scores 4 points.

The value TQI $K$, depending on $K_D$, which describes the internal safety trip by passenger train, and the variable $x_{D7}$ of the product $\overline{Q}_{D7} \cdot x_{D7}$, showing operational state of a hand brake is calculated by the formula:

$$x_{D7} = \frac{\sum_{c=1}^{n_{car}} \left( \frac{HB_f}{HB_{max}} + \frac{EBC_f}{EBC_{max}} \right)_c}{n_{car} \cdot n_b}, \tag{7.10}$$

where $x_{D7}$ is the variable of the 7-th sub-criterion of the key criterion D; $HB_f$ is the factual score condition and action of handbrake in carriage $c$ (0, 1 or 2 points); $HB_{max}$ is the maximum possible score of the condition and action of handbrake (2 points); $EBC_f$ is the factual score condition and action of emergency brake cock in the carriage $c$ (0, 1, 2, 3 or 4 points); $EBC_{max}$ is the maximum possible score of the condition and action of emergency brake cock lever in carriage $c$ (4 points); $n_{car}$ is the number of carriages in the passenger train ($c$ = 1, 2, ..., $n_{car}$); $n_b$ is the number of the types of braking measures in the carriage of the train ($n_b = 2$).

### *7.1.8 Call the Police Option*

Conflicts may arise during the boarding and disembarking process between passengers, outsiders (terrorists, thieves, troublemakers) and train staff, which may require the involvement of law enforcement agencies. The first port of call is the police, who can be called to the train and whose guaranteed arrival affects the safety of the trip.

The international association of government-controlled police organisations, RAILPOL (RAILway POLice), is responsible for the policing of Europe's railways. The aim of the Association is to strengthen and intensify international railway police cooperation, prevent threats, enhance the effectiveness of measures to combat cross-border crime and act as a link between the police and the railway sector. Pickpocketing, vandalism, and graffiti have become international. Organised crime groups involved in illegal migration, trafficking in people, money matters, drugs and firearms use railways as a means of transport. Terrorists also use logistics bases, roads, and railways to travel across Europe, even if their targets are not in Europe [36].

Security concerns and fear of crime (FoC) in a city can affect travel behaviour through changes in travel choices and habits. In this sense, the feeling of being "safe" or "at risk" in public transport depends not only on factors such as lighting, travelling companionship or crowding, but also on latent individual-specific characteristics, among which the fear of crime is a key issue to be considered when formulating a transport security policy [37].

Around 36% of public transport passengers in Mexico City reported feeling unsafe or not very safe. The results of the ordinary logistic regression show that the level of fear of crime is similar for all modes of transport. Those travelling for more than 30 min were more worried. In addition, the fear of crime on public transport is higher among victims of crime, younger passengers, those who have a poor opinion of the local police and those who are fearful of crime in their own neighbourhood. These findings suggest that fear reduction policies are an important partner of crime reduction policies in increasing quality of life, and that public transport is an important, although neglected, dimension of such policies [38].

The focus is on how crime concentrates in and around the public transport system. It discusses the causes of the lack of evidence on crime in public transport. A description of the challenges in identifying and analysing concentrations of crime is provided. International evidence of crime in this area is reviewed. The causes and explanations of such events and their specialisation are examined. Prevention options are foreseen, some emerging trends and future research approaches are discussed [39].

There are two types of police authorities on Korean railways: the Special Judicial Police and the Local Subway Police. The existence of two separate entities has drawbacks for overall rail safety: operational inefficiencies, lack of customer-oriented policing and synergic effects. The purpose of this study [40] is to search for solutions to rail crimes that contribute to rapid spatial movements since they take place on fast moving trains.

The nature and frequency of transit crime and passenger concerns about safety at transit hubs and during the journey are examined through a bibliometric analysis and a systematic literature review [41]. 3137 documents were received, and 245 documents were selected for in-depth analysis. The studies are mostly oriented towards large cities focusing on the safety of the rail-bound environments and examples of interventions to improve the actual and perceived safety of public transport (PT) users. The English-language literature dominates in this area of concern. By highlighting the importance of the transit environment throughout the journey, the article also advocates for greater inclusion of passenger safety needs and multi-stakeholder involvement into PT policy.

The British Transport Police (BTP) is one of the oldest police forces in the world and one of the two national police forces in England, Scotland, and Wales. Fighting terrorism on the rail network is vital to the security of the UK. A survey of public opinion and BTP officials [42] on the effectiveness of existing security measures, particularly in relation to the fear of terrorism, found that armed police are the most effective and reliable security instrument. Respondents ranked the activities of information companies as the least effective. The qualitative analysis showed that an increase in the number of armed police, without communication from the information companies, would have a negative impact on the public by increasing fear.

The purpose of this study [40] is to search for solutions to rail crimes that contribute to rapid spatial movements since they take place on fast moving trains. Security on public transport is essential for mobility, and the fear of crime can be more of a personal issue than the crime itself. It has been found [43] that among the most important parameters affecting passengers with respect to their perceived safety are lighting, surveillance, behaviour of other persons, time of day, and passenger's own gender.

Sexual harassment on trains is also not uncommon. BTP is urging people to report any sexual harassment they experience or see while travelling by train. BTP has launched the *Railway Guardian* app to report incidents. Passengers can also text police officers. Scotland is considering women-only rail carriages. Police records show that 46 sexual assaults against women were recorded on Scottish railways in 2021. There were also 301 reports of assault, threats, or harassment on trains [44]. The number of sexual offences recorded against women in 2021 is still the highest in the last 10 years of available data [45]. Another study focuses on sexual assault cases on commuter trains in Tokyo [46].

The increase in the number of illegal migrants in Lithuania is a premeditated hybrid war waged by the Belarusian regime against Lithuania and, consequently, against the European Union as a whole. There have been cases of illegal migrants being detained by police officers on Lithuanian railway trains. Police officers are often called to trains to deal with undisciplined passengers, such as those who consume or are under the influence of alcoholic beverages and drugs, those who smoke in places other than those designated for that purpose, those who do not wear face masks during the pandemic, and those who disregard the rules governing the transport of passengers and luggage.

Call the police on a train option is scored 0, 1 or 2 points. Where there is no provision for calling the police to the train, the train is awarded 0 points. Score 1 point is awarded when police officers can be summoned to the train by radio via the train driver. When police officers can be called on board the train directly via the nearest railway station (GSM-R radio system), the option is scored 2 points.

The value of TQI $K$, depending on $K_D$, which descraibes the internal safety trip by passenger train, and the variable $x_{D8}$ of the product $\overline{Q}_{D8} \cdot x_{D8}$, showing possibility of calling the police is calculated by the formula:

$$x_{D8} = \frac{CPO_f}{CPO_{max}}, \tag{7.11}$$

where $x_{D8}$ is the variable of the 8-th sub-criterion of the key criterion D; $CPO_f$ is the factual score of the possibility of calling the police in passenger train (0, 1 or 2 points); $CPO_{max}$ is the maximum possible score of possibility of calling the police in train (2 points).

## 7.2 Additive Model for the Sub-criteria of the Key Criterion D

The comprehensive quality index $K_D$, describing the sub-criteria with reference to the intrinsic pasenger safety on train, was calculated as follows:

$$K_D = \overline{Z}_D^* \cdot \left[ \overline{Q}_{D1} \cdot \frac{\sum_{c=1}^{n_{car}} \left( \frac{PFE_f}{PFE_{max}} + \frac{APE_f}{APE_{max}} + \frac{PPE_f}{PPE_{max}} + \frac{FAS_f}{FAS_{max}} + \frac{AFE_f}{AFE_{max}} \right)_c}{n_{car} \cdot n_{ex}} \right.$$
$$+ \overline{Q}_{D2} \cdot \frac{FAK_f}{FAK_{max}} + \overline{Q}_{D3} \cdot \frac{ACF_f}{ACF_{max}} + \overline{Q}_{D4} \cdot \frac{T_{lim} - T_{ABANf}}{T_{lim} - T_{air}}$$
$$+ \overline{Q}_{D5} \cdot \frac{\sum_{c=1}^{n_{car}} \left( \frac{EE_f}{EE_{max}} \right)_c}{n_{car}} + \overline{Q}_{D6} \cdot \frac{\sum_{c=1}^{n_{car}} \left( \frac{RSGDL_f}{RSTGL_{max}} \right)_c}{n_{car}}$$
$$\left. + \overline{Q}_{D7} \cdot \frac{\sum_{c=1}^{n_{car}} \left( \frac{HB_f}{HB_{max}} + \frac{EBC_f}{EBC_{max}} \right)_c}{n_{car} \cdot n_b} + \overline{Q}_{D8} \cdot \frac{CPO_f}{CPO_{max}} \right]. \tag{7.12}$$

In practice, using the mathematical additive model (7.12), the weight $\overline{Z}_D^*$ of the key criterion D is taken from Table 3.16. The local weights $\overline{Q}_{Di}$ of its eight sub-criteria are taken from Fig. 3.7d. For each sub-criterion (D1, D2, ..., D8), the variables $x_{D1}$, $x_{D2}$, ..., $x_{D8}$ shall be calculated by taking the values of the parameters assessing the quality of the passenger train under study, and computed from formulae (7.2)–(7.11).

The element $K_D$ of the CQI of a passenger train, which depends on the intrinsic safety elements and parameters, is ideally (the highest train quality) equal to $\overline{Z}_D^* =$

0.286. Average train quality is met when $K_D = 0.143$, while the worst possible quality is assessed $K_D = 0$.

## 7.3 Testing and Assessment of the Adequacy for the Model of Key Criterion D

Using the formulas suggested in the chapter 7 for determining CQI components (eight sub-criteria) the authors could compute their significances based on the real (factual) data taken from technical specifications, standards, measures, surveys, and other reliable sources. The tests were carried out on IPT No. 6, which consisted of 6 carriages (Fig. 4.8).

***Number and condition of fire extinguishing equipment***. The six carriages of the IPT No. 6 under study have a total of five fire extinguishing facilities: PFE, APE, PPE, FAS ir AFE. One carriage (sleeping car No 1 with 54 berths) is not equipped with FAS, so this carriage $FAS_f = 0$. The variable calculated for the whole train according to formula (7.2) is $x_{D1} = 0.971$.

***First aid kits***. The calculated availability of first aid kits on all international trains "LTG Link" includes: a stretcher, a first aid kit containing various bandages. There are no medicines in the first aid kit, except for a skin burn treatment. The first aid kit is always located in the train manager's compartment. The bandages in the on-board first aid kit would not be sufficient in an emergency, such as a terrorist attack or an act of war. Therefore, for this criterion, ISP No. 6 is $FAK_f = 1$. If the number of bandages were sufficient in an emergency event, then the D2 sub-criterion would be the one with the highest score $FAK_{max} = 2$. The variable $x_{D2} = 0.50$ is computed from formula (7.3).

***Ambulance call facility***. KENWOOD TK-250 portable radios are currently in use on passenger trains, as are mobile phones. "LTG Link" has installed a radio link that was awarded a gold medal in the competition "Lithuanian Product of the Year 2010". The system of GSM-R covers 1563.3 km of railway lines. The system provides for a high quality of service, resulting in more efficient carriage marshalling, shunting and railway track maintenance operations. The system gives the train manager direct access to the nearest station attendant, traffic steward, and passenger services operator. GSM-R replaced the communication facilities previously used by "LTG Link", which were uncertified and did not meet the requirements set by the Communications Regulatory Authority (CRA) and the EU. The ambulance can be summoned to IPT No. 6 by radio via the train driver. This option is therefore assessed $AFC_f = 1$. If an ambulance could be summoned to the train via the nearest station attendant, this option would be scored $ACF_f = ACF_{max} = 2$. The variable computed from formula (7.4) to indicate the quality of IPT No. 6 is $x_{D3} = 0.5$.

***Operational status of axle box overheating and the fire alarm system***. IPT No. 6 under study consists of 7 carriages (Fig. 4.8). The train was stationary for more than 10 min at 5 intermediate stations, so the actual temperature of all axle boxes was measured (Table 7.2). The temperature of the axle boxes $T_{ABANf}$ was measured with an infra-red thermometer. It was measured by the train carriage attendants. Side I of the carriage shows the axle boxes on the platform side and side II shows the axle boxes on the opposite side. The measurements show that the ambient air temperature dropped from + 22 °C to + 3 °C at each of the intermediate stations from $B_1$ to $B_8$ as the train moved from departure station A to arrival station C. Due to heat exchange processes, the colder ambient air cooled the axle boxes and therefore their actual temperature also decreased. Having measured the temperature $T_{ABANf}$ of the axle boxes for 280 times along the route of the train, the maximum difference $\Delta T_{maxl}$ was found in each ambient air temperature zone.

The maximum temperature difference $\Delta T_{maxl} = + 8.1$ °C was measured at intermediate station $B_5$ ($T_{air} = +16$°C) for the sleeping car No 1 with 54 berths, which was $T_{ABANfmax} = 24.1$°C. This axle box was the most heated and therefore the temperature of this axle box was taken for the calculation of the variable $x_{D4}$ in formula (7.5) and for the safety assessment of the whole IPT No. 6. The variable $x_{D4} = 0.816$ was computed.

***Availability of emergency exits on-board***. There are 2 emergency exits on all carriages of the train under investigation (IPT No. 6). The inspection of the train carriages revealed that carriage No 1 (Fig. 4.8) has two emergency exits (Fig. 7.1). However, one of them is faulty, as there is no special hammer attached to it to break the glass. The variable that indicates the safety of the train and is associated with the number and state of emergency exits, has been computed from formula (7.8), and is equal to 0.964.

***Maintaining railings, stairs, gangways, doors, and locks***. The inspection of the seven carriages of passenger train IPT No. 6 revealed minor faults in five technical elements (railings, stairs, gangways, doors, and locks). Carriage No 3 (dining-car) had a broken door lock and carriage No 6 (compartment car) had problematic door opening. The variable $x_{D6}$ that indicates the intrinsic safety of the train has been computed from formula (7.9) and is equal to 0.929.

***Handbrake status and function, condition of breaks***. All IPT No. 6 carriages are equipped with two ($n_b = 2$) two automatic braking systems: hand brakes and emergency brake levers. Therefore, the state and performance of the handbrake on each carriage is assessed $HB_f = 2$, and $HB_{max} = 2$ for the emergency break levers. The variable $x_{D7}$ computed from formula (7.10) scores 1 point and indicates the high quality of the train in relation to this safety indicator.

***Call the police option***. In the numerical example, the quality assessment of the train under study in terms of the calling the police officers is based on the assumption that police officers can be summoned to the IPT No. 6 by radio via the train driver. This option is therefore assessed $CPO_f = 1$ and $CPO_{max} = 2$. The variable $x_{D8} = 0.5$, calculated from formula (7.11), indicates the average quality of the train with respect to this average safety sub-criterion.

The quality level of the criterion D and the respective $x_{Di}$ value, based on the factual data, are presented in Table 7.3. The quality of train represented by the estimate of the variable is equal to 1, when it is based on the highest values of the (real) data, it is

**Table 7.2** The variation of the car axle box temperature $T_{ABANf}$, measured by infrared thermometer

| Carriage no. in Fig. 4.8 a | Carriage side | Intermediate station (Table 5.3) | | | | | | | | | |
|---|---|---|---|---|---|---|---|---|---|---|---|
| | | $B_1$ | | $B_2$ | | $B_5$ | | $B_6$ | | $B_8$ | |
| | | The ambient air temperature $T_{air}$ | | | | | | | | | |
| | | + 22 °C | | + 20 °C | | + 16 °C | | + 9 °C | | + 3 °C | |
| Luxury car 7 18 berths | I | 21.1 | 22.2 | 21.0 | 22.0 | 14.5 | 16.1 | 3.9 | 4.2 | 4.2 | 2.0 |
| | | 21.6 | 21.3 | 21.4 | 21.1 | 17.1 | 14.8 | 3.7 | 3.8 | 2.3 | 1.0 |
| | II | 21.0 | 22.3 | 19.9 | 22.1 | 14.2 | 15.8 | 4.0 | 4.3 | 4.4 | 2.1 |
| | | 21.4 | 21.2 | 21.1 | 19.8 | 16.9 | 15.0 | 3.9 | 3.9 | 2.5 | 1.6 |
| Compartment car 6 36 berths | I | 21.4 | 21.9 | 21.2 | 21.6 | 16.5 | 16.0 | 6.3 | 6.2 | 2.1 | 2.8 |
| | | 20.9 | 22.5 | 20.1 | 22.2 | 16.1 | 18.2 | 8.1 | 5.4 | 8.1 | 3.6 |
| | II | 21.2 | 21.9 | 21.0 | 21.2 | 16.4 | 15.9 | 6.4 | 6.5 | 2.7 | 3.0 |
| | | 20.7 | 22.3 | 20.5 | 22.0 | 15.7 | 18.0 | 8.3 | 5.9 | 7.5 | 3.2 |
| Compartment car 5 36 berths | I | 22.8 | 22.8 | 22.4 | 22.3 | 19.3 | 18.3 | 9.7 | 8.0 | 2.6 | 0.3 |
| | | 21.0 | 21.2 | 19.8 | 21.0 | 19.0 | 15.6 | 9.5 | 9.7 | 4.4 | 2.5 |
| | II | 22.9 | 21.9 | 22.5 | 21.6 | 18.9 | 18.1 | 9.6 | 8.2 | 2.1 | 1.2 |
| | | 21.2 | 21.3 | 19.7 | 19.8 | 18.5 | 16.2 | 9.7 | 9.6 | 3.9 | 2.6 |
| Compartment car 4 36 berths | I | 24.5 | 21.7 | 24.1 | 21.2 | 15.6 | 15.2 | 8.1 | 10.6 | 1.9 | 1.7 |
| | | 22.6 | 22.3 | 22.3 | 22.0 | 16.4 | 18.2 | 10.1 | 9.9 | 2.2 | 1.5 |
| | II | 24.6 | 21.8 | 24.2 | 21.5 | 15.7 | 15.0 | 8.3 | 10.7 | 1.7 | 1.2 |
| | | 23.0 | 22.4 | 22.8 | 22.1 | 15.9 | 17.8 | 10.0 | 10.1 | 2.0 | 1.1 |
| Dining-car 3 | I | 23.2 | 21.8 | 22.8 | 21.3 | 18.3 | 13.4 | 8.4 | 8.9 | 1.9 | 1.5 |
| | | 21.9 | 23.2 | 21.2 | 22.9 | 16.6 | 17.5 | 8.2 | 8.1 | 1.6 | 1.7 |
| | II | 23.3 | 21.9 | 23.1 | 21.4 | 18.5 | 14.0 | 8.6 | 8.8 | 1.6 | 1.2 |
| | | 21.7 | 23.4 | 21.3 | 23.1 | 17.1 | 16.9 | 8.4 | 8.2 | 1.2 | 1.4 |
| Compartment car 2 36 berths | I | 25.3 | 25.1 | 24.9 | 24.8 | 17.4 | 16.3 | 8.3 | 7.2 | 0.7 | 0.5 |
| | | 24.8 | 24.8 | 24.5 | 24.6 | 18.0 | 18.7 | 7.9 | 7.6 | 1.0 | 2.2 |
| | II | 25.7 | 25.0 | 25.4 | 24.7 | 17.1 | 17.0 | 8.5 | 7.4 | 0.9 | 0.7 |
| | | 24.6 | 24.7 | 24.1 | 24.3 | 18.4 | 18.2 | 8.1 | 7.7 | 1.5 | 1.9 |
| Sleeping car 1 54 berths | I | 23.8 | 22.3 | 23.5 | 21.9 | 23.7 | 20.9 | 9.1 | 9.0 | 7.0 | 4.8 |
| | | 23.2 | 22.8 | 22.8 | 21.8 | 20.7 | 22.0 | 8.7 | 8.9 | 2.8 | 5.9 |
| | II | 23.5 | 21.9 | 23.1 | 21.4 | 24.1 | 21.0 | 9.3 | 9.5 | 7.2 | 5.1 |
| | | 23.6 | 22.5 | 23.2 | 22.1 | 21.0 | 21.9 | 8.9 | 9.1 | 3.1 | 5.7 |
| $\Delta T_{maxl}$ | | + 3.7 | | + 5.4 | | + 8.1 | | + 1.7 | | + 5.1 | |

equal to 0.5, when based on the average values, and to 0, when based on the lowest values of the data.

The index $K_D$ of the sub-criteria describing internal safety trip by international passenger train No. 6 was calculated by formula (7.12) as follows:

$$K_D = 0.286(0.1524 \times 0.971 + 0.1008 \times 0.5 + 0.1007 \times 0.5 + 0.1887 \times 0.816 + 0.1208 \times 0.964 + 0.0908 \times 0.929 + 0.1606 \times 1 + 0.0852 \times 0.5) = 0.2307.$$

The sum of the products of the normalised Local weights $\overline{Q}_{Di}$ of the eight-criteria of the key criterion D (Table 3.10 and Fig. 3.7d) and the variables $x_{Di}$ (Table 7.3) indicates that the quality of travel on IPT No. 6 in terms of intrinsic safety is very good due to the fact that it makes more than 80% of maximum quality.

Taking the weight of the key criterion D ($\overline{Z}_D^* = 0.286$) and the sum of the products of all 8 sub-criteria $\overline{Q}_{Di} \times x_{Di}$ which is equal to 0.8067, the impact of this key criterion on the CQI $K_D$ of the IPT No. 6 was computed (formula 7.1). The effect of intrinsic

**Table 7.3** Quality level of the sub-criteria of the key criterion D and the respective value of variables $x_{Di}$ based on real data

| Variable $x_{Di}$ of the key criterion D (formula) | Real (actual) vale of the variable component | The calculated value $x_{Di}$ |
|---|---|---|
| $x_{D1}$ (7.2) | $PFE_{f1} = PFE_{f2} = PFE_{f3} = PFE_{f4} = PFE_{f5} = PFE_{f6} = PFE_{f7} = 2; PFE_{max} = 2;$ $APE_{f1} = APE_{f2} = APE_{f3} = APE_{f4} = APE_{f5} = APE_{f6} = APE_{f7} = 2; APE_{max} = 2;$ $PPE_{f1} = PPE_{f2} = PPE_{f3} = PPE_{f4} = PPE_{f5} = PPE_{f6} = PPE_{f7} = 2; PPE_{max} = 2;$ $FAS_{f1} = 0; FAS_{f2} = FAS_{f3} = FAS_{f4} = FAS_{f5} = FAS_{f6} = FAS_{f7} = 2; FAS_{max} = 2;$ $AFE_{f1} = AFE_{f2} = AFE_{f3} = AFE_{f4} = AFE_{f5} = AFE_{f6} = AFE_{f7} = 2; AFE_{max} = 2;$ $n_{car} = 7; n_{ex} = 5$ | 0.971 |
| $x_{D2}$ (7.3) | $FAK_f = 1; FAK_{max} = 2$ | 0.50 |
| $x_{D3}$ (7.4) | $ACF_f = 1; ACF_{max} = 2$ | 0.50 |
| $x_{D4}$ (7.5) | $T_{air} = 16°C; T_{lim} = 60°C; T_{ABANfmax} = 24.1°C$ | 0.816 |
| $x_{D5}$ (7.8) | $EE_{f1} = 3; EE_{f2-7} = 4; EE_{max} = 4; n_{car} = 7$ | 0.964 |
| $x_{D6}$ (7.9) | $RSGDL_{f1,2,4,5,7} = 4; RSGDL_{f3,6} = 3; RSGDL_{max} = 4; n_{car} = 7$ | 0.929 |
| $x_{D7}$ (7.10) | $HB_{f1-7} = 2; HB_{max} = 2; EBC_{f1-7} = 4; EBC_{max} = 4; n_b = 2;$ $n_{car} = 7;$ | 1 |
| $x_{D8}$ (7.11) | $CPO_f = 1; CPO_{max} = 2;$ | 0.50 |

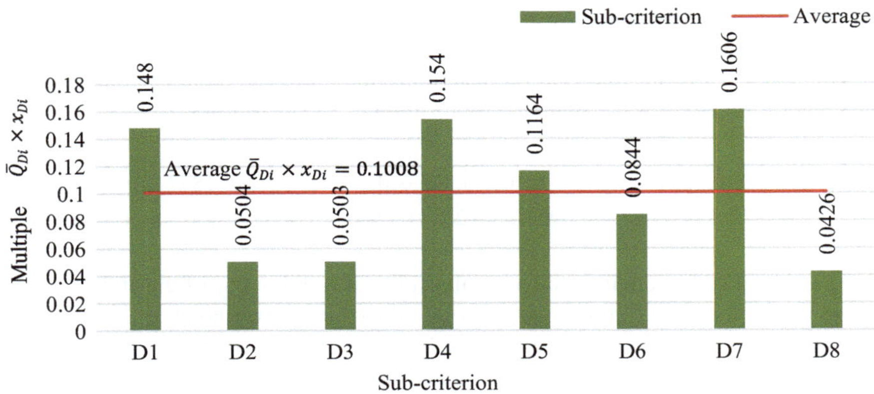


**Fig. 7.4** Findings of the calculations of the impact of the sub-criteria of the key criterion D on the trip by the IPT No. 6

safety factors on the quality of the passenger train under study is indicated by the CQI $K_D = 0.2307$ calculated from real data.

The calculated factual CQI describing the intrinsic safety trip by IPT No. 6, was found to be $K_D = 0.231$. It is approximately 20 percent less than the maximum (indicated the best quality) $K_{Dmax} = 0.286$.

The product of each of the local weight $\overline{Q}_{Di}$ of eight sub-criteria of this Key Criterion D, and of the variable $x_{Di}$ for the international passenger train No. 6 (Fig. 7.4) varies from 0.0426 to 0.1606, i.e., a difference of more than 3.7 times is found. Sub-criteria D1 (number and condition of fire extinguishing equipment), sub-criterion D4 (overheating of the axlebox and the state of operation of the fire alarm system), sub-criterion D7 (the state of operation of the handbrake) all have an impact on the quality of the train journey which is above average ($\overline{Q}_{Di} \times x_{Di} = 0.1008$). Sub-criterion D5 (availability of emergency exits) and sub-criterion D6 (condition of railings, stairs, gangways, doors, and locks) are close to moderate in terms of their impact on the quality of train travel. Sub-criteria D8 (calling the police option), sub-criterion D3 (calling an ambulance) and sub-criterion D2 (availability of first-aid facilities) are rated as of the lowest quality in ISP No. 6.

The numerical example shows that the additive mathematical model (7.12) is correct. It is based on the normalised local weights of the sub-criteria $\overline{Q}_{Di}$ provided by the respondents (train passengers) and the experts (train crew and management staff), as well as on the actual and regulatory parameters of the passenger train under study and provides a realistic result between 0 and $\overline{Z}_D^* = 0.286$. By quantifying the quality of a train, it is easy to compare it with other trains and rank them within the company train fleet.

According to this methodology, the calculated CQI, characterizing a trip by on international passenger train can be used by the railway passenger transport companies' decision maker determining the suitability of individuals trains to carry-out its activities.

# References


1. Hernik J, Mazur R (2018) Innovations and safety in passenger railway transport—the travelers' perspective. In: 33rd International Scientific Conference on Economic and Social Development—"Managerial Issues in Modern Business". Book of Proceedings: Warsaw, 26–27 September 2018, Warsaw, pp 71–81
2. Lv D, Zhang Y, Lin J, Wan P, Hu Y (2020) Multi-view interactive visual exploration of individual association for public transportation passengers. Appl Sci 10(2):628. https://doi.org/10.3390/app10020628
3. Kecklund L, Anderzén I, Petterson S, Haggstrom J, Wahlstrom B (2012) Evacuation in trains—the railway safety challenge. Research Gate, pp 1–9. https://www.researchgate.net/publication/26723174810.1201/b12742-92
4. Capote JA, Alvear D, Abren O, Cuesta A (2012) Analysis of evacuation procedures in high speed trains fires. Fire Saf J 49:35–46. https://doi.org/10.1016/j.firesaf.2011.12.008
5. Xu P, Xiao X, Yao S (2022) A review of train passenger safety—inspiration from passive safety passenger protection technology of automobile. Transp Saf Environ 4(1):1–12. https://doi.org/10.1093/tse/tdab032
6. Bureika G, Komaiško M, Jastremskas V (2017) Modelling the ranking of Lithuanian Railways level crossing by safety level. Transp Probl 12:11–22. https://doi.org/10.20858/tp.2017.12.se.1
7. Vileiniskis M, Remenyte-Prescott R, Rama D (2016) A fault detection method for railway point systems. Proc Inst Mech Eng Part F J Rail Rapid Transit 230(3):852–865. https://doi.org/10.1177/0954409714567487
8. Kuo C-M, Lin C-C (2016) Analysis of derailment criteria. Proc Inst Mech Eng Part F J Rail Rapid Transit 230(4):1158–1163. https://doi.org/10.1177/0954409715583692
9. Maskeliūnaitė L, Sivilevičius H (2020) The model for evaluating criteria describing the internal safety of a railway trip by international train. In: Proceedings of the international conference TRANSBALTICA XI: Transportation science and technology, May 2–3, 2019. Springer, Cham, Vilnius, pp 610–621
10. Lovreglio R, Duan X, Rahouti A, Phipps R, Nilsson D (2021) Comparing the effectiveness of fire extinguisher virtual reality and video training. Virtual Reality 25(5):133–145. https://doi.org/10.1007/s10055-020-00447-5
11. EN 50553:2012. Railway applications—requirements for running capability in case of fire on board of rolling stock. European Standard. https://standards.iteh.ai/catalog/standards/clc/35b51e15-b024-4079-a457-eb9efc61513b/en-50553-2012
12. Stahl P (2017) Fire detection and fire fighting testing and approval on railway vehicles. Environmental Science, Corpus ID: 201721977. https://www.semanticscholar.org/paper/Fire-Detection-and-Fire-Fighting-Testing-and-on-Stahl/2d0505d76a7261722df82278432fc4f529ac758b
13. CEN/TR 17532:2020. Railway applications—fire protection on railway vehicles—Assessment of fire containment and control systems for railway vehicles. https://standards.iteh.ai/catalog/standards/cen/910924f9-44f2-49d8-b8ce-a69c3c637ff7/cen-tr-17532-2020
14. Stacy M (2016) Fire safety on train. Rail Engineer. https://www.railengineer.co.uk/fire-safety-on-trains/
15. Priešgaisrinės saugos normos. Tarptautinio susisiekimo tarp NVS šalių, Estijos, Latvijos ir Lietuvos respublikų keleiviniai vagonai. Priešgaisrinės saugos reikalavimai (39/V-KL) (2003) Patvirtinta dvidešimt penktame Geležinkelių transporto tarybos posėdyje 1999 m. lapkričio 29 d. UAB „Gelspa“,Vilnius, 42 p (in Lithuanian)
16. Atnaujinti vaistinėles vairuotojai turės iki 2023 m. liepos 1 d. (2022) Mano vyriausybė. Lietuvos Respublikos Sveikatos apsaugos ministerija. https://sam.lrv.lt/lt/naujienos/atnaujinti-vaistineles-vairuotojai-tures-iki-2023-m-liepos-1-d (in Lithuanian)
17. Sahaidak-Nikitiuk R, Gretska G, Demchenko N (2019) Study of medical kits from passenger trains. ScienceRise Pharm Sci 1(17):23–31. https://doi.org/10.15587/2519-4852.2019.158002
18. Liew CH, Flaherty GT (2020) Experiences and attitudes of international travelers with cardiovascular disease: a qualitative analysis. Am J Trop Med Hyg 102(3):689–697. https://doi.org/10.4269/ajtmh.19-0793

19. Radijo ryšio tarp traukinio viršininko ir lokomotyvo mašinisto priemonių naudojimo instrukcija (KL/42) (1999) Patvirtinta Riedmenų ūkio direktoriaus 1998 m. gruodžio 28 d. Informacijos ir leidybos centras, Vilnius, 12 p (in Lithuanian)
20. Bosso N, Gugliotta A, Zampieri N (2018) Design of a test rig for railway axle-boxes. Proc Struct Integr 12:344–352. https://doi.org/10.1016/j.prostr.2018.11.082
21. Hao W, Liu F (2020) Axle temperature monitoring and neural network prediction analysis for high-speed train under operation. Symmetry 12(10):1662, 1–13. https://doi.org/10.3390/sym12101662
22. Wang Z, Cheng Y, Allen P, Yin Z, Zou D, Zhang W (2020) Analysis of vibration and temperature on the axle box bearing of a high-speed train. Veh Syst Dyn 58(10):1605–1628. https://doi.org/10.1080/00423114.2019.1645340
23. Techninės vagonų priežiūros instrukcija (66-V) (2011) (in Lithuanian)
24. Passenger Train Emergency Systems: Review of Egress Variables and Egress Simulation Models (2013) Department of Transportation. Federal Railroad Administration. Office of Research and Development, Washington, US. file:///C:/Users/40507/Downloads/dot_9856_DS1.pdf
25. Noonan S (2020) Passenger train doors. Expert article. The Experts Robson Forensic. https://www.robsonforensic.com/articles/passenger-train-door-expert
26. Innovative Product for Train Door Condition Monitoring Based on Vibration Analysis Algorithms (2020) European Commission. Cordis EU research results. https://cordis.europa.eu/project/id/730766
27. Waterson P, Pilcher C, Evans S, Moore J (2012) Developing safety signs for children on board trains. Appl Ergon 43(1):254–265. https://doi.org/10.1016/j.apergo.2011.05.012
28. Ham S, Han S-Y, Kim S, Park HJ, Park K-J, Choi J-H (2019) A comparative study of fault diagnosis for train door system: traditional versus deep learning approaches. Sensors 19(23):5160, 1–15. https://doi.org/10.3390/s19235160
29. Khode S, Shaikh K, Shaikh N, Pawar R (2022) Design and development of train staircase mechanism for easy accessibility of passengers to achieve low level platform. IJARIIE 8(3):1779–1783. https://ijariie.com/AdminUploadPdf/DESIGN_AND_DEVELOPMENT_OF_TRAIN_STAIRCASE_MECHANISM_FOR_EASY_ACCESSIBILITY_OF_PASSENGERS_TO_ACHIEVE_LOW_LEVEL_PLATFORM_ijariie16877.pdf
30. Gujarathi MT, Dhongade M, Sonwane M, Aher G, Gorde G (2022) A paper on design & development of train staircase mechanism for easy accessibility of passengers to achieve low level platform. J Modern Eng Technol Sci 4(5):5530–5534. https://www.irjmets.com/uploadedfiles/paper//issue_5_may_2022/25020/final/fin_irjmets1654188079.pdf
31. Holloway C, Thoreau R, Roan T-R, Boampong D, Clarke T, Watts D, Tyler N (2016) Effect of vertical step height on boarding and alighting time of train passengers. Proc Inst Mech Eng Part F J Rail Rapid Transit 230(4):1234–1241. https://doi.org/10.1177/0954409715590480
32. Yao S, Li X, Xiao X, Xu P, Che Q (2021) Injuries to standing passengers using different handrails in subway train collision accidents. Int J Crashworth 26(1):53–64. https://doi.org/10.1080/13588265.2019.1687152
33. Seriani S, Fernandes VA, Moraga P, Cortes F (2022) Experimental location of the vertical handrail to improve the accessibility of wheelchair passengers boarding and alighting at metro stations—a pilot study. Sustainability 14(15):9224, 1–22. https://doi.org/10.3390/su14159224
34. Techninio geležinkelių naudojimo nuostatai (1996) Lietuvos Respublikos susisiekimo ministerija. Patvirtinta Lietuvos Respublikos susisiekimo ministro 1996 m. rugsėjo 20 d. įsakymu Nr.297. Informacijos ir leidybos centras, Vilnius, 126 p (in Lithuanian)
35. Günay M, Korkmaz ME, Özmen R (2020) An investigation on braking systems used in railway vehicles. Eng Sci Technol Int J 23:421–431. https://doi.org/10.1016/j.jestch.2020.01.009
36. About RAILPOL (2018) RAILPOL. European Association of Railway Police Forces. https://www.railpol.eu/site/about-railpol
37. Soto J, Fontalvo-Orozco M, Useche SA (2022) Public transportation and fear of crime at BRT systems: approaching to the case of Barranquilla (Colombia) through integrated choice and latent variable models. Transp Res Part A Policy Pract 155:142–160. https://doi.org/10.1016/j.tra.2021.11.001

38. Vilalta C (2011) Fear of crime in public transport: research in Mexico city. Crime Prev Commun Saf 13(3):171–186. https://doi.org/10.1057/cpcs.2011.4
39. Newton A (2014) Crime on public transport. In: Bruinsma G, Weisburd D (eds) Encyclopedia of criminology and criminal justice. Springer, New York, NY, pp 709–720. https://doi.org/10.1007/978-1-4614-5690-2_301
40. Tack-Hyun S, Yong-Sang L, Jeong-Seok D (2006) A study on efficient crime prevention in the railway industry—focusing on the Ministry of Construction and Transportation's Railway Police Office. J Korean Soc Railway 9(4). https://www.researchgate.net/publication/263360031_A_study_on_Efficient_Crime_Prevention_in_the_Railway_Industry_-_Focusing_on_the_Ministry_of_Construction_and_Transportation's_Railway_Police_Office_-
41. Ceccato V, Gaudelet N, Graf G (2022) Crime and safety in transit environments: a systematic review of the English and the French literature, 1970–2020. Public Transp 14:105–153. https://doi.org/10.1007/s12469-021-00265-1
42. Power N, McManus MA, Lynch R, Bonworth J (2016) Fear of crime on the rail networks: perceptions of the UK public and British Transport Police. Crime Prev Commun Saf 18(2):91–104. https://doi.org/10.1057/cpcs.2016.2
43. Sundling C, Ceccato V (2022) The impact of rail-based stations on passengers' safety perceptions. A systematic review of international evidence. Transp Res F Traffic Psychol Behav 86(2):99–120. https://doi.org/10.1016/j.trf.2022.02.011
44. Passengers urged to report sexual harassment on trains (2022) BBS. News. https://www.bbc.com/news/uk-scotland-scotland-politics-62029407
45. Merson A (2022) "Women are at risk": sex assaults hit 10-year high on trains. The Press and Journal Evening Express. https://www.pressandjournal.co.uk/fp/politics/scottish-politics/4281728/women-trains-sex-assaults/
46. Chowdhury (2022) Sexual assault on public transport: crowds, nation, and violence in the urban commons Social Cult Geogr, pp 1–17. https://doi.org/10.1080/14649365.2022.2052170

# Chapter 8
# Implementation of the Quality Assessment System for Rail Travel

*Think about the trains you will take, not the trains you missed!*
Mehmet Murat ildan

## 8.1 Improving the Quality of Rail Passenger Services

In the EU, the sector of passenger transport is highly consumer oriented. Transport operators have stated that a satisfied passenger will come back and thus increase the revenue of the transport operator. Over the last decade, efforts have been made to improve rail passenger transport, but in practice this has been very slow. The primary focus is usually on railway lines with great demand, especially high-speed train lines and national train lines. Different methods have been employed to ensure the quality of transport services. Quality has a major impact on passenger services. One of the most useful methods for assessing service quality is surveys. Surveys can be carried out from both passenger and transport operator perspectives. A comprehensive survey has provided a framework for maintaining the current level of service and offered a few recommendations for improving the service [1].

The study examines and assesses the logistics quality of rail passenger transport. It describes the relationship between logistics and transport, as well as the needs and functions of rail transport logistics. Quality indicators for rail passenger transport and a new methodology for the assessment of quality are presented. This methodology has been tested on a selected route and relates to monitoring the quality of traffic routes on the rail network [2].

The analysis of passenger preferences for selected marketing activities carried out by the leading passenger rail companies is summarised. The aim of the analysis was to find out whether the time and price categories (cost categories) used to develop the operator's offer in most of the marketing activities were in line with contemporary passenger preferences [3].



H. Sivilevičius and L. Maskeliūnaitė, *Assessment of the Quality of Passenger Transportation by Train Using Multiple Criteria Decision Making Methods*, Lecture Notes in Intelligent Transportation and Infrastructure,
https://doi.org/10.1007/978-3-031-85877-2_8

The world record for the world's longest passenger train, set by the Swiss company Rhaetion Railway, is helping to boost the popularity of rail transport. The 25 km Alpine Albula-Bernina route was covered in just over an hour by a 1.9-km train made up of 100 carriages. The journey down the slopes of the mountains to highlight Switzerland's engineering achievements and commemorate the 175th anniversary of the country's railways was watched by many curious people [4].

The Service Quality Report 2019 prepared by the Passenger Transport Directorate of Joint-stock Company "Lietuvos geležinkeliai" does not contain the concept of service quality. Joint-stock Company "Lietuvos geležinkeliai" is committed to ensuring the quality of its services: information on rail passenger transport services and ticketing options is provided at railway stations, on trains, by phone, by e-mail and on the Company official websites. The punctuality of the trains is assessed, and passenger complaints are dealt with. Of the 1320 complaints made by passengers in 2019, the majority were about: money charged but tickets not sent on time; technical faults at ticket counters; banking problems; train delays; inconvenient timetables; quality of passenger service [5].

Attention is paid to passengers with special needs. New trains are being procured to accommodate such passengers, infrastructure is being upgraded, and technical measures (lifts, elevators, ramps) are being introduced. Cooperation with organisations uniting people with disabilities (Lithuanian Disability Forum) is promoted and staff training is organised.

Surveys are frequently made use of "LTG Link" has conducted an online passenger survey to find out passengers' needs and expectations regarding the annual train timetables [6]. The Service Quality Report provides statistics on the number of passengers transported per year on both domestic and international trains, the number of bicycles sold, and the number of loyalty card and concessionary tickets sold. The report describes the measures that should ensure the cleanliness of trains and station premises without quantifying the actual quality. It is not clear from the data presented in the report what are the reasons behind passengers' choice to travel by train and what are the most important factors in terms of the quality of the journey for a passenger. There is no information on the passenger's experience of train travel. It is also impossible to establish the quality of an individual route/train.

## 8.2 Putting the New Model for Assessing the Quality of Train Travel in Practice

The multi-criteria integrated assessment model can be successfully used to analyse and compare the quality of international trains on individual routes. The multi-criteria evaluation methodology presented in this monograph can be used to calculate the factual values of the passenger transport quality criteria to identify which of the quality criteria under consideration have the highest and which have the lowest impact on the QTT. The improvement of the quality of rail travel requires a set of measures that would lead to the best possible result, which would allow passengers to choose rail travel as an alternative to other modes of transport.

The value of the CQI $K$ (formula 3.1), calculated from the weights/values of the key criteria and sub-criteria $\overline{Z}_f^*$ and $\overline{\omega}_i$ (or $\overline{\omega}_i$) as well as the variables $x_{fi}$ of passenger train journey quality as determined by expert studies, shall be compared with the recommended levels (Fig. 8.1). These six levels of train quality (excellent, very good, average, poor, very poor) represent different values of the CQI $K$. In a country with a strong economy, the quality requirements for passenger trains are the highest. A train is considered excellent when $K$ is between 0.95 and 1.00, very good when $K$ is between 0.85 and 0.95, good when $K$ is between 0.7 and 0.85, average when $K$ is between 0.5 and 0.7, poor when $K$ is between 0.25 and 0.5 and very poor when $K$ is between 0.00 and 0.25. Passenger trains for countries with average economies have lower quality requirements. The CQI is modified with a reduction of the values of the quality levels by 0.05–0.10. (Fig. 8.1). The lowest quality requirements are for passenger trains in low economy countries. The assignment of a country to one of the three groups of different economies is subjective, as is the correspondence of the quality level thresholds to the calculated $K$ index.

Calculated value of the quality index for the studied IPT No 6:

$$K = K_A + K_B + K_C + K_D = 0.1415 + 0.1157 + 0.1518 + 0.2307 = 0.6397.$$

With $K = 0.64$, the quality of this train is average in strong economies and good in medium and weak economies (Fig. 8.1). The thresholds for quality levels may be modified by the decision-makers, but the general principle is not subject to change.

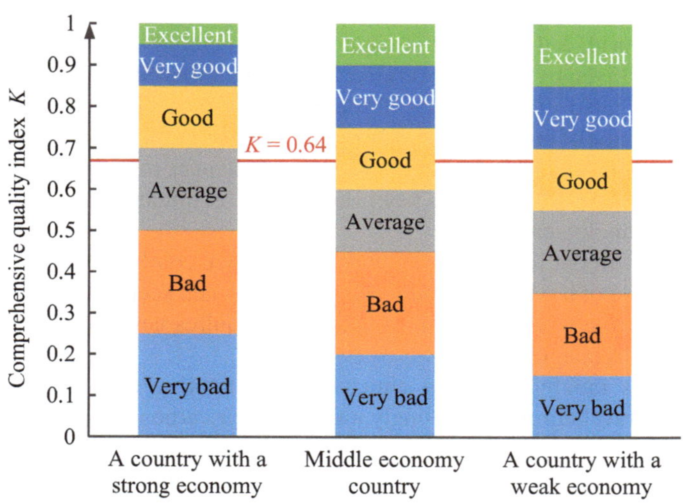


**Fig. 8.1** Impact of the Passenger Train Comprehensive Quality Index $K$ on the quality level of train travel in different economies

The assessment methodology can be applied to international train formations made up of different passenger carriages, with different pricing and services provided. This assessment methodology can easily be applied to local trains. Local trains are not subject to the quality criteria that are applicable to trains on international routes, e.g., cost of visas, expertise of customs and border officials, etc. The mean local weight of the rejected sub-criterion is divided in proportion to the weights of the remaining sub-criteria in the group and the resulting fractions are added to the other sub-criteria. The largest proportion of the weight of the absent sub-criterion is added to the largest (most important) sub-criterion. The sum of the local mean weights of the remaining sub-criteria is equal to one.

A multi-criteria assessment of a single international passenger train requires knowledge regarding the construction and the categories of the carriages, the range of services provided, the ticket fares, the skills of the service staff, etc. The calculated CQI for an international passenger train is expressed as a single number, making it easy to compare with the quality index of another train. A train with a higher estimated $K$ is of higher quality.

The results of the proposed multi-objective integrated assessment model could be successfully used in the nominations for "Best Passenger Train Team", "Best Train Journey" or "Best Train Route". Based on the calculated multi-criteria train quality indicator $K$, which compares one train with another, trains could be classified according to the well-known classification of hotels, i.e., the number of stars.

## 8.3 Rationale for Future Studies

All over the world, rail transport is undergoing something of a renaissance, with demand being driven by a few converging trends. Digital evolution, climate change, rising demand, rapid urbanisation - all these mega-trends have major implications for the future shape of our railways, pulling it in new directions. These trends need to be nurtured so that the rail industry can meet the challenges of tomorrow. Improving passengers' travel experience is a key priority. This is influenced by the arrival of new digital technologies.

Passengers will opt for rail only if their journeys are comfortable, fast, safe, and reliable. The cost of travel is also important. To attract and keep new passengers, rail services need to be high quality. To achieve this requires new thinking and a new level of technical integration.

The planet's rapidly changing climate is recalibrating the risk equation for everyone and travellers are increasingly aware of their carbon footprint. More stringent requirements on greenhouse gases may have an impact on transport planning, with a preference for cleaner vehicles. With increased urbanisation and road congestion, as well as environmental considerations and energy consumption, railways are a strong alternative to road transport both for passengers and freight. Train speeds are rising.

High speed can offer a sustainable, progressive means of managing the economic and environmental costs of the journey. It can also offer fast connectivity, a positive passenger experience, and can result in economic benefits. With every new railway line connecting people and communities, the popularity of rail grows. Demand for rail services is on the rise. Coping with this will require new competences, knowledge, and experience [7].

The future of rail transport in Lithuania is linked to the construction of *Rail Baltica,* a European rail line. *Rail Baltica* is a rail transport infrastructure project that aims to integrate the Baltic States into the European railway network. The project total value is about 24 billion Eur, and it is the largest investment aimed at the enhancement of mobility in the region, development of business, tourism, and trade. The project involves 5 EU countries: Poland, Lithuania, Latvia, Estonia, and Finland which is involved indirectly. The line will connect Helsinki, Tallinn, Pärnu, Riga, Panevėžys, Kaunas, Vilnius, and Warsaw. 870 km of high-speed electrified railway double track will be built in Lithuania, Latvia, and Estonia. The project will have an electrified railway line controlled by ERTMS (European Rail Traffic Management System). The estimated speed of the passenger trains is 249 kms per hour. A modern high-speed rail link is expected to provide a convenient, safe, and environmentally friendly alternative for passenger movement. Passengers in the Baltic States will enjoy the comfort of train travel that has long been available in Western Europe [8]. Future rolling stock must comply with the principles of sustainable mobility.

Critical infrastructure networks, such as transport and electricity networks, are essential for the functioning of society and the economy. Growing transport demand is increasing congestion on rail networks, making them more interdependent and difficult to operate. In addition, more and more disruptions can be expected in the future due to system failures and climate change. Train cancellations and delays hamper passenger mobility. There is now a growing need to quantify the impact of disruptions and the evolution of system functioning [9].

Technology has the potential to provide train passengers with up-to date information and tailored services to improve the rail travel experience. However, there is a lack of knowledge about the types of innovations and services that are preferred by the travelling public [10]. As rail tracks and rolling stock improve, so do the expectations of rail passengers. The supply and quality of services and the competence of train staff are bound to change. It is likely that the future rail passenger will be interested in mobility indicators such as speed, comfort, and reliability. However, we must accept that different people may have different needs. This requires passenger surveys.

Rail passenger transport, acting as an integral mode of transport, has the prerequisites to form the essential axis of an integrated transport system, and thus the core of a sustainable public passenger transport system. To meet the preconditions, rail passenger transport must focus on research and innovation in the coming decades. Research and innovation are to focus on the following key areas: digitalisation and economic efficiency, automation of the rail system and Artificial Intelligence, sustainable and environmentally friendly solutions, cutting edge RTV life cycle management and enhancement of the performance of RTV, the integration of railways into a door-to-door system (improving accessibility of rail transport services for customers) and safety and security (increasing customer safety and asset protection) [11].

In future, it is foreseen to develop a computer programme to apply the mathematical models presented in the monograph to calculate a comprehensive train quality index.

## References

1. Abramović B, Šipuš D (2020) Quality assessment of regional railway passenger transport. Sustainable Rail Transport (pp 83–96). Springer International Publishing. https://www.springerprofessional.de/en/quality-assessment-of-regional-railway-passenger-transport/17008902
2. Dedík M, Gašparik J, Záhumenská Z (2017) Quality assessment in the logistics of rail passenger transport. MATEC Web Conf 134:1–7. https://doi.org/10.1051/matecconf/201713400009
3. Rosa G (2021) Passenger preferences in rail transport in Poland as regards travelling time and cost. Sustainability 13(9):4737. https://doi.org/10.3390/su13094737
4. Switzerland sets record with world's longest passenger train (2022). BBC News. https://www.bbc.com/news/av/world-europe-63442530
5. Teikiamų paslaugų kokybės ataskaita (2019) Vadovaujantis Europos Parlamento ir Tarybos reglamentu (EB) 1371/2007 „Dėl geležinkelių keleivių teisių ir pareigų". AB „Lietuvos geležinkeliai" Keleivių vežimo direkcija, Vilnius, 11 p (in Lithuanian)
6. Traukinių keleiviai kviečiami atsakyti į keletą klausimų apie metinius traukinių tvarkaraščius (2022) Varėnos rajono savivaldybė. https://varena.lt/aktualijos/traukiniu-keleiviai-kvieciami-atsakyti-keleta-klausimu-apie-metinius-traukiniu-tvarkarascius/ (in Lithuanian)
7. Morrell M, Goulding L, Luebkeman Ch (2019) Future of Rail 2050. Arup, London, 60 p. file://C:/Users/40507/Downloads/Arup_FutureofRail2050_2019%20(1).pdf
8. Maskeliūnaitė L (2021) Railways in Lithuania: from tsarist Russia to Rail Baltica. Transport 36(4):364–375. https://doi.org/10.3846/transport.2021.16086
9. Bešinović N (2020) Resilience in railway transport systems: a literature review and research agenda. Transp Rev 40(4):457–478. https://doi.org/10.1080/01441647.2020.1728419
10. Oliveira L, Bruen C, Birrell S, Cain R (2019) What passengers really want: assessing the value of rail innovation to improve experiences. Transp Res Interdiscip Perspect 1(100014):1–11. https://doi.org/10.1016/j.trip.2019.100014
11. Hlavatý J, Ližbetin J (2021) Innovation in rail passenger transport as a basis for the safety of public passenger transport. Transp Res Proc 53:98–105. https://doi.org/10.1016/j.trpro.2021.02.013

# Index




H. Sivilevičius and L. Maskeliūnaitė, *Assessment of the Quality of Passenger Transportation by Train Using Multiple Criteria Decision Making Methods*, Lecture Notes in Intelligent Transportation and Infrastructure,
https://doi.org/10.1007/978-3-031-85877-2

MIX
Papier aus verantwortungsvollen Quellen
Paper from responsible sources
FSC® C105338

If you have any concerns about our products,
you can contact us on
**ProductSafety@springernature.com**

In case Publisher is established outside the EU,
the EU authorized representative is:
**Springer Nature Customer Service Center GmbH**
**Europaplatz 3, 69115 Heidelberg, Germany**

Printed by Libri Plureos GmbH
in Hamburg, Germany